Food Wars

In the years since publication of the first edition of *Food Wars* much has happened in the world of food policy. This new edition brings these developments fully up to date within the original analytical framework of competing paradigms or worldviews shaping the direction and decision-making within food politics and policy.

The key theme of the importance of integrating human and environmental health has become even more pressing. In the first edition the authors set out and brought together the different strands of emerging agendas and competing narratives. The second edition retains the same core structure and includes updated examples, case studies and the new issues which show how these conflicting tendencies have played out in practice over recent years and what this tells us about the way the global food system is heading. Examples of key issues given increased attention include:

- nutrition, including the global rise in obesity, as well as chronic conditions, hunger and under-nutrition
- the environment, particularly the challenges of climate change, biodiversity loss, water stress and food security
- food industry concentration and market power
- volatility and uncertainty food prices and policy responses
- tensions over food, democracy and citizenship
- social and cultural aspects impacting food and nutrition policies.

Tim Lang is Professor of Food Policy at City University London, UK. He has been a consultant to UN agencies, including auditing the Global Top 25 Food Companies on food and health for the WHO. He was a UK Sustainable Development Commissioner, 2006–11, and has worked closely with civil society organisations locally, nationally and internationally. In 2010 he was appointed to the Mayor of London's Food Board. He is a frequent commentator on food policy in the media.

Michael Heasman is Lecturer in Health and Wellbeing, Faculty of Health Studies, University of Bradford, UK. He has held academic positions at universities including Alberta, Aalborg, SOAS, Reading and Harper Adams. He is Research Fellow at the Centre for Food Policy, City University London. He has been a consultant to Agriculture & Agri-Food Canada, VTT in Finland and CSIRO in Australia.

This welcome new edition of *Food Wars* explores the deepening integration of health and environment in food policy. The authors were prescient 15 years ago in identifying two emerging food system trajectories. Since 2008 battle has been explicitly engaged over meanings of food system sustainability and equity. The updated, well documented synthesis of economy, science, power, consumption, and human relations with our earthly home is an even more valuable guide for students, food system innovators, and policymakers.

Harriet Friedmann, *Professor Emeritus of Sociology, University of Toronto, Canada and Visiting Professor, Institute of Social Studies (Erasmus University)*

Will it be possible to achieve a socially just and environmentally sustainable system of global food production and consumption? The first edition of *Food Wars* broke new ground in exploring the competing paradigms that underpin modern food systems. This edition tackles today's most confronting food issues – obesity, hunger, environmental degradation, and corporate domination – providing a critical, nuanced, understanding of contemporary food policy. It is compulsory reading for anyone who cares about the future of food and farming.

Geoffrey Lawrence, *Emeritus Professor of Sociology, University of Queensland, Australia and President, International Rural Sociology Association*

Food Wars, 2nd edition, is essential reading for anyone wanting to understand the battles underpinning global food provisioning. With significantly updated chapters, it intensifies the call and provides a much needed vision to radically change the way our food system is organised and governed.

Damian Maye, *Reader in Agri-Food Studies, University of Gloucestershire, UK*

What's so terrific about this book is its basis in theory applied to real-world, cross-cutting food issues involving government, business, and civil society. The authors emphasize the need for all of us to advocate for healthier and more sustainable food systems, for food peace rather than food wars, and to do so now.

Marion Nestle, *Professor of Nutrition, Food Studies, and Public Health, New York University, USA and author of* Soda Politics: Taking on Big Soda (and Winning)

Food Wars

The global battle for mouths, minds and markets

Second edition

Tim Lang and Michael Heasman

First published 2004 by Earthscan

This edition published 2015
by Routledge
2 Park Square, Milton Park, Abingdon, Oxon OX14 4RN

and by Routledge
711 Third Avenue, New York, NY 10017

Routledge is an imprint of the Taylor & Francis Group, an informa business

British Library Cataloguing-in-Publication Data
A catalogue record for this book is available from the British Library

Library of Congress Cataloging in Publication Data
Lang, Tim.
    Food wars : the global battle for mouths, minds and markets /
    Tim Lang and Michael Heasman. -- 2nd ed.
        pages cm
    Includes bibliographical references and index.
    1. Nutrition policy. 2. Food supply. I. Heasman, Michael, 1958- II. Title.
    TX359.L36 2016
    363.8'561--dc23   2015016997

ISBN: 978-1-138-80258-2 (hbk)
ISBN: 978-1-138-80262-9 (pbk)
ISBN: 978-1-315-75411-6 (ebk)

Typeset in Times New Roman
by HWA Text and Data Management, London
Printed and bound by CPI Group (UK) Ltd, Croydon, CR0 4YY

Contents

Figures

Tables

Abbreviations

AFN	alternative food networks
AICR	American Institute of Cancer Research
AMR	antimicrobial resistance
ANZFA	Australia–New Zealand Food Authority (now FSANZ)
AoA	Agreement on Agriculture (of the GATT)
BMI	body mass index
bn	billion
BSE	bovine spongiform encephalopathy
CAP	Common Agricultural Policy
CAPr	Committee of Advertising Practice (UK)
CDC	Centers for Disease Control and Prevention (USA)
CEO	chief executive officer
CEO	Corporate Europe Observatory (NGO)
CFC	chlorofluorocarbon
CGIAR	Consultative Group on International Agricultural Research (a consortium)
CHD	coronary heart disease
CI	Consumers International (world body of consumer NGOs)
CIAA	Confederation of Food and Drink Industries of Europe (now FDE)
CO_2e	carbon dioxide (CO_2) equivalent
Codex	Codex Alimentarius Commission (joint WHO/FAO body)
COMA	Committee on Medical Aspects of Food Policy (UK, replaced by SACN)
COPD	chronic obstructive pulmonary disease
CSR	corporate social responsibility
CVD	cardiovascular disease
DALY	disability adjusted life year
Defra	Department of Environment, Food and Rural Affairs (UK)
DNA	deoxyribonucleic acid (hereditary material in genes)
DPAS	Diet and Physical Activity Strategy (of the WHO)
EC	European Commission
EFSA	European Food Safety Authority
EU	European Union

FAO	Food and Agriculture Organisation (of the UN)
FDA	Food and Drug Administration (USA)
FDE	Food Drink Europe (industry body of the EU)
FSANZ	Food Standards Australia New Zealand (formerly ANZFA)
G8	group of eight leading economies (now G7)
G20	group of central bank governors from twenty leading economies
GATT	General Agreement on Tariffs and Trade
GBD	Global Burden of Disease (a research study)
GDA	guideline daily amounts (measure of desirable nutrient intake)
GM	genetic modification
GMO	genetically modified organism
GPS	global positioning system (space-based satellite navigation systems)
GVC	global value chain
ha	hectare
HACCP	Hazards Analysis Critical Control Point
IARC	International Agency for Research on Cancer
ICN	International Conference on Nutrition (1992)
ICN2	International Conference on Nutrition (2014)
IFES	integrated food energy system
IFPRI	International Food Policy Research Institute
IMF	International Monetary Fund
IPCC	Intergovernmental Panel on Climate Change (UN advisors)
kcal	kilocalorie (nutritional energy)
kg	kilo (weight)
LFE	local food economy
m	million
MAI	Mediterranean Adequacy Index
MDG	Millennium Development Goals
MEA	Millennium ecosystems assessment (report to UN)
NAFTA	North American Free Trade Agreement
NCD	non-communicable disease
NGO	non-governmental organisation
NIDDM	non-insulin-dependent diabetes mellitus
NP	non-producing country
OECD	Organisation for Economic Cooperation and Development
OIE	World Organisation for Animal Health
PCB	polychlorinated biphenyl
POPs	persistent organic pollutants
PPP	public private partnership
R&D	research and development
SACN	Standing Advisory Committee on Nutrition (UK replaced COMA)
SDG	Sustainable Development Goals (of the UN)
SFSC	short food supply chain
SME	small- or medium-sized enterprise

SPS	Sanitary and Phytosanitary Standards (part of 1994 GATT set by Codex)
TBT	Technical Barriers to Trade (part of the 1994 GATT)
TNCs	transnational corporations
TPP	Trans-Pacific Partnership
TTIP	Transatlantic Trade and Investment Partnership
UK	United Kingdom
UN	United Nations
UNCED	United Nations Conference on Environment and Development (1992)
UNCTAD	United Nations Conference on Trade and Development
UNDESA	UN Department of Economic and Social Affairs
UNEP	United Nations Environment Programme
UNESCO	UN educational, scientific and cultural organisation
UNICEF	UN Children's Fund
US(A)	United States (of America)
USDA	United States Department of Agriculture
WCRF	World Cancer Research Fund
WHO	World Health Organisation (of the United Nations)
WHO-E	World Health Organisation Regional Office for Europe
WIPO	World Intellectual Property Organisation
WRAP	Waste Resources Action Programme (UK)
WTO	World Trade Organisation

Preface

In the years since the first edition of *Food Wars* was published in 2004, much has happened in the world of food policy. This new edition brings these developments up to date within our original analytical framework of competing paradigms or worldviews shaping the direction and decision-making within food politics and policy. Our key theme of the importance of analyzing whether human and environmental health are aligned with social justice and the food economy has become even more pressing.

In the first edition we set out a big theory about there being a clash of policy paradigms, and that the world of food was presently uncertain. An old order was clearly no longer up to the job, but there were tensions between competing narratives to replace it. The book set out to make sense of these different strands and to assess the implications of these emerging agendas and competing narratives. We also asked, and still ask, whether the evidence from science is clear about what policy makers ought to be doing. We think it is. Indeed, one motivation for this second edition is to incorporate new data and analyses that have come into the public sphere since we began to map our theory back in 2000.

This second edition is a completely rewritten book, while keeping the style and approach that our readers appreciated. We have updated examples, added vast amounts of new data, case studies and references. We, like our peer group, are increasingly interested in the tricky task of how to address cross-cutting issues. Do health and environment align with social imperatives and business models? How can policy makers ensure that actions taken for public health improvement do not conflict with environmental or social justice imperatives? If everyone ate like Americans or Europeans, what would that mean for food policy makers and global food business?

As we mused such questions for this second edition, we felt confident that our original core arguments remain pertinent. The food system *is* in trouble, but it is possible to make sense of this situation and to unpick the barriers to progress. Indeed, the arguments we made back in 2004 are, in some respects, even more necessary today. The evidence for system change has grown stronger since this century began.

Nutrition related problems, for instance, continue unabated, not least the global rise of obesity and non-communicable diseases (such as diabetes and heart disease), hand-in hand with continued widespread existence of hunger and malnutrition. These human tragedies remain largely unresolved, despite wide-ranging policy interventions and awareness of the costs of inaction. There is often a mix of policy drift and denial. It is sad that policy strategies are often ineffective, but this should be an incentive to clarify the arguments, analyse the reasons for failure and help spread public understanding of the need for change. The public – consumers, workers, scientists, media, i.e. all of us – is a crucial element in any mapping of change. History suggests that politicians and policy makers only act if they feel under pressure to do so. The public mood gives politicians legitimacy to facilitate change.

We took on the daunting task of rewriting the whole book, rather than just making small changes, because so much has happened in recent years yet so much has stayed the same! Environmental considerations have concentrated the minds of decision-makers, anxious about issues such as oil and commodity price volatility, the impact of climate change, and continuing geopolitical insecurity. Implementation and political will, alas, continue to lag behind the mounting evidence of the crises facing the future security of food supply.

We, like many academics and food policy analysts, are sometimes impatient at the slow progress being made to address real challenges in and on the food system. We have tried hard in this edition to capture the positive side of the situation too. From recognition, action can follow, and evidence can help build up pressure and focus minds. We have tried to highlight the many events and innovations being made around the world to tackle some of the food system's pressing problems. Attention to food matters rocketed throughout the world when global food commodity prices rose in 2007–08. Prices became volatile. Commerce became and remains worried. NGOs became vociferous. Connections between health, environment and an unequal world became legitimate high level policy debates. Politicians momentarily looked engaged.

In 2015, as this book goes to press, that welcome attention has subsided somewhat, though the data and realities suggest this should not be the case. In the world of food policy, serious questions are again being asked about whether governments really can set the required new frameworks needed for a complex 21st century food world. Even giant companies are frustrated sometimes. A gap has opened up between what needs to be done and political will. But, also, there is a gap between what consumers do and what they say they are prepared to do. Unpicking these tensions kept our motivation high, and it spurred us on in the re-writing task throughout 2014–15. Like many academics and analysts working in this area, we are convinced this is one of the biggest issues for humanity and policy makers to sort out.

We'd like to thank our many friends, colleagues and students who have given us so many ideas, feedback and support for the arguments in *Food Wars*. We thank also Tim Hardwick, Ashley Wright, Laurence Eastham and the team at Routledge. We thank all those who gave permissions to use figures and data. And finally we give heartfelt thanks to our families and friends who put up with us in the process, in particular to Liz Castledine, who put up with us over many weeks and weekends in London. We hope readers find this new edition useful.

Tim Lang and Michael Heasman
September 2015

1 Introduction

The food policy problem

Freedom from want of food, therefore, must mean making available for every citizen in every country sufficient of the right kind of food for health. If we are planning food for the people, no lower standard can be accepted.

Sir John Boyd Orr, writing in 1943; later the first Director-General of the Food and Agriculture Organisation of the United Nations (1880–1971)[1]

Core arguments

In this book we argue that humanity has reached a critical juncture in its relationship to food supply and food policy, and that both public and corporate policies are failing to adequately grasp the enormity of the challenge, let alone how to implement much needed change. From this perspective, food policy is in crisis, particularly over health and the environment. Both need to be addressed if society is to be well served. We argue the case for a new vision which links human health with the environment, often termed ecological public health. This cuts across the whole spectrum of the working of the global food system – from food culture and the way food business operates to agriculture and land use. The case for public and policy engagement to enable a secure and sustainable food supply for the future is pressing. Food policy needs to provide solutions to the worldwide burdens of diet-related ill health, food-related environmental damage and the social inequalities associated with these. At present, food policy is in a phase we term the 'Food Wars', with competing and sometimes contradictory analyses and solutions, and with different actors and vested interests putting forward divergent food policy agendas. These activities taken together represent a critical struggle over the future of food and the shaping of minds, markets and mouths. This will force us to change our relationship with what and how we eat and how the world's food is grown, processed and sold.

Food is an intimate part of people's daily lives. It is a biological necessity but it also shapes and is a vehicle for the way people interact with friends, family, work colleagues and ourselves. It is associated with pleasure, seduction, pain, power, sharing and caring. As people eat their daily food, bought in the shops that they know, buying brands that they are familiar with, it is hard to imagine that there is such a thing as a global food system, stretching from the local corner store to the giant food conglomerate, from the small farmer to the marketing mogul, and from the consumer to a web of business interactions. This book is partly about that, trying to bring to attention the complexity of the way food is produced and processed. We also consider its impact on long-term public health and well-being. In addition, there is overwhelming evidence about food's impact on the environment. The natural resources used to grow, process and distribute food are a major factor, for example, in climate change, energy and water use. And there is a strong social dimension to the Food Wars; social factors such as class, gender and culture shape and are influenced by the material and biological realities of food systems. As data on these issues has improved, policy-makers and increasingly the consuming public and food businesses have become aware of tensions over the shape of the food system, in relation to both human health and the pressures on the environmental resources needed to sustain our food supply. What seems an obviously good thing – food – has become problematic with layers of complexity making it hard to grasp an overview to tackle the global challenges ahead. In this book we set out a framing narrative based around three competing 'paradigms' to help business, policy-makers and civil society situate problems and solutions facing the global food system – this is set out in detail in Chapter 2.

Our particular interest here is in food policy: the decision-making that shapes the way the world of food operates and is controlled. In the past, food policy was often seen as the responsibility and terrain of governments. They set food policies, if they wanted. It was a 'top down' process. Today, however, that view of food policy is obsolete. While the role of government remains important, such as in negotiations involving international forums over trade agreements and setting standards, corporations also set food policies, both explicitly and implicitly, and may be more important in shaping food systems than governments. As we will argue later, food governance has become 'hollowed out' in its oversight of food systems. Civil society organisations, too, have policies on food. They champion their interests and voice them to get leverage over change, if they can. The modern view of food policy in this book, therefore, is that food policy is itself contested terrain, fought over and created through complex processes of stakeholder engagement.

The need for a 'big picture' perspective of global food system challenges moved into sharp focus for policy and business in 2007–2008 when, alongside the global financial crisis, the world experienced a food price crisis. This crisis was unusual in a number of respects. For example, while food commodity prices historically tend to rise and fall, in the 2007 food price crisis the prices of most of the world's major commodities – rice, wheat, maize (corn) and soy – in unison saw a sharp upward price spike. While food prices dipped and then rose more gently over subsequent years, they remain stubbornly high – creating costs and uncertainties for business and rising food prices for consumers.[2 3]

Added to this, the strong evidence for the environmental unsustainability of many parts of the global food system in the face of climate change or the overuse of natural resources such as the exploitation of global fisheries or of freshwater, has forced governments and business to become much more proactive.[4] They are responding to the threats global food markets face such as the ability for food markets to operate 'efficiently' or the inability to maintain 'food security' due to environmental degradation.[5] Global policy-makers also continue to be confronted with the double burden of disease – namely, persistent hunger and malnutrition at the same time as the global human health tragedy of escalating numbers of people experiencing non-communicable diseases (NCDs) such as diabetes, cardiovascular diseases (CVDs) and certain cancers, as well as the ongoing and spiralling rates of overweight and obesity in populations.[6]

We thus see the world of food policy as one experiencing ongoing conflicts. Consensus may be arrived at regarding the extent of the problems, but it should not be assumed there is agreement on solutions or even the nature or scope of the actions required. The principal argument of this book is that food policies are formed and fractured by these series of conflicts – what we term the 'Food Wars thesis' – structured around what we present as the three dominant narratives or 'paradigms' we explain in Chapter 2. These paradigmatic narratives offer different conceptions of the relationship between food, health and environment. They offer distinct and sometimes competing choices for public policy, the corporate sector and civil society. We argue that health in particular has often been somewhat marginalised in policy, although that is changing with rising obesity in particular, and that the Food Wars can be analysed, in part, as the manoeuvring or positioning by different interest groups seeking to influence the future of food.

Addressing what we regard as the linked challenges of diet-related health, environmental sustainability and corporate and civil society policy objectives will require better processes for making food policies and reform of the institutions of food governance; they need to be shaped in an integrated way. Unless this is done, we believe that food supply chains will lose public trust. If they are to achieve popular support and legitimacy, they will need to be infused with what we call 'food democracy', a notion we explore towards the end of the book. This goes beyond mere consumerism, the exercise of choice at the check-out or selecting between brands.

Our focus therefore is on the wider policy choices that shape how humanity orders its food economy and on urging public policy to play a positive role in promoting the public good. To this end, we explore key elements of the world of food that we consider to be crucial. These are:

- *health*: the relationships between diet, disease, nutrition and public health;
- *business*: the way food is produced and handled, from farm inputs to consumption;
- *consumer culture*: how, why and where people consume food;
- *society*: how food is framed by values, norms, roles and social divisions;

- *the environment*: the use and misuse of land, sea and other natural resources when producing food; and
- *food governance*: how the food economy is regulated and how food policy choices are made and implemented.

These issues are often studied in isolation, and at times deserve such micro-attention. But the scale of the pressures and challenges in the context of the global food supply now suggests that this 'compartmental' approach is no longer a viable way of handling food policy-making. The book calls for a new framework for making food policy choices. In the 2000s – particularly when a global commodity and food price crisis emerged in 2007–08 – minds were focused on the vulnerability of future food supplies. The developed world suddenly became worried that its security, not just Africa's or Asia's, was uncertain. This brought to wider attention a concern that had been growing among scientists and policy analysts around the food system's sustainability and whether the 20th-century legacy and definition of progress were still appropriate.

While today's food economy is grounded in a long history of production, experimentation and technological change, the industrial food supply is still relatively young in human history – a little more than 150 to 200 years old. Since World War II the food economy has undergone remarkable commercial and technological expansion in order to provide food for an unprecedented growth in human population – to more than 7 billion in the 2010s and forecast to reach 8 billion by 2020 and 9.6 billion by 2050.[7] Yet the commodity price crisis raised old questions about whether enough food can realistically be produced for 9 billion plus or so by mid-21st century. To put it starkly, could the world feed itself with this explosion of mouths at a time of planetary and societal stress and strain? Or feed it in a way and with diets (particularly high in meat and dairy) that came to dominate in the late 20th century? The means to address these questions have become highly contested, with technological solutions being promoted from some quarters, while others offer the argument that there is enough food but it is mal-distributed and/or wasted, and that plant-based diets would be beneficial for health and environment.

In the first edition of this book (published in 2004), it seemed radical to be arguing for a new integrated approach to food policy, one which took health, environment and the public interest equally seriously, and outlining a paradigm in which these goals were at the heart of future food systems. Yet as we produced the second edition of the book over a decade later, the assurance that the 20th-century model we call Productionism had worked perfectly was gone. The issue was now whether to unleash a new industrial food revolution or to go for a softer route. One thing is certain: the sustainability of food production systems and the quality of foodstuffs in the developed and developing worlds are being challenged as never before. The current food system appears to lurch from crisis to crisis: from earlier serious health scares such as BSE ('mad cow disease') to continuing major outbreaks of foodborne disease, from a world in which food corporations' power seemed unstoppable to one where their investments and sure-footedness could be questioned. In this new era, global food supply faces new challenges: a continuing surge in population growth in some parts of the world and an increasingly aged population in others; a global population where the majority now live in urban centres yet rely on food from the rural; the introduction of radical new technologies such as genetic modification and nanotechnology yet consumers saying they want authenticity and integrity; a new global scale and scope of corporate control and influence; a fragility in consumer trust in food governance and institutions; and persistent health problems associated with inadequate diet such as heart disease, obesity and diabetes which, alongside hunger and famine, affect hundreds of millions of people. This is a complex, troubling picture for food policy analysts to attempt to clarify, let alone politicians with their eyes on short-term issues.

This book argues that such challenges cannot be met in a piecemeal fashion. There has to be a clear commitment to framing food around public health and sustainability. Our concern is to make the links across discrete policy areas, from the way food is produced to the management of consumption and the healthiness of foodstuffs. We argue that the future viability of the food economy depends upon policy-makers articulating this necessity and delivering a new reality.

But is this easier said than done? Difficult questions loom. How does this new vision fit demands from developing countries? How can population health goals be reconciled with the way people want to live their lives? Can consumers realistically continue to expect ever cheaper food? Have consumers in the affluent world reached the limits of cheapness? Can consumers in the developing world expect the same trajectory of dietary change, or could they do things differently? What sort of intensification in production is best for human and environmental health? How can patterns of food trade benefit more people? What are the acceptable limits to the continuing concentration of market share by giant food companies? To what extent should public money support food production, if at all? How can changing consumption patterns be built around lower impact food systems? How does this all fit a divided world where rich and poor live and eat differently and have very different extremes of choices?

Why Food Wars?

Every day, millions of men, women and children are direct or indirect casualties of failures of food policy to deliver safe, nutritious and life-enhancing diets. People raised in the developed world since World War II may think that the damage is felt only in areas of the world that suffer famine, malnutrition or other deficiencies; but in the rich world too there is a huge toll. While Western societies have increased the caloric content of their diets and boosted the sheer quantity of food, they have at the same time introduced methods of production, distribution and consumption that threaten the future of the food system that delivers those calories. Even in developed countries there are defined problems associated with low income and deprivation. The argument has been made that, despite tens of thousands of food products being on hypermarket shelves, the production methods that delivered them have reduced the quality and nutritional value of some foodstuffs (such as the loss of essential bioactive components like vitamins and minerals). A policy assumption that health will follow wealth generation and sufficiency of supply now looks over-simplistic in the face of the coincidence of under-, over-, and mal-nutrition. The UN's World Health Organisation (WHO) and Food and Agriculture Organisation (FAO) suggest that there are around 0.9 billion hungry, 1.4 billion overweight (of whom 500 million are obese) and 2 billion mal-nourished.[8]

All around us, food culture is divided. On the one hand, the pursuit of cheap food through mass industrialisation has brought a greater range of foods to vast numbers of people. On the other, there are now questions about how much is needed and pressures on quality. Again, on the one hand, we have 'celebrity chefs' with top-rated TV shows, cookery and diet books on the best-seller lists, and popular media concerns about food quality, safety and availability. On the other hand, a crisis of food supply still dominates great tracts of the world. In the USA, one of the world's richest countries and superpowers, obesity has become a chronic problem. One estimate is that the US spent $190bn on obesity-related healthcare expenses in 2005, double previous estimates.[9]

This book is partly about these dichotomies: over- and under-consumption; over- and underproduction; over- and under-availability; intensification versus extensification; sustainable and unsustainable food systems; and hi-tech solutions versus traditional, culturally based ones, knowledge- and skill-rich food systems versus de-skilled and knowledge-poor ones; affluent world modes of eating versus simpler dietary patterns. The book summarises and discusses these.

In the Food Wars, there are numerous conflicts over the quality of food; food safety; nutrition; trade in foodstuffs; corporate control of food supply; food poverty and supply insecurity; the coexistence of the overfed and underfed; the unprecedented environmental damage from food production and the role and purpose of technology. How are organisations, policy-makers, businesses, farmers, non-governmental organisations (NGOs) and even individuals to tackle the enormity of the global and local challenges now confronting the food system? Despite apparent food abundance, the security of the food supply cannot be taken for granted.

We set out to write this book in the early 2000s because we were frustrated that key policy-makers who ought to be getting a grip on this complex world of policy appeared to be skirting around major problems rather than facing them, or too often dealing with the challenges separately in neat policy boxes rather than holistically. Academics were not helping much either; mostly they remained in discourses set by their academic disciplines, in silos, rather than helping create bigger pictures to offer to policy-makers. This silo mentality remains a challenge across the food system, not just within academia. Sectoral interests have their comfort zones, their reflex rationales for what they do. The food industry, and government, tend to continue to see the responsibility for food as lying with the individual consumer, and any 'liberal'-minded intervention in food supply, it is argued, is condescending: treating individuals as victims rather than intelligent food consumers. From that perspective, any leadership by government can be painted as a 'nanny-state' which has no right to tell people what they should or should not eat. Such an approach, we argue, ignores the realities and the scale of the crisis in food, health, environment and society, where food cultural rules have been turned upside down in a few decades; an 'eat anything anytime' world is environmentally and health irresponsible, says others. They see the challenges facing the world of food as beyond the scope of individuals or single companies to resolve. It will need clearer direction and frameworks either facilitated or created by democratically formed and accountable governments to address structural problems such as food-related inequalities, social injustice and problems of access, affordability and availability of food, and the unsustainability of local, national and global food systems. Also, to put responsibility onto individual consumers or families shows a profound misunderstanding of the power relationships shaping food supply. Collectively consumers might be more influential, of course, but consumers don't operate in concert.

Much of this book is our attempt to resolve the complex battles over what the 'food policy' problem really is and what to do about it.

Are radical options in food and health feasible or even possible?

The good news is that across the world of food policy there is a recognition that radical solutions are needed. We use the word 'radical' in its original Latin sense of getting to the roots of problems. In that respect the growth of big overview reports of the world's food systems since the first edition of this book are most helpful.[5 10 11] Distinct policy choices are emerging which will frame business and consumer opportunities. In the nutrition sciences, for example, a new 'ecological nutrition' is being developed along evolutionary principles – seeking diets that suit humans' evolutionary legacy. Such thinking is being offered as a radical way of linking food and health but the critics see this as excessively individualised and losing the public sphere. Personalised health also lends itself to the narrow 'technical fix' approach of many solutions being offered to feed the world. The revolution in 'life sciences', based upon an understanding that genes predispose people to diseases and that diet may trigger genetic predispositions – the so-called nutrigenomic understanding – could have profound implications for the 'personalisation' or 'individualisation' trend.

We argue that beneath the apparently calm surface of the 20th-century food supply chain (which, for all its scandals and monetary crises, has increased output and fed more people than ever before in human history) there are powerful undercurrents. Food businesses now know that they face structural uncertainties not least from climate change, energy and volatile markets. The food business world changed in 2007–8. A new uncertainty crept through the West's capitals and finance houses, and searches for new directions began to emerge.[12 13] The long drive down in food commodity prices appears to have stopped. Volatility became accepted as a possible new norm.[14] In this context, there will inevitably be winners and losers in the conventional business stakes. Even giant food companies can get into trouble – as happened in the UK when major retailers and branded food processors were revealed to have sold supposedly beef products (such as beef burgers) that had significant quantities of meat protein derived from horsemeat.[15] Some critics saw this scandal as an inevitable consequence of powerful players such as retailers squeezing their supply chains to lower costs in increasingly competitive markets. And when oil prices halved in 2014–15, this altered the carbon (energy) reduction strategies that had begun to be normalised following the oil price hike of 2007–8. Doing the right thing, like doing the wrong thing, can be a risky business. Certainties can become uncertainties, and then reverse or go fluid.

In this new, more volatile world, the map of consumer behaviour is also being redrawn. Most obviously, some rapidly developing countries' consumers have entered food markets and are making choices which follow patterns Westerners did before them, in a process known as the nutrition transition.[16] This is the new food and health problem for policy-makers in countries like China, India and Brazil. These countries, in turn are starting to redraw the geographies of food supply chains as their demand or supply for commodities such as meat or for animal feed are impacting global food flows to meet fast changing consumer demand. Apparently, this further proves the power of consumers in markets. Given higher incomes consumers will buy what markets offer them. Western food companies have not surprisingly rushed into developing markets with their products. But this has raised questions about how real consumer power is in markets. Are markets shaped by powerful marketing? What is meant by a 'market' in a globalizing world where influences, tastes and food products cross borders so rapidly? And what are real needs?

Recognizing the structural changes in the food world, some leading food magnates have seen nutrition as a potentially useful ally rather than threatening their interests. From the mid-1990s, nutrition science has been used (and abused) for new product development, food marketing and business strategy. In 2003, for example, the Chief Executive Officer of Nestlé, the planet's largest food company, stated that Nestlé aimed to become the world's leading nutrition and wellness company within five years.[17] In this respect nutrition itself is becoming increasingly commoditised – a process that has been referred to as 'nutritionism' whereby consumer perceptions of food quality and manufactured food products are increasingly being perceived as 'healthful' depending on the content of 'good' or 'bad' nutrients.[18] Yet where are the politicians to champion and articulate a public policy vision for food, nutrition and health to over-ride this commercial dominance? Indeed, what should such a vision look like? What should it include or exclude? How can it embrace the whole food chain, from growing food to final consumption? World conferences which ought to set out and debate the new visions such as the UN's 2014 second International Conference on Nutrition (ICN2), a recall of the first ICN in 1992 parallel to the Rio UN Conference on Environment and Development (UNCED), struggled to provide what was needed.[19 20]

Another radical conflict, in a polarised form, can be seen in the global battle between a 'GM' (genetically modified) future or an 'organic' future – with both camps making claims of enhanced health and environmental benefits as the rationale for their competing ways of carrying out agriculture. Through examples such as these we try in this book, where possible, to explore where there are 'radical' options in food and health policy and whether these are feasible, what their scope might be, and how our conceptual model of three competing paradigms helps to understand the choices being made, their consequences and the most appropriate ways forward.

In recent years there has been a wave of major international reports about food and about agriculture in particular. These share the deep concern about future uncertainties but vary in their solutions and approaches.[13 21-25] There have been many international meetings looking at future food policy, e.g. the United Nations in Rio de Janeiro at the Rio+20 in 2012,[26] or the ICN2 in 2014 mentioned above, or hosted by the G8 in 2008 and 2013,[26 27] or within global think-tanks.[13 28] But political change has been slower and messier. Despite the growth of interest in the broad area covered by this book, we are concerned about the timidity of formal political, public and policy engagement with the complex issues this book outlines. The stakes are high, of course, and some reluctance to face facts is understandable and human. But the excuses must stop. Like many others, we think it is time to wake up and 'face the music', not least to understand the tensions between different positions across commerce, states and civil society about possible and desirable futures for food systems. These are vying for dominance. To understand the Food Wars, we need to unpick what the motives and implications of the various positions are.

The case for a systemic rethink of the global food system – our motives back in the early 2000s for the first edition of this book, and today again in the mid-2010s, may be summarised thus·

- The model of food and agriculture put in place in the mid 20th century has been very successful in raising output and lowering prices but it has put quantity before quality and assumed expanding consumer choice would automatically improve public health.[29]
- Humanity has moved from an agricultural/rural to a hypermarket/urban food culture in a remarkably short time, a process that continues to roll out fast in the developing world.[30]
- While policy attention has traditionally been on agriculture, it is what happens *off* the farm in terms of processing, retail and food service that is in effect changing the shape and dynamics of the food economy.[31]
- Food has a deep impact on the environment, health and society. The 21st century will have to address this crisis of sustainability in the broadest sense of sustainability.[4 32]
- Throughout the world, diets are changing in ways that carry huge health implications and challenges.[16] This is in part due to trade liberalisation and in part to consumer aspirations; in this respect, there is both a 'push' and a 'pull' in the food system.
- Food, nutrition and health challenges are global. Countries like Brazil, India and China are already in the grip of a double burden of food-related disease: non-communicable diseases (such as heart disease, some cancers,

diabetes and obesity) take a heavy toll in all countries. Even sub-Saharan Africa has an obesity problem. At the same time mass hunger persists as a global problem.[33]

- There are limits to how far an individualised medical model of food and nutrition, which is becoming influential on policy-makers and investors, can resolve the scale and range of eating and health challenges, let alone diet's environmental impacts.[34]
- The environmental pressures on food production are reaching crisis scale.[35] This includes: over-fishing, damage to and loss of soil to grow food in, water stress and greenhouse gas emissions. The 20th-century food system achieved much of its success by 'mining' the environment, notably using oil, primarily for fertilisers.
- Consumer confidence and trust can be fragile.[36] Consumers are unclear about what a good diet is and experience competing messages about what to do, eat and responsibilities. This tension is coming to a head over how to define what a sustainable diet is.[37]
- The world has many diverse food cultures and traditions but these are being altered by globalisation of tastes, commercial reach and consumer opportunities via the internet and global media.[38] Assumptions about food are altering.
- Demographic pressures loom large. The 20th century managed to produce vastly more food (at a cost) but resources, land and eco-systems will be squeezed even harder by the anticipated extra 2 billion (bn) people by 2050. The 7 bn in 2000 will be 9bn by 2050 and possibly 11bn by 2100.[39]
- The food system suffers continual social injustice and inequalities. This includes: mal-distribution of food despite plentiful supplies, poor overall access to a good diet, inequities in the labour process, poor and even no wages in some sectors, poor working conditions, vast inequalities in health outcomes, and unfair returns for key suppliers along the food chain.[40]
- While few people disagree that the world faces a crisis of 'food (in)security', the term food security can mean different things. To some it is just a matter of raising production; to others it is a matter of access, affordability, utilisation and appropriateness, too.[41]
- An old tension about who has power and control over food has taken on new urgency, as already giant food corporations extend their global reach, influence culture, and shape how the commodified food system works. The development of intellectual property rights to seeds and brand imagery sits alongside very old concerns about ownership of land and the emergence of 'land grabs'.[42 43]
- New and unexpected risks and shocks to food systems can have major consequences on food for human consumption. Examples include subsidies for the production of biofuels at the expense of crops for food supply, the increasing amount of speculative capital potentially distorting commodity markets, or the prospect of new outbreaks of avian flu or other pandemics relating to animals and human health.[44]

An outline of the book

This is a book about ideas of how the future of food is to be shaped and conceived. We address this task by setting up a conceptual framework in Chapter 2 of three paradigms. There we discuss in detail the character of the Food Wars thesis, the assumptions of the paradigms that inform this book and what we mean by a paradigm. We argue that food policy often has a troubled relationship with evidence – sometimes lagging, sometimes leading it. How much evidence is needed to change policy?

Chapters 3 to 8 deal in detail with the evidence in support of our conceptual framework as it relates to health, food policy and the dynamics of the food system. We start by looking at the evidence of how the world's diet is changing and facing the problems of both under-consumption and over-consumption, often within the same countries. There is the mythology that the rich world suffers heart disease while the poor world suffers hunger. Diet-related diseases such as heart disease are becoming rapidly more prevalent in low- and medium-income countries. We show how diet- and health-related problems are growing in scale, not diminishing, as might be assumed with better food supply. Concerns such as obesity are dominating nutrition policy discourses, while equally troubling trends like the incredible rise in global cases of diabetes are proving very problematic to manage or contain.

Chapter 4 turns to another war zone: essentially a conflict over sustainability. By this we mean battles over quality, environmental impact and how to audit this. Our case is that the food supply chain is committed to producing a range of foodstuffs in environmentally unsustainable and wasteful ways that militate against human health. Today's food supply chain, while seemingly appropriate for the past, is now shown to damage and threaten the environment. Food, a means for life, is threatening its own continued production. Too many policy-makers still believe that they can merely 'bolt on' an eco-friendly niche market to the crisis of food and the environment. A reorientation of the entire food supply chain is needed if both human and environmental health are to be delivered.

In Chapter 5 we look at how public policy has responded to evidence about diet and disease. We give a short historical overview of changing conceptions of public health and the importance of nutrition, arguing that nutrition is a battleground between those who see it as framed by social objectives and those who believe that targeting only 'at risk' individuals is a more effective intervention. We review how governments have tended to rely upon health education as the mechanism for improving public health, setting dietary goals and offering guidelines which put responsibility upon individuals for their own health. We question this food policy strategy. We discuss how governments are wary of governing and even if they are not, the internationalisation of food trades means intergovernmental negotiations make negotiations complex.

The success or failure of food policy will be dependent on how it relates to the workings of the food economy and affects particular food business interests. In Chapter 6 we present an overview of what is meant by the food system/economy, arguing that, while consolidation and concentration of the power of the food industry is a long-running trend, the scale and pace of this change are new. The food industry is relying on a twin strategy to take it into the future: first, relying on technology and 'technical fixes' to resolve most problems; and second, aligning itself with the interests of consumers.

While most food companies today will describe themselves and their activities as 'consumer-led', we argue that this phrase is too superficial. We propose that a better grasp of food and consumer culture would help public policy analysts face what is happening in modern food markets. In Chapter 7 we map out what we see as the new consumer culture and landscape. Even at the basic market-led level, the rich-world consumer is developing a very different conception of food: convenience, snacking, ready meals, an eating-out culture and a food lifestyle that meet time constraints, and that recognise women's role in food and society.

In the commercial context, there is no respite in the tragedies continually hitting rural and farming communities. While farmers and the land are being squeezed, oligopolies from agribusiness to food processing, retailing and even food service dictate the workings of the food supply chain. We suggest that much public policy response to date has been at best reactive rather than proactive; and in many instances, NGOs and the business and scientific communities, albeit differently, have been more in tune with wider societal trends about food and health than policy-makers and government. But, as we argue in Chapters 8 and 9, future food and health choices must ultimately be resolved in public discourse: designing and reworking the institutional 'architecture' of food policy to deliver public goods is a pressing challenge. There are many public forums in which food is discussed and policies are made, but the problem is their integration and whether the collective picture is appropriate to the challenge of humanity and eco-systems. There is a crisis of institutions and of governance (that curious English word that refers to the science and practice of government) at all levels – local, national, regional and global. The processes of government are too often trapped in 'boxes' of responsibility, with no one retaining overall responsibility across the different compartments. We describe this as a long-term battle of food democracy versus food control.

Our objective in this book is to contribute to the debate and to suggest that there is already available a wide range of policy options and alternative voices. In Chapter 9 we argue that policy-makers too often assume that they have little choice and consequently discourage the alternatives. But we think that there is a new era of experimentation underway and through our 'paradigms' we show that there are different ways of assessing and making choices.

A new conception of health – linking human and ecological health – has to be at the heart of any food policy.[32 45] To deliver a healthy and environmentally appropriate pattern of consumption requires a reorientation of priorities. New knowledge and debate about food, health and the environment are needed across and within societies. Food is a common challenge for all humanity and must not be left to tiny elites deciding what is good for the rest of us. Light needs to be shone into some murky quarters of the food policy world. Globalisation has meant new high level policy processes which can lack accountability. International meetings of big food business and governments warrant more transparency. Even in those quarters, it is interesting to note the development of a range of alternative food scenarios. We detect the growth of 'insurance policies' against unforeseen crises. But ultimately, the public must be engaged too, not least to tackle the unacceptable legacy of disease, ill health and environmental damage. In that respect, there is a welcome burgeoning of vibrant civil society campaigns. This is sometimes seen as an emerging 'food movement' but whether that, too, is integrated and coherent remains to be seen, because it embraces a diversity of issues such as world hunger, corporate power, local food policies, disadvantaged communities, fair trade, diet and health, and much more. Will this gel to push for system change?

This book offers a panorama of the contemporary, complex world of food. Food continues to pose problems in public and corporate policy and, vice versa, public and corporate policy causes problems for the world of food. Crises can spark change, as they did over food safety in the 1990s and price spikes in the 2000s. But, in our view, the framework of public policy on food is too fragmented and piecemeal. Problems are addressed too often in an *ad hoc* or interim manner when what is required is a revised and systematic framework for addressing food policy, integrating core drivers such as health, business, environmental impact, consumer experience and policy management. There is recognition in some countries that this is needed but internationally it remains in embryonic stages and is subject to considerable conflict. There is some way to go in the Food Wars before there is Food Peace.

References

1   Boyd Orr SJ. *Food and the People*. London: Pilot Press, 1943.
2   OECD, FAO. *Agricultural Outlook 2011–20*. Paris and Rome: Organisation for Economic Cooperation and Development and Food and Agriculture Organisation, 2011.
3   Clapp J, Cohen M, eds. *The Global Food Crisis: Governance Challenges and Opportunities*. Waterloo, Canada: Wilfred Laurier University Press, 2009.
4   Sage C. *Environment and Food*. Abingdon: Routledge, 2012.
5   Foresight. *The Future of Food and Farming: Challenges and Choices for Global Sustainability*. Final Report. London: Government Office for Science, 2011: 211.
6   Chronic Diseases and Development. *The Lancet*, 2011: 377: 1438–1868. <http://www.thelancet.com/series/chronic-diseases-and-development>
7   UN Department of Economic and Social Affairs Population Division. *World Population Prospects: The 2012 Revision – Key Findings and Advance Tables*. New York: United Nations, 2013.
8   FAO. What we do. <http://www.fao.org/about/what-we-do/so1/en> Rome: Food and Agriculture Organisation, 2013.
9   Cawley J, Meyerhoefer C. The medical care costs of obesity: an instrumental variables approach. *Journal of Health Economics*, 31(1) (2012): 219–30.
10   Foresight. *Tackling Obesities: Future Choices—Project Report*. London: The Stationery Office, 2007.
11   Paillard S, Treyer S, Dorin B, eds. *Agrimonde: Scenarios and Challenges for Feeding the World in 2050*. Paris: Editions Quae, 2011.

12   World Economic Forum, Accenture. *More with Less: Scaling Sustainable Consumption with Resource Efficiency*. Geneva: World Economic Forum, 2012.

13   World Economic Forum, McKinsey & Co. *Realizing a New Vision for Agriculture: A Roadmap for Stakeholders*. Davos: World Economic Forum, 2010.

14   FAO. *The State of Food Insecurity in the World: How does International Price Volatility Affect Domestic Economies and Food Security?* Rome: Food and Agriculture Organisation, 2011.

15   NAO. *Food Safety and Authenticity in the Processed Meat Supply Chain.* <http://www.nao.org.uk/report/food-safety-and-authenticity-in-the-processed-meat-supply-chain> London: National Audit Office, 2013.

16   Popkin B. *The World is Fat: The Fads, Trends, Policies and Products that are Fattening the Human Race*. New York: Avery/Penguin, 2009.

17   Brabeck P. Quoted in Food Navigator, Nutrition and Health, 2003. <http://foodnavigator.com/news.news.asp?id=7700> [accessed May 2003].

18   Scrinis G. *Nutritionism: The Science and Politics of Dietary Advice*. New York: Columbia University Press, 2013.

19   ICN2. *Conference Outcome Document: Framework for Action – From Commitments to Action*. Rome: Food and Agriculture Organisation of the United Nations, 2014.

20   ICN2. *Conference Outcome Document: Rome Declaration on Nutrition*. Rome: Food and Agriculture Organisation, 2014.

21   IAASTD. *Global Report and Synthesis Report*. London: International Assessment of Agricultural Science and Technology Development Knowledge, 2008.

22   World Bank. *World Development Report 2008: Agriculture for Development*. Washington, DC: World Bank, 2007.

23   FAO, IFAD, IMF, *et al. Price Volatility in Food and Agricultural Markets: Policy Responses*. Rome: Food and Agriculture Organisation, IFAD, IMF, OECD, UNCTAD, WFP, the World Bank, the WTO, IFPRI, and the UN High Level Task Force (HLTF), 2011.

24   FAO, Bioversity International. *Final Document: International Scientific Symposium: Biodiversity and Sustainable Diets – United against Hunger. 3–5 November 2010, Rome.*   <http://www.eurofir.net/sites/default/files/9th%20IFDC/FAO_Symposium_final_121110.pdf> Rome: Food and Agriculture Organisation, 2010.

25   UNEP. *Avoiding Future Famines: Strengthening the Ecological Basis of Food Security through Sustainable Food Systems*. Nairobi: United Nations Environment Programme, 2012.

26   United Nations. *Report of the United Nations Conference on Sustainable Development, Rio de Janeiro, Brazil, 20–22 June 2012*. New York: United Nations, 2012.

27   G8. 'L'Aquila' Joint Statement on Global Food Security L'Aquila Food Security Initiative (AFSI), 10 July 2009. <http://www.g8italia2009.it/static/G8_Allegato/LAquila_Joint_Statement_on_Global_Food_Security%5B1%5D,0.pdf> Rome: G8 Leaders, 2009.

28   IFPRI. *Global Food Policy Report 2011*. Washington, DC: International Food Policy Research Institute, 2011.

29   Patel R. *Stuffed and Starved: Markets, Power and the Hidden Battle for the World Food System*. London: Portobello, 2007.

30   Steel C. *Hungry City: How Food Shapes our Lives*. London: Chatto & Windus, 2008.

31   Morgan K, Marsden T, Murdoch J. *Worlds of Food: Place, Power and Provenance in the Food Chain*. Oxford: Oxford University Press, 2006.

32   Lang T, Barling D, Caraher M. *Food Policy: Integrating Health, Environment and Society*. Oxford: Oxford University Press, 2009.

33   Alwin DA, ed. *Global Status Report on Noncommunicable Diseases 2010*. <http://www.who.int/nmh/publications/ncd_report2010/en> Geneva: World Health Organisation, 2011.

34   House of Lords Science and Technology Select Committee. *Behaviour Change: Report of the House of Lords Science and Technology Select Committee*. 2nd Report of Session 2010–12. London: The Stationery Office, 2011.

35   UNEP, Nellemann C, MacDevette M, *et al*. *The Environmental Food Crisis: The Environment's Role in Averting Future Food Crises*. A UNEP rapid response assessment. Arendal, Norway: United Nations Environment Programme/GRID-Arendal 2009.

36   Kjaernes U, Harvey M, Warde A. *Trust in Food: An Institutional and Comparative Analysis*. Basingstoke: Macmillan/Palgrave, 2007.

37   Lang T. Sustainable diets: Hairshirts or a better food future? *Development*, 57(4) (2014): 1–17.

38   Fernandez-Armesto F. *Food: A History*. London: Macmillan, 2001.

39   UN Population Division. *World Population Prospects: The 2012 Revision*. Washington, DC: United Nations, 2013.

40   Lamb H. *Fighting the Banana Wars and Other Fairtrade Battles*. London: Rider/Ebury, 2008.

41   Lang T, Barling D. Food security and food sustainability: reformulating the debate. *The Geographical Journal*, 178(4) (2012): 313–26.

42   GRAIN. *Seized! The 2008 Land Grab for Food and Financial Security*. Barcelona: GRAIN, 2008.

43   Pearce F. *The Land Grabbers: The New Fight over Who Owns the Earth*. London: Transworld Publishers, 2012.

44   O'Riordan T, Lenton T, eds. *Addressing Tipping Points for a Precarious Future*. Oxford: Oxford University Press/British Academy, 2013.

45   Rayner G, Lang T. *Ecological Public Health: Reshaping the Conditions for Good Health*. Abingdon: Routledge/Earthscan, 2012.

2   The Food Wars thesis

If you know before you look, you cannot see for knowing.

Sir Terry Frost RA (British artist 1915–2003)[1]

Core arguments

Different visions for the future of food are shaping the potential for how food will be produced and marketed. Inevitably, there are policy choices – for the state, the corporate sector and civil society. Human and environmental health need to be at the heart of these choices. Three broad conceptual frameworks or 'paradigms' compete to understand pressures on the future of food, and propose different ways forward for food policy, the food economy and how people live. The three paradigms each claim to deliver production to satisfy human needs, and to deliver health and other public benefits. The challenge for policy-makers is how to sift through the evidence and to give a fair hearing to a range of choices. This process is sometimes difficult because the relationship between evidence and policy is not what it seems. The world of food is on the cusp of a far-reaching transition. The three paradigms outlined in this chapter provide a way of understanding the challenges.

Introduction

The world is producing more food than ever to feed more mouths than ever.[2] For the better off there are more food and beverage product choices than it is possible to imagine. In the USA, for example, around 20,000 new food products are launched every year.[3] A US supermarket will have around 38,000 items on sale.[4] This is true in Europe and many rich societies across the world. It's a model of food choice which is being universalised. Yet for many people their food availability is very different, one of restricted range and choice. This vast disparity of experience is symbolised by the great differences in life expectancy. The Australian average is 82 years. Afghanistan's is 60. While there are lots of factors shaping life expectancy, food is one of them. How much we get, what quality, at what cost ... all affect our food intake. While this direct impact is well-known and evidenced – as is shown throughout this book – nevertheless throughout the world a general feeling of unease and mistrust about the future of our food supply began to be voiced in the late 20th century. Scandals about food quality and safety affected rich societies not just poorer ones. Food problems and threats became regular fodder for media coverage. Although the 20th century had witnessed remarkable success in producing more food, famine and hunger continued. In 2013, the FAO estimated that 12.5 per cent of the world's population (868 million people) are undernourished in terms of energy intake. 26 per cent of the world's children are stunted, 2 billion people suffer from one or more micronutrient deficiencies, 1.4 billion people are overweight, of whom 500 million are obese.[2 5] The optimism of the mid-20th-century food policy planners that, with good management and science, the problem of food shortages would disappear, has not been fulfilled. Food's capacity to cause problems has not lessened. It seems just to have got more complicated. Over-, under- and mal-consumption coexist, sometimes in the same country, town and families.

This book explores how new relationships are already apparent throughout the entire food supply chain, from the way the food is produced to its consumption. Alongside the mainstream model of the food economy symbolised by the supermarket, the big intensive farm, the mass product factory and vast consumer choice – alternative models are also mooted – of shorter food supply chains, localised systems, less processed foods. This contrast between mass and alternative food networks has been much studied by social scientists.[6] No wonder there are such arguments about food. The pace and scale of change witnessed in the 20th century engendered such reactions. New foods, new processes, new styles, all poured out onto the public. Quickly, however, studies from medical and other scientists began to cast doubt on whether simply having more products guaranteed good health. Arguments about quality, control over diet, and long-term impacts surfaced. This book explores those arguments, making the case that food is one of the most important and thorny issues facing policy-makers.

We see the world of food supply currently in the throes of yet another period of change, the latest in a series of long-term transitions from settled agriculture to a food world dominated by farming and agriculture to one where agribusiness and commodity producers vie with food service and retail giant companies to control access to consumption. Brands and brand loyalties are part of this battle for control. Identifying and studying these battles is why we describe the current food policy situation as one of Food Wars.

In this chapter we offer our core conceptual model, amplified and explored throughout the book. It can be put simply. We see food policy as a territory in which three main frameworks or 'paradigms' compete for dominance. The first is what we call the Productionist paradigm. This is historically the dominant of the three and has been, throughout the 20th century, the model of what is meant by food policy. It proposes that the purpose of policy is to produce more food and that doing this will deliver progress, health and well-being. More food keeps hunger at bay, lowers prices, makes food more affordable and is thus accessible. There is a problem, however. This model began to look a bit simplistic by the 1970s, as evidence that population health did not automatically follow from there being more food, and also from early evidence about food's environmental impact. And that is why the second paradigm began to be articulated. This is what we term a Life Sciences Integrated paradigm. This sees the scientific possibilities of genetics and molecular biology as the best way to deliver more food and thus health and to prevent negative consequences. Ever more sophisticated science and technology will unleash the productive capacity to feed ever more mouths better and with fewer costly and damaging inputs. Against this, however, we posit an Ecologically Integrated paradigm which articulates the view that food systems are inextricably bound to biological eco-systems and nature, and that a good food system maintains eco-systems and builds food culture around them. In this paradigm, good food policy is about balancing the interests of humans and eco-systems.

Before we explore these paradigms in more detail, we need to set out some basic assumptions about food policy and the food supply chain that informs our theory.

Food policy choices

Throughout this book, we use the terms 'food policy', 'food and health policy' and 'food and farming policy'. Since the turn of the millennium, there has been a welcome spate of books articulating the challenges facing food policy, some public facing,[7-9] others more academic.[6 10 11] Alongside these many books, there are countless heavy reports giving substance to the view that the food policy challenge – far from being resolved by Productionist or the Life Sciences Integration paradigms – is deepening and needs a radical overhaul.[12-15] In addition many non-governmental organisations (NGOs) have produced accounts of the food system in crisis whether viewed through the lens of social justice or environment.[16-18] But we retain here our simple definition of food policy as referring to those policies and policy-making processes that shape the outcome of the food supply chain, food culture and who eats what, when and how, and with what consequences. Our task here is to unravel the strands of competing interests, decision-making and policy objectives and strategies. In our view, it would be wrong to talk of there being one food policy or one food policy-maker, whether at local, national or global level. As an earlier analyst noted back in 1980, there is not one food policy but many food poli*cies* and policy-maker*s*; the combination shapes the overall dynamics of the food system.[19]

Food policy-making is essentially a social process, with actors pursuing goals. The shape of the food supply chain is the outcome of myriad decisions and actions from production to consumption; it can involve people and organisations who may not even call themselves policy-makers. For example, the food industry, when it sets specifications for food products, is in part determining the nutritional intake of consumers; healthcare planners, when facing the burgeoning costs of managing the rise of certain diseases (such as diabetes and some cancers) are making decisions that shape food policies, dealing with the results of how food is produced and consumed. Equally, competition authorities or town planners, when making decisions about retail market share or the siting of supermarkets, are determining issues as diverse as prices, access to food shops and local culture. The value of adopting this broad conception of food policy is that it helps to make sense of what otherwise can appear just a political 'bun-fight' or, worse, just an anarchic, disparate, inchoate jumble. It is true that food policy is contested terrain: a battle of interests, knowledge and beliefs. But it is quite possible to make sense of what appears a mess. Themes and continuities emerge: the fear of hunger, the desire to make food affordable, the problem of public health, competition for land.

The sort of food economy that exists is the result of a set of conscious policy choices made in the past, including both state and corporate decisions, involving funding for particular types of food production and processing, the setting of research priorities and national and strategic objectives, the provision of education and information, the creation of rules for trade and safe food and law enforcement and sanctions when things go wrong. And also a bundle of decisions *not* to do things.

To look at food policy as a set of processes and decisions to shape the food system requires us to look along the whole of the food supply chain. Even today, too many writers and experts see food policy as primarily a concern about agriculture and primary production. In a majority urbanised world, this cannot still be true, not least since power off the land frequently shapes what happens on the land.[20] Our approach requires us to look at food as the outcome of a system of relationships between sectors and interests. Figure 2.1 gives a simplified version of what is meant by the food system or food supply chain, a term originally promoted by agricultural economists who now use a different term – 'value chain' – to analyse how, from farm to consumer, raw commodities get value added to them. The important point to note is that analysis from a food value chain perspective assumes that change in one part of the chain, intentionally or not, has an impact on other parts. Increasingly, analysis from a food-chain perspective is used to understand trends and the global restructuring of the food supply.

Key characteristics of the food supply chain

The model of the food supply chain in Figure 2.1 allows us to note some key characteristics of the modern supply system. We can summarise these under four main arguments.

Pressures 'off the farm' dominate the food system

Although farming is still the biggest employer on the planet (with 1.4 billion people engaged), it is no longer the main power in the food system. In the past, agriculture dominated food policy thinking and still plays a significant role in international trade and budgetary debates, notably about the rights and wrongs of subsidies, and what to do about maintenance of food supply. The food supply

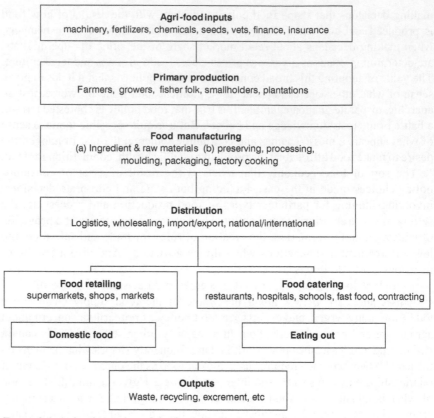

Figure 2.1 A simple version of the food supply chain

Source: Adapted from WHO

chain today is increasingly driven by forces away from the farm. In recent years, the 'financialisation' of food commodity markets has been seen as a worrying international issue, particularly with new volatility in food prices after decades of relative stability, despite price rises and drops.[21 22] Another hot topic has been 'land grabs' with pressure on developing countries to allow richer countries to buy or control land for food.[23] Pressures off the land are more important in framing the food economy than politicians often like to admit. Most money and value is extracted from consumers by economic sectors after the farm gate.[24] Today, the main drivers of the food supply chain are the powerful forces of processors, traders, retailers, caterers and financial speculators in agribusiness (as emerged after the 2007–8 price rise). This generates a paradox. Rhetoric in contemporary food policy focuses on the farm (especially the family farm in US Farm Bill politics or the smallholder in developing countries or the concern of large food manufacturers for their farm suppliers) when the reality is that power lies off the farm and money goes off the farm.

Control over consumption is a key to food system dynamics

Power in the modern food economy is increasingly driven by concerns about the consumption end of the food supply chain. There are critical fights between processors, retailers and caterers/foodservice over which of them is to make most money from capturing consumer tastes and purchasing.[25] Retailers increasingly offer their own brands in competition with existing manufacturers' brands. Most analysts agree, however, that in the late 20th century the food retailer became the power broker between primary producers and consumption;[26] they define the connections in the value chain depicted in Figure 2.1. In developed economies, and increasingly in developing countries, the retailer is the most powerful actor in the corridors of power and in shaping what consumers get access to eat. Individual consumers are diverse and usually unconscious of their collective influence; they can be badly organised and they carry most of the health costs of current food supply, yet they are often made responsible by politicians for their own diet-related (ill) health, using the argument that they are ultimately answerable for what they eat, a position fiercely contested by the consumer movement.

Public, state and corporate interests do not necessarily correspond

The pace of development and the structure of the food chain is being increasingly shaped by a limited number of very powerful food conglomerates. This power is manifest in how they set standards, logistics, planning, financial commitments, and contracts. While this has been an evolving process, consolidation in the food industry has now reached a new level of influence in key markets.[27] These corporate interests see food policy-making as part of their business strategy and are often well represented in the state policy arenas. This has led to the criticism that the state is not functioning as a protector of the consumer. In recent decades there has been a strong push towards self-regulatory models of governance such as public–private partnerships (PPPs) especially in relation to tackling diet-related ill health.[28] This raises a problem for food governance – the role of public democratic control, accountability and public responsibility – an issue raised throughout this book but particularly addressed in Chapter 8.

Health and eco-systems have been marginalised in the food economy but are now increasingly a threat

Although the food supply chain model in Figure 2.1 is a simplified description of the food economy, it can imply support for the view that human and environmental health are an outcome of the smooth running of the food supply chain. In fact, both public health and environmental protection have been seriously damaged by apparent progress in the 20th century.[29-31] Food is one of the major sources of environmental degradation and causes of morbidity and premature death – hunger in poorer countries and non-communicable diseases in the richer world. The role of the food system in shaping human and environmental health has become an increasingly fraught battleground in the food wars.[32 33]

As has already been intimated in the book so far, the simple version of the food supply chain (Figure 2.1) is in reality much more complex. The food chain draws upon inputs and resources from nature, culture and social mores transmitted through families, and wider economic drivers. How food ingredients and materials flow through the chain is also affected by shaping forces which surround the food chain. These include, on the one hand, deliberate policies and interventions (such as government policies, laws and regulations) and, on the other hand, social interventions from other industries (e.g. finance), science and research, civil society, and the consciousness industries (advertising, media, marketing). This dynamic mix has further consequences as the food chain creates impacts on health, the environment, culture and society. Figure 2.2 presents a schematic representation of this complex web of cross-cutting flows. At the top are the inputs to the food chain flow. On the right are shaping factors. On the left are at least four levels of food government. At the bottom are some key consequences of the dynamic mix. Even this figure over-simplifies what in reality may be even messier! But Figure 2.2 helps present the kind of perspective that this book proposes for modern food policy analysis. Food policy is no longer what it was from the mid-19th century to the mid-20th century. It is no longer dominated by the state with 'top-down' powers. Indeed much academic research has unpicked that simple view of government as governing. There always were blurred boundaries between government, industry and the public, but the power relations have shifted importantly with governments in the West subscribing to neo-liberal influenced ideologies which try to restrict the functions of government and the state. One consequence of this ideological change is uncertainty about responsibilities and control over public benefits.

The war of paradigms: time for a new framework?

Our concern in this book is to draw attention to the overall framework of food policy thinking. Our Food Wars thesis, and the paradigms that we offer and explore, come from our observation of a number of key conflict zones in the Food Wars. When evidence of poor diet's impact on population health is so strong,[34] why are policy-makers reluctant to place public health at the heart of how the food system is organised? When supposedly modern, efficient food production has such a massive impact on the environment,[29] why are consumers still largely kept in ignorance of the consequences of their personal choices? When food is so obviously mal-distributed on the planet, with billions over-eating while others under or mal-consume, why is there such a focus on ever increasing production? The 21st century, in many academic researchers' views, requires an overhaul, not least in what we want from policy-makers.

Throughout the book, we try to set out the evidence about the food system. Let us be clear, there have been many gains in the last two centuries – more people fed, greater varieties in diet, beneficial technical innovation. Nevertheless, big

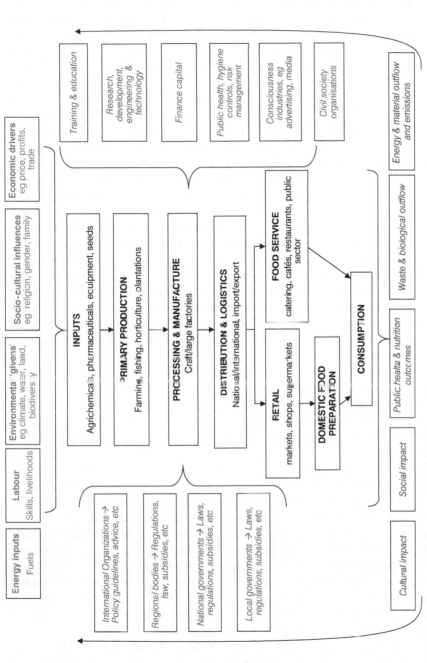

Figure 2.2 The food system, its inputs, outputs and influences

structural changes are almost certainly needed due to the consequences of past decisions and practices. While to propose the case for structural change used to be seen as politically outlandish, since the 2000s the case for change has become more central, helped by evidence of the enormity of food's impact on climate change, water use, waste, obesity, the persistence of hunger and food inequalities within not just between societies. This recognition increased in and after the 2007–8 banking and commodity crisis. The adherence to productivism (see below) – which marginalizes food policy as mostly about producing more food and enhancing market efficiencies – was shaken. The West's politicians felt the squeeze themselves. Prices rocketed, then fell, then rose, and now even proponents of 'business as usual' accept that price volatility is the new normality, and that the long drop in prices is probably over.[35] The argument that we and other academics had been making for some time that fundamental choices about food systems loomed became more mainstream. Reports emerged from governments and scientists suggesting ongoing food crises,[13 36-38] and contributed to the respectability of the argument proposed in this book and by many others that a paradigm shift is not only necessary but underway.

What is meant by this word 'paradigm'? A paradigm is a way of thinking, a set of assumptions from which new knowledge is generated, a way of seeing the world which shapes intellectual beliefs and actions. We use the term 'food paradigm' to indicate a set of shared understandings, common rules and ways of conceiving problems and solutions about food. A paradigm for us is an underlying, fundamental set of framing assumptions that shape the way a body of knowledge is thought of.

The take-up by policy-makers of the concept 'paradigm' is usually traced back to the work of Thomas Kuhn (1922–96), the US philosopher of science who first popularised the term.[39] In fact, he merely built on a concept spelled out by Ludwig Wittgenstein (1889–1951), the Austrian-born philosopher. Kuhn took Wittgenstein's concept of paradigms and applied it to science as a process of making ideas: a set of 'universally recognised scientific achievements that for a time provide model problems and solutions to a community of practitioners'. Kuhn was interested in how scientific understanding went through momentous crisis points and what determined why one accepted framework of thinking fell by the wayside while another triumphed in its place. For example, the work of Isaac Newton transformed how humans thought of the physical world; nearly three centuries later the new physics of relativity which Albert Einstein and others introduced created another paradigm shift which replaced the Newtonian worldview, transforming and superseding its tenets. Kuhn himself was said to have used the word paradigm with at least 21 different shades of meaning,[40] and academics and policy-makers today use the term 'paradigm' more fluidly or metaphorically than even Kuhn originally intended.

The food system that developed rapidly after World War II exemplified a way of thinking that we call the Productionist paradigm. This remains the dominant worldview in food policy, but it is, as we have already sketched, now under strain for not being able to address the depth and complexity of problems. Modifications

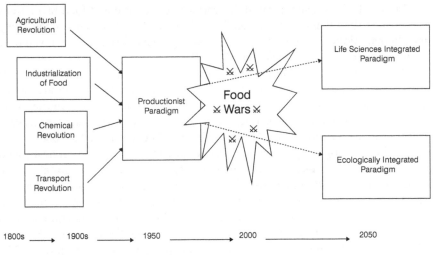

LEGEND:
✕ = Key battlegrounds in the Food Wars. These include:
• Diet, health and disease prevention
• Environmental crisis
• Capturing the consumer
• Controlling food supply
• What sort of food business
• Competing visions and ideologies

Figure 2.3 The era of the Food Wars

to the Productionist paradigm have constantly been and are being made. We argue here that there are also two emerging frameworks, which we call the Life Sciences Integrated paradigm and the Ecologically Integrated paradigm. These challenge some if not all of the basic precepts of the Productionist paradigm. Both these new paradigms are grounded in the science of biology, but each interprets biological and societal systems in ways that offer differing choices for the future of food: how food is produced, who produces it and how it is processed and sold; questions of social justice and fairness; the geographies and spaces for food production and consumption (global versus local sourcing); and not least the place of diet and food in public health. Figure 2.3 illustrates the competition and tensions between the three paradigms from the last two centuries to the present. The world is now in the era of fully fledged Food Wars.

The Productionist paradigm

Although the focus of this book is on modern food policy choices, and most of our framework is about relatively recent history – the last 200 years – and looking ahead at this century, it is essential to remember that the relationship of humans to food production is very old indeed. Most historians think that settled agriculture began around 10,000 or so years ago. The relationship between humans, the land, food and production, and its impact on health and the environment, can be seen as going through certain stages. There are different analyses of what those stages are. Table 2.1 gives a summary of the various agricultural and food revolutions over the last 10 millennia.[41] It should be noted that while this seems and is a long time, human biology and food-related physiology have themselves evolved over an even longer period, perhaps 400–500 millennia.[42]

While acknowledging the long view illustrated in Table 2.1, our interest here is about where policy-makers take humanity and food next. The Productionist paradigm has been all about the industrialisation of food and farming, and applying the principles and logic of industrial production to increase food output, in order to improve human welfare. Thus industrial thinking has been applied to control what is essentially biological processes of plants, growth, water, photosynthesis, land and nutrient flows. This has been called the attempt to 'refashion nature'.[43] The hi-tech approach to Productionism in farming now looks to 'precision farming' (see Table 2.1), for example, and to ever larger-scale and more micro-managed control systems, using robotics and GIS navigation systems, and computerised knowledge. It should be noted however that there are counter-arguments about the role of skills and knowledge. These are not the preserve just of the hi-tech. The 'low-tech + hi-skills' approach of agro-ecology, for example, and of low impact food systems suggest a different version of knowledge intensity.[6 44] There are also arguments for how, if public health is at the centre of production, the food system is extensively remodelled.[45-47]

The origins of the Productionist paradigm lie in the industrialisation of food over the last 200 years (see Figure 2.3). It drew upon advances in chemical, transport and agricultural technologies, as well as plant and animal breeding and husbandry techniques. Over this period food supply in many parts of the world has moved from often local, small-scale production to concentrated production and mass distribution of foodstuffs. Such a shift is a defining characteristic of the Productionist paradigm (even though it should be noted that much global food production is still local or regionally based).

With the arrival of industrialisation from the 19th century and the explosion of urban populations in the last two centuries, the social division of food became even more politically sensitive. Reliance on trade in agricultural commodities, such as spices and sugar,[48] already considerable for some foods, grew; pressures to intensify production accelerated, increasing rural poverty; increased output from the land reduced the actual labour required on the land; new energy sources and machinery replaced labour and animal haulage. The features of this slow agricultural revolution include: the increased use of inputs and of plant and animal breeding, the growth of fewer but larger farms, mechanisation and a reliance on fossil fuels.[49 50]

For us the Productionist paradigm goes far beyond the farm: it typified the whole 20th-century outlook – gaining serious policy support from the 1930s and 1940s onwards – in which the food supply chain became production-led in order to increase the quantity of food and to feed the growing opportunities for food processing. It developed a science base to further the goals of increasing output. Universities, colleges of agriculture, extension services and a panoply of support were gradually incorporated into this paradigm, which came to dominate food

Table 2.1 Periods of technical change and their impact on farming, culture and health

Era/revolution	Date	Impact on Farming	Culture	Implications for food-related health
Settled agriculture	From 8500 BCE on	Decline of hunter-gathering; greater control over food supply but new skills needed	Fixed human habitats; division between 'wild' and 'cultivated'	Risks of crop failures dependent on local conditions and cultivation and storage skills; diet entirely local and subject to self-reliance; food safety subject to herbal skills
Iron age	5000–6000 BCE	Tougher implements (ploughs, saws)	Emergence of technology; spread of artistic expression	New techniques for preparing food for domestic consumption (pots and pans); food still overwhelmingly local, but trade in some preservable foods (e.g. oil, spices)
Feudal and peasant agriculture (not in some regions, e.g. North America)	Variable, by region/ continent	Spread of enclosed land (parcelling up of formerly common land by private landowners); use of animals as motive power; marginalisation of nomadic practices	Division of labour; settlement around land-based production and village systems	Food insecurity subject to climate, wars, location; peasant uprisings against oppression and hunger
Industrial and agricultural revolution in Europe and US	Mid-18th century	Land enclosure; rotation systems; rural labour leaves for towns; emergence of mechanisation	Growth of towns; emergence of industrial working class with no access to land; rise of democratic demands	Transport and energy revolutions dramatically raise output and spread foods; improved range of foods available to more people; emergence of commodity trading on significant scale; emergence of industrial working-class diets

Continued

Table 2.1 continued

Era/revolution	Date	Impact on Farming	Culture	Implications for food-related health
Mendelian genetics	1860s; applied in early 20th century	Plant breeding gives new varieties with 'hybrid vigour'	Beginnings of biological science in everyday life, e.g. enzymes	Plant availability extends beyond original 'Vavilov' area; increased potential for variety in the diet, in turn increases chances of diet providing all essential nutrients for a healthy life
The oil era	20th century	Animal traction replaced by the tractor; spread of modern, intensive agricultural techniques	Car use and supermarkets rise; emergence of large-scale food processors; modern mass consumerist food culture and brands take off	Less land used to grow feed for animals as motive power; rise of impact of excess calorie intake leading to diet-related chronic diseases; discovery of vitamins stresses importance of micronutrients; increase in food trade gives ever wider food choice
Green Revolution in developing countries	1960s and after	Systematic plant breeding programs on key regional crops (rice, potatoes) to raise yields	Concentration of farming in larger holdings and more commercialised, intensive agriculture	Transition from underproduction to global surplus with continued mal-distribution; over-consumption continues to rise
Modern livestock revolution	1980s and after	Growth of meat consumption creates 'pull' in agriculture; increased use of cereals to produce meat	Rising incomes as more low-income countries achieve affluence; meat consumption rises (in meat-eating cultures); food suitable for humans (e.g. soya) is redirected to animals	Rise in meat consumption associated with Nutrition Transition; global evidence of simultaneous under-, over-, and mal-consumption; beginning of the end of the 1940s production-focused policy consensus that increased output will, if guided by science and if distributed fairly, end most food-related health problems.

| Precision farming | End of 20th century/start of 21st | New generation of industrial crops; emergence of 'biological era': crop protection, genetic modification, genomics. Application of off-farm technology eg nanotechnology, robotics, information technology, GPS. | Debate about drivers of progress, patent ownership; consumer information becomes central to management in 'risk society'. Possible disconnection of consumers from nature | Uncertain as yet; focus is upon output; debates about safety and human health impacts and whether these technologies will deliver food security gains to whole populations; unverified promises to provide technical solutions to degenerative diseases (e.g., nutrigenomics) coincide uncertainties about the environment and eco-systems (e.g. 'superweeds'). |

Source: adapted from Lang[41]

policy in the mid-20th century. Our analysis of the Productionist paradigm is that it was built not just upon the agricultural revolution from the 18th century onwards (and of the chemical and transport revolutions too), but on the capacity of food processors to invent, preserve, store and distribute food en masse, and of investment in distribution.[51 52]

The triumph of the Productionist paradigm was cemented in the experience of mid-20th-century starvation, food shortages, and mal-distribution which affected many countries in the 1930s.[53 54] Throughout the world, governments began to accept the case for (if not reality) of new national and international policies designed to increase production by applying large-scale industrial techniques. World War II was the defining moment that witnessed the triumph of Productionist thinking for post-war reconstruction. The overarching goal of the paradigm was to increase output and efficiencies of labour and capital for increasingly urbanised populations.

Now, over half a century on from the consolidation of the Productionist paradigm, it is under strain and showing up major limitations. Although Productionism has been successful in raising output in line with an unprecedented rise in world population from 3 billion in the 1950s to nigh 7 billion at the end of the century, 1.9 to 2.2 billion people are estimated to remain untouched by the full impact of modern agricultural technology.[55] Moreover health and environmental strains now threaten the continuation of the Productionist model. It is too crude just to produce more food. Modern policy-making is troubled by matters such as oil security, climate change, water depletion, soil pollution as well as public concerns for example about animal welfare and the nature of plant and animal breeding.[56 57] There is also a serious battle over power and ownership, not just in terms of companies (which we discuss in Chapter 6) but also of the intellectual property and even the genetic basis of foods.[58]

To achieve its objectives, industrialised production historically focused on monocultures (single crops in a field) rather than diversity, but this created a reliance on artificial inputs (agrichemicals such as herbicides, pesticides and fertilisers) and energy-intensive engineering both on and off the farm. The future sustainability and commercial viability of the Productionist paradigm, its suitability as a business model, not just its profitability, is now far from certain, with agribusiness and politicians, as well as markets and consumers, now questioning how food is produced.[59 60]

In this book, the term 'productionism' is used to convey the big picture we have been drawing above. In the academic literature, a not dissimilar term 'productivism' is used.[61-63] These two terms are different. Productivism is used broadly to indicate the pursuit of capital and labour efficiency on farms, a pursuit of efficiency in primary production. 'Productive farmers are seen as better farmers with greater yields displaying the virtuousness of hard work.'[63: 262] We use productionism to go wider and to refer to the entire food system being focused on production per se. It is a policy to deliver societal goals such as health and food security, not just developing the efficiency of individual farm units. The 1940s Productionist thinkers' goal was more than just technical. Theirs was a clear social vision to adddress public health failures exposed by industrialisation and the boom-bust of whole economies.[64 65] Productionism has been a progressive force, whereas the academic literature on productivism presents it as a narrower, purely economic perspective.

Two new paradigms vying to replace Productionism

The Productionist paradigm, as has already suggested above, came under attack for its over-use of inputs (such as agrichemicals, veterinary drugs), environmental damage,[56] and for facilitating cheapened industrial ingredients for processed foods (such as high fructose corn syrup, artificial additives, and trans fatty acids).[66] As the evidence for these criticisms emerged, first in the 1970s and increasingly thereafter, two strong alternative paradigms emerged, competing for the future of food.

Both claim human and environmental health benefits. Both are grounded in new understanding of the science of biology, one genetics in particular, the other ecology. Both claim to be science-informed and both seek to transform the Productionist paradigm and change the business model. One is what we call the Life Sciences Integrated paradigm and the other the Ecologically Integrated paradigm. Figure 2.2 situates both paradigms historically in the context of the Food Wars. Both derive from a common root: the argument that the 21st century will be the century of biology, seen through a highly technical lens to be a 'bio+techno' century, in one case, and 'Bio+ecology' in the other.[67] 'Bio' is sometimes represented as the driver of innovation. If the 20th century was characterised by the emergence of post-industrialisation and 'information', then the 21st century is said to be the age of biological science. The battle is on, however, for which version of biology triumphs. One sees this as reducing to genetics and mining the life sciences. The other emphasises working within and for eco-systems. There are radical differences between the bio visions. One seeks control and improvement of nature, while the other vision seeks to minimise the use of such chemicals and to 'work with nature'. The word 'ecology' (or the shorthand 'eco') is being bent to each purpose.[68] The battle over bio has already created controversies, notably over genetically modified crops and cloning. Languages in many tongues are being forced to inject new 'bio'-words into their lexicon: there is now bioprocessing, bioprospecting, bioprivacy, bioextinction, biodiversty, bioscience, bioinformatics, biovigilance, biosafety, bioterrorism, biosecurity and, of course, biotechnologies.

The Life Sciences Integrated paradigm

The Life Sciences Integrated paradigm describes the rapidly emerging scientific framework that is heralding the application of new biological technologies to food production. We propose this paradigm as a way of capturing a body of thought that has as its core a mechanistic and fairly medicalised interpretation of human and environmental health. In this, food is perceived as almost like a drug, a solution to diseased conditions, part of a planned, personalised, controllable and systemic manipulation of the determinants of health and ill health. This highly sophisticated thinking about food and health is at the heart of the application of biotechnology to food production, and its application on an industrial scale is at the core of the Life Sciences Integrated paradigm. Techniques in biotechnology are already delivering many advances in food production methods, food handling and consumer products.

We should stress here that this paradigm means more than genetic modification (GM) alone, and includes the whole spectrum of biotechnology: that is, the use and manipulation of living materials in the manufacture and processing of foodstuffs. Enzymes, for example, are a key processing aid within biotechnology that receive only negligible publicity or adverse publicity.[69] It is GM that has become the central defining image of the Life Sciences Integrated paradigm and the focus of media, consumer, and policy attention. GM seeds and the chemical inputs they require are reshaping the biological base of agricultural production at a speed that is unprecedented in human food production. Despite the relatively crude state of the technology, GM has been introduced into world food systems at a rate that some see as irreversible. The long-term implications for agricultural environments, and for the structure and power relationships in the food chain, are unknown.

The novelty of the science – it can mean taking the genes from unrelated species and inserting them into another to forge a new plant or animal that would not be possible in nature (known technically as 'recombinant DNA biotechnology') – represents a revolutionary technological shift. This has changed the economics of whole input industries already, particularly the seeds and agrichemicals sectors which have consolidated.[70] One of the attractions of the paradigm is that in many respects it relies on a simple reinterpretation of the existing Productionist paradigm but claims to remedy a number of its limitations: from lessening environmental impacts, through improving human health from greater food production, to creating new products with enhanced, yet sometimes contested, health benefits.

From an agricultural point of view, the spread of commercialised GM crops has been extensive in certain crops – cotton, maize and soya – and claims are made that it heralds the new Green Revolution.[71] As a result, the Life Sciences Integrated paradigm is well placed to become the dominant paradigm of the early 21st century. Plantings of GM crops have risen from zero in the mid-1990s to 1.7 million hectares in 1996 to 170 million hectares by 2012, according to industry sources.[72] The USA is the main country adopting GM, followed by Brazil.

Nutrigenomics is another line of research being pursued within the emerging Life Sciences Integrated paradigm and, to some extent, it typifies what that paradigm offers. Nutrigenomics seeks to understand how nutrition and particular dietary intakes interact with the structure and expression of genes and with genetic pre-potential. Why can one person eat a diet high in fats and not get cancer or heart disease, when another cannot?[73 74] If it were possible to unravel the interaction of genes, diet, ingredients and lifestyle, the promise of delivering an individualised or personalised approach to food and health might be realised. Nutrigenomics is the application of this new understanding of genes and how they operate in plants, animals and microorganisms. Researchers try to unravel the mechanisms involved – which foods and which ingredients have an impact on which genes and which diseases. This would have been inconceivable without the completion of the mapping of the human genome by US and European geneticists.

Nutrigenomics promises a targeted fix to the diet and health policy problem, based on an acceptance that both micronutrients and macronutrients alter the metabolic programming of cells, and on an understanding of how diet is a key factor in disease.[75] The commercial as well as academic search is on for bio-active ingredients which could be exploited for health. Anticipating this area to grow, one US biotech company working in this area, stated that:

> by being able to elucidate genetic profiles of individuals, diets will be formulated from crop to fork to confer prevention or retard disease progression. As basic science advances converge with e-commerce, new opportunities will emerge to deliver to consumers, whose genetic susceptibility to specific diets and diseases are known, products tailored to individual dietary needs.[76]

Although researchers are attracting funds to this work, realizing health gains is probably some time off. Even if nutrigenomics does yield more precise understanding of the diet–gene–health connection, many observers consider that existing population dietary advice still stands. Even if some people are more likely to trigger degenerative diseases from eating a particular balance of nutrients, the population as a whole would benefit from attaining already known dietary goals such as restricting consumption of saturated and total fats and increasing intake of vitamins and trace elements from fruit and vegetables.

Nutrigenomics, argue the sceptics, may offer commercial wealth by selling to the 'worried well' and rich consumers, but it is probably of little relevance to global public health. Already concerns have been raised. One concern is that big business logic is outstripping the public's ability to make informed choices. Other ethical dilemmas relate to the problem of privacy and cost. Nevertheless, nutrigenomics suits the more individualised policy approach of looking after one's own health. It says little about the need to alter the environment that reduces the chance of whole populations taking exercise or consuming a wholesome diet.

Nanotechnology is another avenue of hi-tech science feeding into the food industry. Nanotechnology is the manipulation of matter at the atomic molecular level and its application in food for defined purposes. Nano refers to scale rather than process. A nanometre (nm) is one billionth of a metre. A human hair is about 80,000 nm thick. Nanotechnology has been heavily invested in by science and the food industry to create new materials which can be included in foods.[77] It is being introduced, as the US FDA phrased it, 'to allow scientists to work on the scale of molecules to create, explore, and manipulate the biological and material worlds measured in nanometers'.[78] It is being used, for example, in packaging to improve barrier properties, better temperature performance and for thinner films.[79] Nanotechnology is already being applied right across the spectrum of food technology in agriculture, processing, packaging and food supplements. Examples include:[80]

- Agriculture: nanocapsules to deliver pesticides and other agrichemicals, growth hormones and vaccines; nanochips to identify preservation and tracking;
- Food processing: nanocapsules to improve bio-availability of nutriceuticals for flavour enhancement; nanoparticles to bind and remove chemicals or pathogens;
- Food packaging: antibodies attached to fluorescent nanoparticles to detect chemicals of foodborne pathogens; lighter, stronger and more heat-resistant films with silicate nanoparticles;
- Food supplements: nano-sized powders to increase absorption of nutrients.

The quiet introduction of nanotechnology has been viewed with some concern by some civil society watchdogs, with one respected Canadian NGO calling for a moratorium and far wider public consultation and regulatory oversight.[81]

An ongoing trend in food processing within the Life Sciences Integrated paradigm is what has been termed 'substitutionism' by Goodman and Redclift.[43] They described the tendency within food manufacturing to find substitutes for 'inorganic' ingredients for the 'organic' – a classic example being the replacement of sugar (an 'organic' ingredient) with chemically derived 'artificial' sweeteners. There have been ongoing attempts over the past three decades to find such magic bullet substitutes for many ingredients from 'fat substitutes' (a technology that largely failed) to a range of food processing additives or ingredients. As technologies evolve there is continuing investment interest in finding the next 'magic bullet' such as mayonnaise without eggs or burgers without meat or 'liquid meals' rather than a 'real' dinner. These are all real examples from business start-ups in the early 2000s and part of their promotion is that these technologies address sustainability issues, not least are less wasteful on the planet's resources and cut out the exploitation of animals.

One the most high-profile examples was the media frenzy around the cooking and eating of the first lab-grown burger in August 2013.[82] The burger was 'grown' in a laboratory in the Netherlands using the cells derived from a cow. Professor Mark Post of Maastricht University is the scientist behind the technology but interestingly it was Sergey Brin, the co-founder of Google, who put up the cash – $330,000 – to enable the research. But this is not the only Silicon Valley entrepreneur behind the next generation of scientists looking to change the food we eat. An article in the *Financial Times* in October 2013 focused on three such start-ups in Silicon Valley. One of these, Impossible Foods, is working on making meat 'in a better way' using biomass rather than lab-grown 'meat' as in the Netherlands. Investors were reported as already putting in $75m into the company – but Impossible Foods is not alone, other meat substitute start-ups include Beyond Meat and New York-based Modern Meadow.[83]

Another company mentioned was Hampton Creek which is working on an ingredient that works as a substitute for eggs based on a variety of yellow pea, while Soylent is a company trying to replace meals with a liquid-based product. All these hi-tech companies situate their activities in terms of helping to produce a small food environmental footprint and about providing healthier consumer choices.[83]

The Life Sciences Integrated paradigm is being supported by considerable investment, mostly private but also public, and while the US was the initial home of much research, it has spread throughout the OECD member states. This is mostly investment for the health market with much less for agriculture and food and beverage. In Australia, 18 per cent of firms operating in biotechnology were food, agriculture and beverage related, whereas in Belgium the figure was 23 per cent and Canada 27 per cent.[84]

The Ecologically Integrated paradigm

Since the 1930s, there has been persistent criticism of industrial agribusiness which argues that agriculture and food supply are crucially integrated to the goals of improving nutrition and health; and that the Productionist paradigm distorts that connection by putting the focus on a particular version of agricultural production.[85] Ecological integration suggests that for public problems such as diet-related ill health and environmental damage to be prevented, a holistic perspective is required. One cannot resolve obesity, for example, by a drug or surgery. Obesity is the result of a distorted relationship between humans, food culture, food availability, pricing and supply. Environments need to be changed to right wrongs. In a world where people use cars to be mobile, their food needs ought to go down but instead eating continues, and at a population level, exhortation to take more exercise can fall on deaf ears. Obesity is systemic problem, thus a new set of relationships between factors needs to be orchestrated. That's the meaning of ecological integration – getting the factors to work in a common direction.

There are many examples of Ecological Integration thinking emerging in policy such as: agroforestry (where tree growing enhances food production at ground level), agroecology (see below), organic food systems, and some low impact, resource-conserving short food supply chains. Key factors in Ecologically Integrated thinking are a high level of social inclusiveness, engaging with small-scale producers and a general commitment to low-impact living and food systems. Sustainability is seen through a filter that combines social, environmental and economic criteria. Knowledge is broad-based, inter-disciplinary and combines scientific insights with indigenous knowledge. In theory, ecological integrated knowledge aspires to be more participatory rather than hierarchical.[86]

The Ecologically Integrated paradigm is grounded firmly in the science of biology, but takes an integrative and less industrial approach to nature. Its core assumption recognises mutual dependencies, symbiotic relationships and the complexity of interactions. It aims to preserve ecological diversity, taking a more holistic view of health and society than the 'medicalised' Life Sciences paradigm. Agroecology is one example of farming within the Ecologically Integrated paradigm.[87] Certainly, agroecology has gained support particularly among specialists working with farmers in the developing world who lack financial capital but have rich social capital. For them, ecological integration is not just about growing food within environmental limits but actively seeking to harness the protective and synergistic capacities of complex eco-systems to increase and improve food resilience.[12] The world's poor farmers and citizens facing food crisis can rely only on self-reliance and small-scale farming; for them agroecological technologies offer one of the few viable alternatives. This is because a guiding principle of the Ecologically Integrated paradigm is that diverse natural communities are productive and should be supported.[88] A hurdle to overcome in this respect is the specificity of regional eco-systems and the need for specialist local knowledge. This paradigm therefore contrasts with the homogeneous technological packages characteristic of both the Productionist paradigm and the Life Sciences Integrated paradigm, relying upon bio-pesticides technologies to combat insect pests and develop resistant plant varieties and crop rotations; on microbial antagonists to combat plant pathogens and produce better rotations; and on cover cropping to suppress weeds, replacing synthetic fertilisers with bio-fertilisers. There is an increasing emphasis on skills and knowledge management in contrast to the single technician managing thousands of hectares on a 'recipe' basis; it would relink the people with the land, encourage small-scale management units and return alienated farm workers to the land.

Agroecology is emerging as the discipline that provides the basic ecological principles of the study, design and management of agro-ecosystems with a view to productivity conserving natural resources, and to systems that are culturally sensitive, socially just and economically viable. Such technologies include organic matter accumulation, nutrient cycling, soil biological activity, natural control mechanisms (disease suppression, biocontrol of insects and weeds), resource conservation and regeneration (to include soil, water and germplasm), general enhancement of agrobiodiversity and synergisms between components. The agroecology model, however, has yet to demonstrate its widespread applicability, not only in the developing, but also in the developed world.

The Ecologically Integrated paradigm is being particularly influential in 'grassroots' movements and at the local and sub-national level of food policy, where there is more urgent attention to the need to fuse social, environmental and health goals through food policy. In the Western world, there is a vibrant sustainable food cities network which broadly subscribes and there are dozens of food policy councils, charters and community-based food initiatives. Often these are positioned as responses to perceived negativities of the industrialised food system,[89] with local attempts better to integrate diverse policy goals.

These 'bottom up' approaches and interests in ecological thinking are now accompanied by an interest in Ecologically Integrated thinking from – perhaps surprisingly to some – big companies and think-tanks. These became increasingly concerned about the fragility of current food systems, not least the risks to their supply chains, from eco-systems failures. Climate change and water shortage, let alone soil or fishery depletion, threaten some food companies directly. In the 2010s, following the business shock of the commodity price spike, a steady flow of reports and corporate commitments emerged suggesting that some big food and drink companies now took the environment seriously within their supply chain management. PepsiCo UK, for instance, made a commitment in 2010 to reduce its greenhouse gas emissions by 50 per cent in five years, achieving 27 per cent in four years, no mean feat.[90] Unilever, another of the world's food giants, introduced a Sustainable Living Plan in 2010.[91] The World Economic Forum, known as a big business forum, published a McKinsey-written Roadmap to sustainable agriculture.[59] This came a decade after some of the global giant food companies joined forces to create a Sustainable Agriculture Initiative.[92] Such projects suggest that Ecologically Integrative thinking is now being taken more seriously across the food system. Indeed, some of the same companies are investing or 'backing' both opposing emerging paradigms, hedging their bets and/ or trying to take advantage from advances in both. The backing of large-scale organic food production has been noted, with critics arguing that organics as an ecologically integrated approach is being subverted by going large-scale and industrial.[93]

The three paradigms summarised

The three paradigms, although explored throughout this book, are so central to our thinking and arguments that we now provide a summary of key features of each: Table 2.2 for the Productionist paradigm, Table 2.3 for the Life Sciences paradigm, and Table 2.4 for the Ecologically Integrated paradigm. All three paradigms share some features. Although we characterise the Productionist paradigm as more focused on quantity than quality, in fact all paradigms take a position on quality but approach it differently. Similarly, all paradigms assume some kind of market economy. Throughout the food supply chain, adherents of each of the paradigms know the importance of macro-economic frameworks such as the rules for trade and the need to deliver food safety.

It should also be noted that the paradigms can be interpreted differently in political terms: early proponents of the Ecologically Integrated paradigm, for instance, were linked to both far Right and more democratic Left political movements, the former interpreting 'nature' in authoritarian terms and arguing that hierarchies and top-down rules are essential to allow intrinsic values to be asserted.[85] The left position, in contrast, argued that ecology needed champions to protect it from the depredations of advanced capitalism.[85] Although Tables 2.2, 2.3 and 2.4 highlight differences, it should be noted that all the paradigms claim a strong health orientation; the role of the food supply chain is to deliver health and satisfy needs. But as the rest of this book explores, what is meant by health, and what the determinants of health are, are matters of debate.

Table 2.2 Features of the Productionist paradigm

Drivers	Commitment to raise output; immediate gains sought through intensification
Key food sector	Commodity markets; high-input agriculture; mass processing for mass markets
Industry approach	Homogeneous products; pursuit of quantity and productivity (throughput); quality defined mostly in cosmetic terms
Scientific focus	Chemistry + pharmaceuticals (antibiotics) + traditional plant breeding
Policy framework	Largely set by agriculture ministries; reliance on subsidies
Consumer focus	Cheapness; appearance of food; homogeneous products; convenience for women; assumes safety of foods
Market focus	Global and national markets; emergence of consumer choice; shift to branding
Environmental assumptions	Cheap energy for inputs and transport; limitless resources natural resources; monoculture; externalisation of waste/pollution
Political support	Historically strong but declining; grounded in landed interests; battles over subsidies
Role of knowledge	Agroeconomists as important as scientists; the State as gatekeeper
Health approach	Health gains assumed to follow from sufficiency of supply and lower prices
Ownership	Technocratic and landed élite

Which will dominate?

As we have suggested already, there are supporters of each paradigm, even raw Productionism. Some believe the Productionist paradigm is still fit for purpose with modifications – a 'business-as-usual plus' approach. Others recognise the weight of evidence that significant changes are imminent. Moreover, within each emerging paradigm, there can be variations in emphasis. For example, aspects of the Life Sciences Integration paradigm are being presented as a new Green Revolution to feed the world with some emphasising the 'technological fixes' on offer from within the bio-sciences,[94] while others (in our view rightly) emphasise the need for technical change to be accompanied by societal improvements.[71] Within the Ecologically Integrated paradigm, too, while there are some who take a 'deep green' approach to the future, for instance arguing that deep ecology necessitates a drastic reduction in human population, there are others who see it as essentially also a vision of social change.[95 96] Interestingly, subscribers to both emerging paradigms are critical of – yet claim the mantle of – the 1970s Green Revolution.[97 98]

Some present the emerging paradigms as favoured by big business (Life Sciences Integrated) versus small business (Ecologically Integrated). There is a glimmer of truth here but the reality is more blurred with small and big business to be found in both camps. It would be wrong to dub the Ecologically Integrated paradigm as reactionary, anti-science or anti-big business: it offers a particular

Table 2.3 Features of the Life Sciences Integrated paradigm

Drivers	Capital-intensive use of Life Sciences (agrofood); commodity production; tight managerial control; mass scale
Key food sector	Commodity traders, food retailers, processors and foodservice vie for domination of supply chains; rise of logistics
Industry approach	Hi-tech; industrial-scale application of biotechnology primarily in agriculture but increasingly in manufacturing (enzymes not just GM); sophisticated use of mass media to shape food markets
Scientific focus	Engineering at molecular level to link genetics, biology, engineering, nutrition; control from laboratory to field and factory; science presented as neutral but tailored by industry-led/oriented funding; big data; farm management technologies such as drones
Policy framework	Big Science expertise but nervousness about consumer reactions; blurred regulatory and policy responsibilities between State and companies
Consumer focus	Consumer sovereignty rhetoric; language of choice; personalised appeal
Market focus	Global ambitions; large companies dominate;
Environmental assumptions	intensive use of biological inputs; claims to deliver environmental and health benefits
Political support	Dominant position in R&D; divisions among rich and poor countries about how to interpret Life Sciences paradigm
Role of knowledge	top-down; expert-led; hi-tech skills; laboratory science base
Health approach	Centres on maintaining mass food output, but recognises new health problems from overconsumption; thinks health can be technical fixed preferably by an individualised basis; seeks to improve beneficial traits of crops for human health
Ownership	Highly capitalised

view of science, business and consumption, aiming to prove it can be 'big' in the sense of ensuring a viable food economy in developed world markets.[99] Indeed, there is a strong line of reports which presents this as a rounded case.[100] In contrast, one of the strengths of the Life Sciences Integrated paradigm is that it has immense influence in the corridors of power and also among the decision-makers in many large food companies. It also builds on the structures of the Productionist paradigm which has proven remarkably resilient in food output and winning consumer support. Meanwhile, the Ecologically Integrated paradigm, once marginal in mainstream food business and portrayed as quirky or backward-looking, is now winning some support. An example is how large food companies now recognise the carbon footprint of meat, for example, or the fragility of fish stocks. So the image of the Ecologically Integrated paradigm is changing. It has some real consumer appeal – integrity, eco-friendliness, etc. – but can be seen as expensive and elitist.

We see a mix of polarisation and fusion ahead, with a general consensus that the world faces huge health, environmental and development challenges, but

Table 2.4 Features of the Ecologically Integrated paradigm

Drivers	Integrative; health at heart of food system; environmental, energy & waste impact reduction; resource conservation; diversity on and off the field; eco-systems resilience
Key food sector	Whole-chain systems approach (from land to consumer); sub-national and regionalised food economies
Industry approach	Traditional; shorter food supply chains; authenticity; minimal processing; select use of biotechnology (fermentation, not GM)
Scientific focus	Interdisciplinary; ecological integration; social and eco-systems resilience
Policy framework	Partnership of ministries; collaborative institutional structures; promotes advantages of decentralisation and team-work
Consumer focus	Citizens not consumers; improved links between the land and consumption; greater transparency
Market focus	Regional and local focus – 'bio-regionalism'; nervous about export-led agriculture; favours smaller companies but increasingly adopted by larger ones
Environmental assumptions	Resources are finite; need to move away from extensive monoculture and reliance on fossil fuels; need to integrate environmental, nature and conservation policy with industrial and social policy
Political support	Weak but growing; strengthening in some countries; some merging of social and land-based movements
Role of knowledge	Knowledge-intensive, rather than input-intensive; skills needed across whole supply chain; knowledge as empowerment
Health approach	Ecological public health approach; promotes diet diversity
Ownership	Varied with some community rhetoric; mix of 'old' landed interests and new businesses

at the same time considerable effort to keep the Western food model on track. At a conceptual level, we predict continued political differences between the paradigms, although companies and consumers might be 'promiscuous'. A key battleground is already the regulatory arena – which products can get approved, with what scientific evidence and credibility? Who is to make the judgements – state, companies or consumers? There is also a battle ahead for access to public monies and political credibility to support the development of different paradigms.

One scenario is that there will not be a period of mutual tolerance between the paradigms but an era of serious conflict, with proponents seeing little middle ground. If the Life Sciences Integrated paradigm becomes well ensconced in the corridors of power, the Ecologically Integrated paradigm may be forced into more strident opposition to win public acceptance (Chapters 5 and 7 explore the drivers of actual behaviour more closely), and vice versa.

The place of food and health in the 'paradigm' framework

We now turn to one of the salient features of the Food Wars – the battle over public health and diet's role within that. In the Productionist paradigm (Figure 2.4), health is portrayed as being enhanced, above all, by increasing production, which required investment in both monetary and scientific terms. Agriculture, the prophets of Productionism argued, deserved massive support if it was to move away from small-scale, low-yield systems.[101] This, incidentally, was their rationale for the now much-derided subsidy system throughout the West: farmers needed support if they were to raise production. As long as food could be adequately and equitably distributed, health benefits would surely result. This Productionist view of health saw the main problems as under-consumption, under-production, poor distribution and waste (mostly on or near farm). The health goal of public policy, therefore, should be to increase production of key health-enhancing ingredients such as milk, meat, wheat, and other 'big' agricultural commodities.[102] Figure 2.4 shows how this policy relationship connects inputs and outputs in health.

The health assumptions on which the Productionist paradigm was built were based on what today would be regarded as a very narrow understanding of nutrition and health. For example, the observation in the 1800s that animal protein aided human growth led to massive resources in countries such as the US and Europe being invested in the development of the dairy and meat industries.[103] The agricultural and agribusiness focus of the Productionist paradigm has also been weakened by the shift of power and finance down the food supply chain to the retailing, trade, food service and other consumer industries. It's towards that end of supply chains that most food money is made (a feature spelled out in Chapter 6). In the US, for example, about half of all food expenditure is on consumption outside the home, with other countries (like the UK) also heading this way.

This change in consumption patterns suggests just how out of touch the Productionist paradigm is with health needs. The explosion of diet-related non-communicable diseases – many of which are heavily affected by food – is the proof, and known since the 1990s what is more.[104 105] Public policy responses from within the Productionist paradigm look increasingly like rearguard actions, when just half a century ago it promoted a proactive approach – policy intervention, new initiatives and new ways of farming. Today, the paradigm is increasingly reliant on doctor-prescribed drugs to combat disease effects (e.g. statins), or foods specially marketed as having health benefits, and fringe crops to try to gain a foothold in 'health consumer' markets, from pomegranate juice to quinoa. No longer does it drive and structure the food supply chain in a way that makes sense for health. (We expand on this theory in Chapter 5.) Even in developing countries, the paradigm has problems, as diet-related diseases have taken hold.[106] The Green Revolution, while delivering more energy-rich macronutrients, has at the same time exacerbated rates of maternal anaemia and childhood deficiencies in iron, zinc and betacarotene because the higher-yielding strains of wheat and rice contain relatively fewer micronutrients.[42]

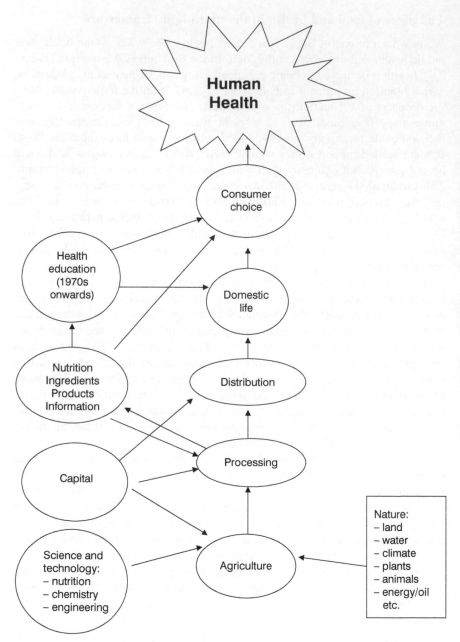

Figure 2.4 Productionist approach to health (1950s to present, with 'health education' included post 1970s)

The World Bank and International Monetary Fund policies supported the production of cash crops in order to increase income.[107] Health education was bolted on in the face of growing evidence about the impact of diet on cardiovascular diseases.[108 109] Rather than rethink the paradigm, governments and corporate interests remained wedded to dietary guidance and public advice.[110] In other words, the policy was to educate consumers to eat more sensibly and to look after themselves. They were bombarded with leaflets and public education programmes, with only mixed results. Other areas of health education, such as nutritional labelling, became hotly contested areas within food policy, rather than being a channel of information to alter the relationship between supply and demand.[111] The paradigm's policy solutions thus emphasised the responsibility of individuals to implement self-change. Nutrition became an increasingly politically charged issue (this is discussed further in Chapter 5).[112 113]

The Productionist paradigm encouraged a reductionist view of the relationship between food and health: that food's contribution to health comes from the 'right' ingestion of the 'correct' balance of ingredients. There is no good or bad food, according to this ethos, only good or bad diets. The onus of responsibility for diet-related health is thereby put on consumers – mainly through what they were expected to read and interpret from food labels. Health was defined as the absence of disease. If consumers wanted to improve their life-expectancy chances, they should eat healthy foods. The old Roman statement of consumer responsibility – *caveat emptor*: 'buyer beware' – prevailed.

The majority of public health opinion takes issue with this position. As we show in Chapter 3, much of diet-related health, like public health itself, is socially determined, particularly by social inequalities and status, which shape life chances.[114-116] Although we have many reservations about the Productionist model – and note too that the failure to ensure equitable distribution of food is not entirely its fault – it has been spectacularly successful in its own key terms. Outputs and total yields have increased, albeit more in some developed areas of the world due to land ownership and macro-economic policies.[117 118] A major success of Productionism, as world population has increased (from 3 billion in 1950s to 7 billion by 2010), is that it has produced increasing amount of food to enable the percentage of the world's population in chronic hunger to decrease from 19 per cent in 1990–92 to 12 per cent in 2011.[5] The absolute numbers have remained stubbornly high (around 1 billion), but the proportion of the total population has dropped.

Why then do we believe the Productionist paradigm has run its course? Producing ever-more food is not the answer to the range of food-related challenges ahead. The current era of the Food Wars (see Figure 2.3) is characterised by heavy doubts about future sustainability as well as being dogged by diet-related health burdens. Evidence mounts as to the financial costs to the taxpayer, with one study calculating that poor diet accounts for 13 per cent of all European Union healthcare costs.[119 120] Within our paradigm-based analysis, it is important to note that neither emerging paradigm replicates the elitist bias of the Productionist paradigm which favours wealthy individuals and developed nations over their poorer counterparts.

The Life Sciences and Ecologically Integrated paradigms' approaches to health

Increasing the food supply is still a critical policy concern, under almost all scenarios. One argument, however, is that if the rich world stopped over-consuming, if less food was wasted, or if animal production stopped relying on grains (nearly half of all grains grown on the planet are fed to animals), the world could feed everyone reasonably. This is an equity-led policy approach. The Life Sciences paradigm's approach to health, however, has a focus on individualised

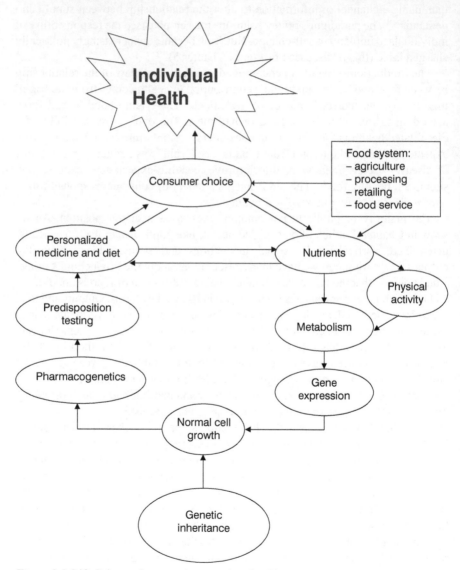

Figure 2.5 Life Sciences Integrated approach to health

health (Figure 2.5). It offers an updated almost medical model of health in that it promises the capacity to understand the constituent parts of disease and the human's capacity to fall prey to particular diseases, and then offers long-term personalised dietary solutions, such as nutrigenomics, implying a highly sophisticated understanding of the minutiae of the biological and genetic 'cogs' in the human 'machine'. The Life Sciences approach disaggregates the complexities of the food–disease–health nexus into discrete parts and offers food or food-derived ingredients as potential aids: health is delivered by science.

A fundamental difference between the Productionist and the Life Sciences paradigms on one hand, and the Ecologically Integrated paradigm (Figure 2.6) on the other hand, is that the first two paradigms conceive of health as an outcome of a long linear process (the food chain or bio-food chain), while the latter conceives of health as something that is an outcome of even more complex interactions including within society. The Life Sciences paradigm pursues a model of health that uses sophisticated science to disaggregate foodstuffs and nutrients to seek their particular characteristics which may then be recombined in new products and new technologically controlled health strategies.

The health approach of the Ecologically Integrated paradigm, as the name implies, is centred on ecology: understanding the interactions of systems and

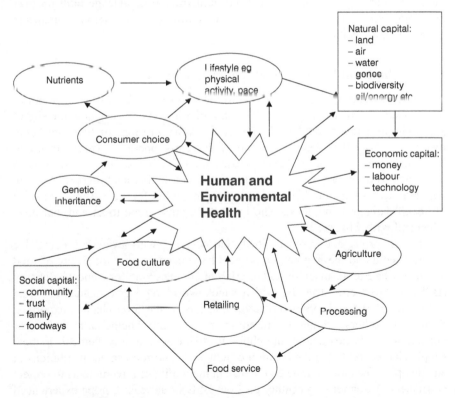

Figure 2.6 Ecologically Integrated approach to health

cycles that are characteristic of biological systems in nature. The emphasis is on process – notably feedback loops, cycles, symbiosis and interconnections. It proposes that the goal of food policy should be to understand these processes and to work with them, rather than to engineer, constrict or fragment them. The Ecological Integrated approach sees monoculture, whether in the field or in diet, as anathema, whereas the other two paradigms see monoculture as a matter of business reality and efficiency; specialism and economies of scale being the route to successful enterprise and output.

Both emerging paradigms claim to deliver environmental and health benefits that they say are not being delivered by the Productionist paradigm: reduced chemical inputs, foods with better nutrient profiles, and food security.[12 13] Both believe that the Productionist approach will be unable to deliver enough food for burgeoning world populations, or not without major environmental shocks.

Where the paradigms differ, however, is on how to deliver the increase in food and its assumed benefits to population health. The Life Sciences paradigm is espoused by the agrichemical and pharmaceutical end of the food system while the Ecologically Integrated paradigm looks to learn from and modernise more traditional farming knowledge. Both embrace concepts of intellectual property, the Life Sciences through the patenting of genetic materials, the Ecologically Integrated paradigm by building on and modernizing knowledge built up over many generations. Both paradigms argue that its approach will deliver human and environmental health.

Ending the Food Wars through policy and evidence

In Figure 2.3, the depiction of the era of the Food Wars features the symbol of the crossed swords used by historians and map-makers to designate the site of a battlefield. The war analogy is apt: unnecessary millions of lives are being lost by the way that the food economy currently operates. The wars might be orchestrated sometimes behind closed doors in boardrooms as well as in parliaments and intergovernmental discourse, but the battles for supremacy are very real. Lives hang on their outcome. People get hurt in the clashes. The scale of health damage is immense. The Food Wars must come to an end, but how, when and with what result?

In wars, propaganda prevails. In recent decades, with awareness rising due to new media and the internet, consumers have become only too aware of food scandals. The European Union had its crisis of bovine spongiform encephalopathy (BSE) or 'mad cow disease', which shook world food governance and trust in the late 1980s and 1990s. Powerful interests in the food chain – exporters, farmers and others – fought to present the risks as either being under control or not serious.[121] It took a crisis in public trust to clean up murky elements in meat supply chains and led to the creation of new laws and food safety institutions in an attempt to recapture public confidence in the ability of governments to protect consumers. A quarter of a century on from BSE's outbreak, Europe experienced another meat scandal in 2013, this time over unlabelled illegal use of horsemeat in processed foods. Horse was passed off as beef. Again, scandal engendered consumer cynicism and jokes but exposed lax standards.[122]

The rise of antibiotic/antimicrobial resistance is another example of intensification leading to health damage. Antibiotics have been one of Productionism's greatest aids. Routine use of antibiotics in intensive animal production – particularly poultry and pigs – was a key factor in spreading antimicrobial resistance – recognised as early as 1956[123] – and putting one of medicine's greatest life-savers onto the threatened lists.[124 125] Despite this being long-known, the food industry and farming resisted controls.

Capturing the consumer

A healthy diet – such as one low in saturated fats (from dairy and meats), sugars and refined carbohydrates, and high in fruits and vegetables and a diversity of foods – also has benefits on environmental grounds.[47 126] The trend towards the Westernised diet with its high content of meat and dairy, and larger land use, and a reliance on a narrow range of crops is warping and distorting eco-systems.[29 127] Neo-liberal economics professes that markets give consumers primacy. Consumers are sovereign and said to be the keystone of how and why market economies deliver efficiencies. Yet in practice consumer organisations complain that consumers tend to be under-informed, are sometimes patronised and heavily targeted by marketing and sponsorship.[128] High calorific foods such as burgers and soft drinks are relatively cheap; labelling does not compensate for price signals; health is distorted.[129] For example, nutritional labelling has been a contested issue between consumers, government and the food industry for decades. Choices – particularly by children – are easily influenced by heavy advertising and cause-related marketing.[130 131] Image is everything.

The focus on the consumption end of the food chain has led to a reinvention of industries dedicated to the marketing and branding of foodstuffs. Major food brands mostly emanating from the West have consolidated their position and are being promoted worldwide, entering Eastern and Southern markets and changing dietary patterns.[132-135] However, in developed countries, health has become a marketing battleground with increasing awareness of the links between diet and disease in a demographically ageing population. Consumers are demanding products with health benefits, a demand which the food industry now attempts to satisfy.

The key issue here for policy is whether a proliferation of particular products with presumed health benefits is the right solution to diet-related ill health, as is often claimed by the food producers. After decades of being highly resistant to the evidence that our food has an impact on ill health,[103 112] large sectors of the food industry now, ironically, see health as a potential growth area. At first sight, this is welcome but there is a real danger that this will either be a short-lived technical fix or merely create a niche – 'health' – in already saturated markets.

Another fundamental concern is what sort of food culture is emerging. Many food traditions and common eating behaviours are altering rapidly in the face of new products, marketing and lifestyles. There is rampant 'burgerisation', for one thing – the domination of US-style fast food, often washed down by soft drinks, across the whole developed world, and increasingly spreading throughout the newly industrializing countries such as China. By the 2000s, Americans spent more per annum on fast food than on higher education, personal computers, computer software or new cars.[136] Marketing departments of food companies often celebrate their role in offering consumers an astonishing range of choice. But this choice comes at considerable environmental cost – long-distance trucking, excessive energy use, monocropping on intensive farms, packaging and waste (see Chapter 4) – as well as the health costs of obesity, diabetes and other degenerative diseases.

Evidence-based policy?

Why is policy not being changed and what is stopping such change? Why has the Productionist paradigm not responded, given increasing weight of evidence as to its shortcomings? In policy analysis, there is often talk of 'evidence-based policies' as the gold standard. But in reality, policies and processes have a problematic relationship with evidence. Political expediency or ideology may be more important. Political will is essential to get change in line with evidence. If policy were based on evidence we would see, for example, immediate action utilizing all available levers to reduce the incidence of non-communicable diseases in line with world data.[137-139] Instead, there have been decades of delay and obfuscation. The diet-related epidemics of heart disease, cancers, obesity and diabetes have spread to the developing world, even to low-income countries. Of deaths from all chronic diseases, 72 per cent occur in low and lower-middle income countries.[105] Obesity has rocketed worldwide (see Chapter 3).

The *evidence* about environmental damage is similarly overwhelming.[29 32 140] Besides the deep data on climate change, there are other environmental impacts associated with food production and the spread of Westernised diets. These include water stress, land use, soil degradation, pollution, waste, energy depletion, and biodiversity loss. Cumulatively, this now alarms growing numbers of scientists. [100 141 142] These data and its implications are summarised later (see Chapter 4).

We see an important tension between policy, practice and evidence across the food system. In medicine, practice aspires to be based solely on solid evidence. This so-called Cochrane gold-standard approach argues that any intervention – and healthcare too – should be based on an excellent quality of evidence. The Cochrane Collaboration is a network of many review groups around the world, made up of health specialists committed to systematic reviews of the effects of healthcare intervention. Such reviews should be rigorous, and based on peer-reviewed journals (rather than the 'grey' literature that has not gone through the anonymous process of peer review), and take note of first-rate evidence, ideally from randomised, double blind studies.[143] There are good grounds for arguing that a Cochrane-type approach could be applied to public health nutrition interventions.[144] The ideal relationship between policy and evidence requires an increase in the availability of evidence to which policy-makers should listen.

In practice, the relationship between food policy and evidence falls into a number of discrete possibilities which include the following:

- Policy with evidence
- Policy without evidence
- Policy despite evidence
- Policy burying evidence
- Policy claiming evidence
- Evidence searching for a policy
- Evidence but no policy
- Evidence despite or in the face of policy
- Evidence in line with policy

In our view there is enough evidence to state that the post-World War II Productionist paradigm of food and farming is no longer credible to meet future needs. It might have worked in the past but evidence today is both broader and stronger than simple Productionism can address.[33]

What is missing is a more innovative and organised health lobby to pursue and ensure a change in policy-making in relation to food and health. One of the challenges made in this book and by other academics is that food policy requires breadth of evidence from many sources and bodies of knowledge which are too often kept away from policy-making about food in general and health in particular. Figure 2.7 illustrates the breadth of thinking and policy input now required for decent decision-making processes and for optimum management and delivery of food policy. The ministries which affect food supply include health, trade, environment, agriculture and food, fisheries, consumer protection, development, foreign affairs and industry, but rarely is there a constructive dialogue across these ministries. Hence there has been a growing debate within policy circles about whether the 21st century needs

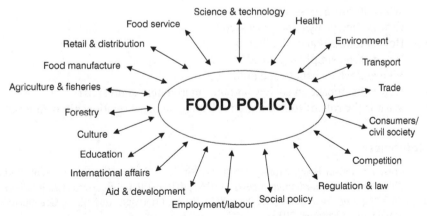

Figure 2.7 The food policy web

to be serviced by food policy councils at different levels of governance to integrate evidence and involve stakeholders, and to come up with feasible solutions.

Too often public policy-making is *ad hoc* or gets distracted by special pleas and particularities when what is needed is some long-term strategic thinking. Ironically, commercial interests do tend to take much longer-term strategic views. Governments come and go, and work to short-term horizons. The mismatch between public and commercial policy is particularly worrying. It is now time for reform. Long-term health must be the key driver of any local, national, regional and global food supply chain (see Chapters 8 and 9).

One thing is already certain. Food policy is moving from a prime concern with raising productivity at any cost to addressing sustainable production and consumption. There is a real danger that the Food Wars may themselves become the dominant ethos of the age, rather than the pursuit of resolution – a new food peace. Food is highly sensitive and can be a weapon of control. The public health, environmental, social justice, ethical and consumers movements – those conglomerations of professions, institutions, organisations and members actively pursuing their interests – have begun to be engaged in the debate about the future of food and the future direction of food policy. New arguments and a new preparedness to stand up against the easy ideological route of individual choice and market forces are being voiced. This is one good result from the alarm at the rise of obesity and overweight, and the growing evidence about environmental degradation, and the shocking exposes of animal mistreatment, and this is the increasing awareness that these matter on a global scale. There is a willingness to consider that different bodies of evidence have tensions, but also that there are overlaps and consistencies.[145] We see this as the pursuit of what is called ecological public health as applied to food systems.[146]

To conclude, in this chapter we have set out our main thesis about the paradigms and have begun to show how it links to big questions on the future of food. These include:

- Who eats and what?
- Who controls access to food?
- How is food grown and processed?
- How is it traded and distributed?
- How is food supply regulated?
- Who and what shapes food policies?
- How is its impact on society and the environment addressed?
- What is the role of food culture in shaping human's relationship with food?

References

1   Frost T. cited in Brian Morley, Obituary: Sir Terry Frost. *Guardian*. 3 Sept. 2003. <http://www.theguardian.com/news/2003/sep/03/guardianobituaries.artsobituaries1>

2   FAO. *State of Food and Agriculture 2013*. Rome: Food and Agriculture Organisation of the United Nations, 2013.

3   USDA ERS. New products. <http://www.ers.usda.gov/topics/food-markets-prices/
    processing-marketing/new-products.aspx#.UbyNEpy67QA>   Washington,   DC:
    Economic Research Service of the US Dept of Agriculture, 2013.
4   Food Marketing Institute (USA). *Supermarket Facts: Industry Overview 2010.*
    Arlington, VA: Food Marketing Institute, 2012.
5   FAO. *State of Food Insecurity 2012.* Rome: Food and Agriculture Organisation,
    2012.
6   Goodman D, DuPuis EM, Goodman MK. *Alternative Food Networks: Knowledge,
    Place and Politics.* London: Routledge, 2012.
7   Pollan M. *The Omnivore's Dilemma: The Search for the Perfect Meal in a Fast-
    Food World.* London: Bloomsbury, 2006.
8   Lawrence F. *Not on the Label: What Really Goes into the Food on your Plate.* 2nd
    edn. London: Penguin, 2013.
9   Lymbery P, Oakeshott I. *Farmageddon: The True Cost of Cheap Meat.* London:
    Bloomsbury, 2014.
10 Clapp J. *Food.* Cambridge: Polity, 2012.
11 Morgan K, Marsden T, Murdoch J. *Worlds of Food: Place, Power and Provenance
    in the Food Chain.* Oxford: Oxford University Press, 2006.
12 IAASTD. *Global Report and Synthesis Report.* London: International Assessment
    of Agricultural Science and Technology Development Knowledge, 2008.
13 Foresight. *The Future of Food and Farming: Challenges and Choices for Global
    Sustainability. Final Report.* London: Government Office for Science, 2011: 211.
14 UNCTAD. Wake up before it is too late. *Trade and Environment Review 2013.*
    Geneva: UN Conference on Trade and Development, 2013: 341.
15 IFPRI. *Global Food Policy Report 2011.* Washington, DC: International Food
    Policy Research Institute, 2011.
16 Lamb H. *Fighting the Banana Wars and Other Fairtrade Battles.* London: Rider/
    Ebury, 2008.
17 Greenpeace International. *Eating up the Amazon.* Amsterdam: Greenpeace
    International, 2006.
18 Wise TA, Murphy S. *Resolving the Food Crisis: Assessing Global Policy Reforms
    since 2007.* <http://ase.tufts.edu/gdae/policy_research/resolving_food_crisis.html>
    Medford, MA, and Minneapolis, MN: Tufts University Global Development and
    Environment Institute, and the Institute for Agricultural and Trade Policy, 2012.
19 Tarrant JR. *Food Policies.* Chichester and New York: J. Wiley & Sons, 1980.
20 Burch D, Lawrence G. Supermarket own brands, supply chains and the transformation
    of the agri-food system. *International Journal of Sociology of Agriculture and
    Food,* 13(1) (2005): 1–18.
21 Oxfam International. *Not a Game: Speculation vs Food Security: Regulating
    Financial Markets to Grow a Better Future.* Oxford: Oxfam International, 2011.
22 Clapp J, Helleiner E. Troubled futures? The global food crisis and the politics of
    agricultural derivatives regulation. *Review of International Political Economy,*
    19(2) (2012): 181–207.
23 De Schutter O. *Reports of the UN Special Rapporteur on the Right to Food.* <http://
    www.srfood.org> Louvain/Geneva: UN Economic and Social Council, 2014.
24 Weiss T. *The Global Food Economy: The Battle for the Future of Farming.* London:
    Zed Books, 2007.
25 Belasco WJ. *Appetite for Change: How the Counterculture Took on the food
    industry.* 2nd edn. Ithaca, NY: Cornell University Press, 2006.

26   Burch D, Lawrence G, eds. *Supermarkets and Agri-food Supply Chains*. Cheltenham: Edward Elgar, 2007.
27   Dalle Mulle E, Ruppanner V. *Exploring the Global Food Supply Chain: Markets, Companies, Systems*. Geneva: 3D, Trade, Human Rights, Equitable Economy, 2010.
28   Kraak VI, Harrigan PB, Lawrence M, *et al*. Balancing the benefits and risks of public-private partnerships to address the global double burden of malnutrition. *Public Health Nutrition* (2011): 1–15.
29   UNEP, Nellemann C, MacDevette M, *et al. The Environmental Food Crisis: The Environment's Role in Averting Future Food Crises*. A UNEP rapid response assessment. Arendal, Norway: United Nations Environment Programme/GRID-Arendal, 2009.
30   WCRF/AICR. *Food, Nutrition, Physical Activity and the Prevention of Cancer: A Global Perspective*. Washington, DC/London: World Cancer Research Fund/ American Institute for Cancer Research, 2007.
31   *Chronic Diseases and Development*. The Lancet 2011: 377: 1438–1868. <http://www.thelancet.com/series/chronic-diseases-and-development>
32   UNEP. *Avoiding Future Famines: Strengthening the Ecological Basis of Food Security through Sustainable Food Systems*. Nairobi: United Nations Environment Programme, 2012.
33   Lang T, Barling D, Caraher M. *Food Policy: Integrating Health, Environment and Society*. Oxford: Oxford University Press, 2009.
34   Alwin DA, ed. *Global Status Report on Noncommunicable Diseases 2010*. <http://www.who.int/nmh/publications/ncd_report2010/en> Geneva: World Health Organisation, 2011.
35   OECD. *OECD-FAO Agricultural Outlook*, 2012.
36   PMSEIC (Australia). *Australia and Food Security in a Changing World*. Canberra: Science, Engineering and Innovation Council of Australia, 2010.
37   Cribb J. *The Coming Famine: The Global Food Crisis and What we can Do to Avoid it*. Berkeley, CA: University of California Press, 2010.
38   Paillard S, Treyer S, Dorin B, eds. *Agrimonde: Scenarios and Challenges for Feeding the World in 2050*. Paris: Editions Quae, 2011.
39   Kuhn TS. *The Structure of Scientific Revolutions*. Chicago, IL: University of Chicago Press, 1970.
40   Masterman M. The nature of a paradigm. In: Lakatos I, Musgrave A, eds. *Criticism and the Growth of Knowledge*. Cambridge: Cambridge University Press, 1970: 59–89.
41   Lang T. Agriculture, food, and health: Perspectives on a long relationship. Brief 2. In: Ruel M, Hawkes C, eds. *Understanding the Links between Agriculture and Health Focus 13*. Washington, DC: International Food Policy Research Institute, 2006: 5–6.
42   McMichael AJ. *Human Frontiers, Environment and Disease*. Cambridge: Cambridge University Press, 2001.
43   Goodman D, Redclift M. *Refashioning Nature: Food, Ecology and Culture*. London: Routledge, 1991.
44   Altieri MA. *Agroecology: The Scientific Basis of Alternative Agriculture*. Boulder, CO: Westview, 1987.
45   Lang T, Rayner G. *Why Health is the Key to Farming and Food*. Report to the Commission on the Future of Farming and Food chaired by Sir Don Curry. London: UK Public Health Association, Chartered Institute of Environmental Health, Faculty

of Public Health Medicine, National Heart Forum and Health Development Agency, 2002.

46   Union of Concerned Scientists. *Food and Agriculture: Plant the Plate.* <http://www.ucsusa.org/assets/documents/food_and_agriculture/plant-the-plate.pdf> Cambridge, MA: Union of Concerned Scientists, 2012.

47   Burlingame B, Dernini S, eds. *Sustainable Diets and Biodiversity: Directions and Solutions for Policy, Research and Action. Proceedings of the International Scientific Symposium 'Biodiversity and Sustainable Diets United against Hunger', 3–5 November 2010, Rome.* Rome: FAO and Bioversity International, 2012.

48   Mintz SW. *Sweetness and Power: The Place of Sugar in Modern History.* Harmondsworth: Penguin Books, 1985.

49   Seddon Q. *The Silent Revolution.* London: BBC Books, 1989.

50   Clunies-Ross T, Hildyard N. *The Politics of Industrial Agriculture.* London: Earthscan, 1992.

51   Hall RH. *Food for Nought: The Decline in Nutrition.* Hagerstown, MD: Harper & Row, 1984.

52   Pyke M. *Industrial Nutrition.* London: MacDonald & Evans, 1950.

53   Brandt K. *The Reconstruction of World Agriculture.* London: George Allen & Unwin, 1945.

54   Vernon J. *Hunger: A Modern History.* Cambridge, MA: Harvard University Press, 2007.

55   Pretty J. *The Living Land.* London: Earthscan, 1998.

56   UNEP, Nellemann C, MacDevette M, Manders T, Eickhout B, Svihus B, Prins AG, Kaltenborn, BP, eds. *The Environmental Food Crisis: The Environment's Role in Averting Future Food Crises.* A UNEP rapid response assessment. Arendal, Norway: United Nations Environment Programme/GRID-Arendal 2009.

57   FoE, CIWF. Eating the planet? How can we feed the world without trashing it. <http://www.ciwf.org.uk/what_we_do/factory_farming/eating_the_planet.aspx> London and Petersfield: Friends of the Earth and Compassion in World Farming, 2009.

58   Tansey G, Rajotte T, eds. *The Future Control of Food: A Guide to International Negotiations and Rules on Intellectual Property, Biodiversity and Food Security.* London and Ottawa: Earthscan and IDRC, 2008.

59   World Economic Forum, McKinsey & Co. *Realizing a New Vision for Agriculture: A Roadmap for Stakeholders.* Davos: World Economic Forum, 2010.

60   FAO. *How to Feed the World in 2050: Background Documents for High Level Forum, Rome, 12-13 October 2009.* Rome: Food and Agriculture Organisation, 2009.

61   Le Heron R. *Globalized Agriculture: Political Choice.* Oxford: Pergamon Press, 1993.

62   McMichael P. *The Global Restructuring of Agro-Food Systems.* Ithaca, NY, and London: Cornell University Press, 1994.

63   Goodman D, Watts MJ, eds. *Globalising Food: Agrarian Questions and Global Restructuring.* London: Routledge, 1997.

64   Boyd Orr J. *As I Recall: The 1880s to the 1960s.* London: MacGibbon & Kee, 1966.

65   Boyd Orr SJ. *Food and the People.* London: Pilot Press, 1943.

66   Nestle M. *What to Eat.* New York: North Point Press (Farrar, Straus & Giroux), 2006.

67   Rifkin J. *The Biotech Century: Harnessing the Gene and Remaking the World.* London: Phoenix, 1999.

68   Rayner G, Lang T. *Ecological Public Health: Reshaping the Conditions of Good Health*. London: Routledge, 2012.

69   Barling D, Vriend HD, Cornelese JA, *et al*. The social aspects of food biotechnology: a European view. *Environmental Toxicology and Pharmacology,* 7 (1999): 85–93.

70   ETC Group. *Putting the Cartel Before the Horse: Farm, Seeds, Soil, Peasants*. Communique 111. Ottawa: ETC Group, 2013.

71   Conway G. *One Billion Hungry: Can we Feed the World?* Ithaca, NY: Cornell University Press, 2012.

72   ISAAA. *ISAAA Update 44*. Ithaca, NY: International Service for the Acquisition of Agri-biotech Applications, 2012.

73   Kaput J, Rodriguez RL. Nutritional genomics: the next frontier in the postgenomic era. *Physiology and Genomics*, 16 (2004): 166–77.

74   Ordovas JM, Mooser V. Nutrigenomics and nutrigenetics. *Current Opinion in Lipidology*, 15 (2004): 101–8.

75   Muller M, Kersten S. Nutrigenomics: goals and strategies. *Nature Reviews of Genetics*, 4 (2003): 315–22.

76   Fogg-Johnson M, Meroli A. Nutrigenomics: the next wave in nutrition research. *Nutraceuticals World* (2004). <http://www.nutraceuticalsworld.com>

77   Weiss J, Takhistov P, Clements JM. Functional materials in food nanotechnology. *Journal of Food Science,* 71(9) (2006): R107–R116.

78   US Food and Drug Administration. *Nanotechnology: A Report of the US Food and Drug Administration Nanotechnology Task Force*. <http://www.fda.gov/ScienceResearch/SpecialTopics/Nanotechnology/ucm301093.htm> Atlanta, GA: US Food and Drug Administration, 2007.

79   Food Standards Agency. *Report on FSA Regulatory Review of the Use of Nanotechnology in Relation to Food*. London: Food Standards Agency, 2008.

80   Berger M. Nanotechnology in agriculture. *Nanowerk.* <http://www.nanowerk.com/spotlight/spotid=37064.php> 2014.

81   ETC. ETC Group has called for a moratorium on the environmental release or commercial use of nanomaterials, a ban on self-assembling nanomaterials and on patents on nanoscale technologies. <http://www.etcgroup.org/content/etc-group-has-called-moratorium-environmental-release-or-commercial-use-nanomaterials-ban>. Ottawa, 2010.

82   Ghosh P. World's first lab-grown burger to be cooked and eaten. <http://www.bbc.co.uk/news/science-environment-22885969> London: BBC News, 2013.

83   Bradshaw T. Food 2.0: the future of what we eat. *Financial Times,* 31 Oct. 2014. <http://www.ft.com/cms/s/2/bfa6fca0-5fbb-11e4-8c27-00144feabdc0.html> [accessed 11/1/15].

84   OECD. *Key Biotechnolgy Indicators.* <http://oe.cd/kbi> Paris: Organisation for Economic Cooperation and Development, 2013.

85   Conford P. *The Origins of the Organic Movement*. Edinburgh: Floris Books, 2001.

86   De Schutter O, van Loqueren G. The new Green Revolution: how 21st century science can feed the world. *Solutions*, 2(4) (2011): 33–44.

87   Altieri MA, Farrell J. *Agroecology: The Science of Sustainable Agriculture*. 2nd edn. Boulder, CO/London: Westview/IT Publications, 1995.

88   Mannion AM. *Agriculture and Environmental Change: Temporal and Spatial Dimensions*. Chichester: Wiley, 1995.

89   Harper A, Shattuck A, Holt-Giménez E, *et al. Food Policy Councils: Lessons Learned*. Oakland, CA: Food First Inc., 2009.

90 PepsiCo UK. 50 in 5 commitment: we plan to reduce our water use and carbon emissions by 50% in 5 years. <http://www.pepsico.co.uk/purpose/environment/reports-and-updates/2010-environment-report/passionate-about-growing/50-in-5> Richmond, Surrey: PepsiCo UK, 2010.

91 Unilever. *Sustainable Living Plan 2010*. <http://www.sustainable-living.unilever.com/the-plan> [accessed Dec. 2010]. London: Unilever plc, 2010.

92 SAI. *Sustainable Agriculture Initiative Platform*. <http://www.saiplatform.org> Brussels: Sustainable Agriculture Initiative, 2014.

93 Guthman J. *Agrarian Dreams: The Paradox of Organic Farming in California.* Berkeley, CA/London: University of California Press, 2004.

94 Cressey D. A new breed? *Nature*, 497 (2 May 2013): 27–9.

95 Porritt J. *The World we Made*. London: Phaidon, 2013.

96 Brown AD. *Feed or Feedback: Agriculture, Population Dynamics and the State of the Planet*. Utrecht: International Books, 2003.

97 Patel R. The long green revolution. *Journal of Peasant Studies*, 40(1) (2013): 1–63.

98 Conway G. *The Doubly Green Revolution: Food for All in the 21st Century*. London: Penguin, 1997.

99 Holt-Giménez E, Altieri M. Agroecology, food sovereignty and the new green revolution. *Agroecology and Sustainable Food Systems*, 37(1) (2012): 90–102.

100 Royal Society. *People and the Planet*. London: Royal Society, 2012.

101 Stapledon SG. *The Land: Now and Tomorrow*. London: Faber & Faber, 1935.

102. Boyd Orr SJ. *Food and the People: Target for Tomorrow No. 3*. London: Pilot Press, 1943.

103 Nestle M. *Food Politics*. Berkeley, CA: University of California Press, 2002.

104 WHO. *Diet, Nutrition and the Prevention of Chronic Diseases*. Geneva: World Health Organisation, 1990.

105 WHO, Global strategy on diet, physical activity and health. 57th World Health Assembly. WHA 57.17, agenda item 12.6. Geneva. World Health Assembly, 2004

106 Popkin BM. The nutrition transition in the developing world. *Development Policy Review*, 21 (2003): 581–97.

107 Bello WF. *The Food Wars*. London: Verso, 2009.

108 Keys A. Coronary heart disease in seven countries. *Circulation*, 41(suppl. 1) (1970): 1–211.

109 Cannon G. *Food and Health: The Experts Agree*. London: Consumers' Association, 1992: 230.

110 Trusswell AS. The evolution of dietary recommendations, goals and guidelines. *Amercan Journal of Clinical Nutrition*, 45 (1987): 1060–72.

111 Gribben C, Gitsham M. *Food Labelling: Understanding Consumer Attitudes and Behaviour*. Berkhamstead (Herts): Ashridge Business School, 2007.

112 Cannon G. *The Politics of Food*. London: Century, 1987.

113 Walker C, Cannon G. *The Food Scandal*. London: Century, 1983.

114 Marmot M. *The Social Determinants of Health*. Oxford: Oxford University Press, 1999.

115 Marmot MG. *Status Syndrome: How your Social Standing Directly Affects your health and Life Expectancy*. London: Bloomsbury Pub., 2004.

116 Wilkinson RG, Pickett K. *The Spirit Level: Why More Equal Societies Almost Always Do Better*. London: Allen Lane, 2009.

117 Smil V. *Feeding the World: A Challenge for the Twenty-First Century*. Cambridge, MA/London: MIT, 2000.

118 Dyson T. *Population and Food: Global Trends and Future Prospects*. London: Routledge, 1996.

119 Scarborough P, Bhatnagar P, Wickramasinghe KK, *et al*. The economic burden of ill health due to diet, physical inactivity, smoking, alcohol and obesity in the UK: an update to 2006-07 NHS costs. *Journal of Public Health*, 33(4) (2011): 527–35.

120 British Heart Foundation. *BHF 2012 European Cardiovascular Disease Statistics*. <http://www.bhf.org.uk/publications/view-publication.aspx?ps=1002098> Oxford and London: British Heart Foundation Health Promotion Research Group, University of Oxford, 2012.

121 Van Zwanenberg P, Millstone E. *BSE: Risk, Science, and Governance*. Oxford and New York: Oxford University Press, 2005.

122 Elliott C. *Elliott Review into the Integrity and Assurance of Food Supply Networks – Interim Report*. <https://www.gov.uk/government/policy-advisory-groups/review-into-the-integrity-and-assurance-of-food-supply-networks> London: HM Government, 2013.

123 National Academies of Science (USA), National Research Council (USA). *First International Conference on the Use of Antibiotics in Agriculture*. Washington, DC: National Academies of Science/National Research Council, 1956.

124 Centers for Disease Control and Prevention. Ten great public health achievements – United States, 1900–1999. *Journal of the American Medical Association*, 281(16) (1999): 1481–3.

125 FAO, WHO, OIE. *Joint FAO/WHO/OIE Expert Meeting on Critically Important Antimicrobials: Report of the FAO/WHO/OIE Expert meeting, FAO, Rome, Italy, 26–30 November 2007*. Rome: Food & Agriculture Organisation, World Health Organisation, World Organisation for Animal Health, 2007.

126 Jacobson MF. *Six Arguments for a Greener Diet: How a More Plant-Based Diet Could Save your Health and the Environment*. Washington, DC: Center for Science in the Public Interest, 2006.

127 Smith P. Delivering food security without increasing pressure on land. *Global Food Security*, 2(1) (2013): 18–23. <http://dx.doi.org/10.1016/j.gfs.2012.11.008>.

128 Lawrence F. *Eat your Heart Out: Why the Food Business is Bad for the Planet and your Health*. London: Penguin, 2008.

129 Consumers Union. *Out of Balance: Marketing of Soda, Candy, Snacks and Fast Foods Drowns Out Healthful Messages*. San Francisco, CA: California Pan-Ethnic Health Network & Consumers Union, 2005.

130 Hastings G, Stead M, Macdermott L, *et al*. *Review of Research on the Effects of Food Promotion to Children: Final Report to the Food Standards Agency by the Centre for Social Marketing, University of Strathclyde*. London: Food Standards Agency, 2004.

131 Hawkes C. *Marketing Food to Children: The Global Regulatory Environment*. Geneva: World Health Organisation, 2004.

132 Norberg-Hodge H. *Ancient Futures: Learning from Ladakh*. San Francisco, CA: Sierra Club Books, 1991.

133 Popkin BM. Urbanisation, lifestyle changes and the nutrition transition. *World Development*, 27(11) (1999): 1905–15.

134 Popkin BM, Nielsen SJ. The sweetening of the world's diet. *Obesity Research*, 11(11) (2003): 1–8.

135 Rayner G, Hawkes C, Lang T, *et al*. Trade liberalisation and the diet transition: a public health response. *Health Promotion International*, 21(suppl. 1) (2006): 67–74.

136 Critser G. *Fatland: How Americans Became the Fattest People in the World*. London: Penguin, 2003.

137 WHO. Top 10 causes of death. Factsheet 310. <http://www.who.int/mediacentre/factsheets/fs310/en> Geneva: World Health Organisation, 2013.

138 WHO/FAO. *Diet, Nutrition and the Prevention of Chronic Diseases*. Report of the joint WHO/FAO expert consultation. WHO Technical Report Series, 916 (TRS 916). Geneva: World Health Organisation & Food and Agriculture Organisation, 2003.

139 WCRF/AICR. *Policy and Action for Cancer Prevention: Food, Nutrition, and Physical Activity: a Global Perspective*. London/Washington, DC: World Cancer Research Fund/American Institute for Cancer Research, 2009.

140 UNEP, World Water Assessment Programme. *Water: A Shared Responsibility*. The United Nations World Water Development Report, 2. Geneva: United Nations Environment Programme, 2006.

141 Rockström J, Steffen W, Noone K, *et al*. Planetary boundaries: exploring the safe operating space for humanity. *Ecology and Society*, 14(2) (2009): 32. <http://www.ecologyandsociety.org/vol14/iss2/art32>

142 McMichael A. *Planetary Overload: Global Environmental Change and the Health of the Human Species*. Cambridge: Cambridge University Press, 1993.

143 Cochrane Collaboration. About the Cochrane collaboration. <www.cochrane.org> Oxford: The Cochrane Collaboration, 2014.

144 Brunner E, Rayner M, Thorogood M, *et al*. Making public health nutrition relevant to evidence-based action. *Public Health Nutrition*, 4(6) (2001): 1297–9.

145 Sustainable Development Commission. *Setting the Table: Advice to Government on Priority Elements of Sustainable Diets*. London: Sustainable Development Commission, 2009.

146 Rayner G, Lang T. *Ecological Public Health: Reshaping the Conditions for Good Health*. Abingdon: Routledge/Earthscan, 2012.

3   Diet, health and disease

Let Reason rule in man, and he dares not trespass against his fellow-creature, but will do as he would be done unto. For Reason tells him, is thy neighbour hungry and naked today, do thou feed him and clothe him, it may be thy case tomorrow, and then he will be ready to help thee.

Gerrard Winstanley, English Leveller, 1609–1676[1]

Core arguments

This chapter explores the evidence of food's impact on human health. The world now has a complex picture of diet-related non-communicable diseases from under-, mal- and over-consumption. These are a major and rising burden to society and economies, and have left policy-makers uncertain what to do (discussed in Chapter 5). This burden of disease emerged slowly but devastatingly in the 20th century, and is now threatening low- and middle-income countries which can barely afford the healthcare consequences. This impact adds to the questioning of Productionism as a policy formula for progress. A key, if not the key, purpose of the Productionist paradigm was to raise food production to contribute to the improvement of population health. Its logic was simple: increasing food output will lower prices which will make food more affordable and available, and thus health will improve. By the 21st century the evidence that this policy logic was critically flawed had become overwhelming. No one anticipated a world where food would be so cheap or markets so flooded with unhealthy foods that dietary patterns could be so skewed. While successfully raising the calorific value of world food supply, Productionism has serviced the over-production of cheap commodities which characterise a phenomenon known as the 'nutrition transition'. This is a shift from simpler diets to the excessive consumption of highly processed foods delivering excess salt, sugar and fat. Patterns of eating which decades ago were associated with the affluent West have appeared in developing countries. The food system has thus snatched defeat from the jaws of victory and is a (and often the) key factor in non-communicable diseases such as heart disease, diabetes, obesity and some cancers. To add insult to injury, problems of under-consumption due to poverty and of food poisoning due to poor sanitation and hygiene remain. In this chapter we set out the scope and scale of this major 21st century food and health problem. The two emerging paradigms take different positions on how to tackle this, one tending to seek technical fixes, the other solutions at the societal and cultural level.

Introduction

Epidemiologists categorise diseases into two broad groups: communicable and non-communicable disease. The distinction is important. Communicable diseases are those carried from person to person or via some intermediary factor. These include diseases such as malaria, food poisoning or SARS. Non-communicable diseases (NCDs) are acquired by lifestyle or other mismatches between humans and their environment. They include many big 'killers' such as cardiovascular disease, strokes, cancers. Their impact may be quick – a sudden fatal heart attack – or slower – diabetes leading to dementia. In the developed world, deaths through infectious and parasitic diseases are very low, whereas in developing countries they are higher. In the past, diet-related non-communicable diseases like coronary heart disease (CHD), obesity, diabetes and cancers were high in developed countries and low in the developing world, but now that dividing line has been blurred, and to some extent erased. Degenerative disease rates are already high in the developing world. For some time, WHO and FAO reports have been warning that world health is in transition with non-communicable diseases now taking a higher toll than communicable diseases. (Mental ill health is also rising worldwide.) Factors in this health transition include diet, demographic change (such as an ageing population) and cultural factors related to globalisation.

This shift in health patterns and the evidence that diet and exercise are key factors is why, in the last quarter of the 20th century, nutrition became such a 'hot' topic again, as it had been in the 1930s. What is to blame for this rise in nutrition-related problems? Are consumers responsible or is the marketing of foodstuffs by commerce the key factor in distorting health? Can public health campaigns persuade consumers to improve their diets? Such questions mattered as evidence grew of the rapidity of the change from restricted to more abundant diets and as food choice was expanded by availability of tens of thousands of food products in supermarkets in the process known as the 'nutrition transition'.[2 3] This threat to health had been observed in a major way first in the West from the 1980s (with early warnings in the 1970s). In the 1990s the World Bank and WHO developed a methodology to calculate how many 'life years' are affected by disease, an indicator known as 'disability adjusted life years' (DALYs). The first *Global Burden of Disease* (GBD) study in 1996 showed that deficient diets accounted for 60 per cent of years of life lost in the established market economies.[4] The second GBD study in 2012 again showed that food dominated the risk factors of ill health. Fourteen of the top 20 risk factors for DALYs were diet-related.[5] Productionism is associated with this health burden, a mix of improvement, distortion and failure. There is more than food, if measured in calories, to feed the world adequately (albeit at an environmental cost – see Chapter 4), but nearly a billion people remain malnourished,[6] despite the over-production of food in global terms. And 1.4 billion people were estimated to be overweight or obese in 2013.[7]

In this chapter, we explore the relationship between diet and disease. We discuss the gross inequalities within and between countries – food deficiencies and poverty amidst food abundance and wealth. The chapter provides a basis for policy questions raised in this book: what is a good food system for the 21st century? A sub-theme of this chapter is why, given the weight and decades of evidence, have policy-makers not responded to this evidence? Chapter 5 will consider what has and has not been done, after conducting a similar review of the evidence of food's impact on the environment.

How does food affect health?

From the 1970s, there has been a growing strand of public health reports alerting policy-makers and the consuming public that diet is a major factor in the causes of death and morbidity. First it was a trickle; now it is a flood. Although deeply unpalatable to some sections of the food industry (see Chapter 6), these reports have been constant reminders of the enormity and scale of the public health crisis.

In the 1970s individual scientific studies began to question the shift to 'modern' diet. Professor Ancel Keys's study of heart disease in seven countries arguably was the first to raise alarm.[8] Others foresaw this likelihood: McCarrison through studies of North India,[9] and also Burkitt studying refined carbohydrates,[10] Cleave on sugars,[11] and many others. But these voices only found institutional, i.e. official hearings, when in 1990 the WHO published its first major report to chronicle food's impact on health.[12] This is probably the first high level report to raise the policy stakes; its authors knew the situation was serious.[13] *Diet, Nutrition and the Prevention of Chronic Disease* documented the high prevalence of diseases which could be prevented by better nutrition. The evidence has grown ever since in a phalanx of WHO reports, some produced with the FAO, the UN body which above all championed productionism.[14-17] The diet-related diseases such reports identified include:

- obesity
- diabetes
- cardiovascular diseases
- cancers
- osteoporosis and bone fractures
- dental disease.

These diseases are not solely exacerbated by poor diet; lack of physical activity matters too, highlighted by studies such as those by Jerry Morris on London busmen.[18] These found that conductors who moved up and down the bus collecting the money and dispensing tickets were healthier than the sedentary drivers in their cabs.[19] We humans do not just have a physiological need or desire to eat, we also have bodies that need – even if we don't desire it! – to be physically active.

This new research highlighting the 'more food in and less exercise out' pattern shaping health is a core reason public health advocates now question Productionism. Launching the joint WHO-FAO *Diet, Nutrition and the Prevention of Chronic Disease* report in 2003, Dr Gro-Harlem Brundtland, then the Director-General of the WHO, stated: '[w]hat is new is that we are laying down the foundation for a global policy response'. That was her hope, certainly. WHO had planned a rapid succession of reports which first outlined the evidence in 2002, then shared agreement as to this evidence in 2003, and finally crafted a global policy response in 2004.[14 16 17] Public health and medical specialists were frustrated at the unnecessary toll of deaths and lost quality of life. Over a decade later, the burden of disease ill health has continued.[5] The evidence was bogged down in an ideological quagmire: more evidence being called for, consumers being blamed, poorer consumers being blamed and other delaying features. How much evidence is needed for action? Even in 1990, with the first major WHO report cited above, there was sufficient evidence to begin to chart change of course, but policy-makers tended to be focused on healthcare rather than prevention; and there were and are strong commercial and cultural barriers to change.

In 2002, the WHO list of the top ten risk factors associated with non-communicable diseases, included eight associated with food and drink (the two not associated were tobacco and unsafe sex):[14]

- blood pressure
- cholesterol
- underweight
- fruit and vegetable intake
- high body mass index
- physical inactivity
- alcohol
- unsafe water, sanitation and hygiene.

In 2003 the *World Cancer Report*, the most comprehensive global examination of the disease then to date, calculated that cancer rates could further increase by 50 per cent to 15 million new cases by 2020.[20] To stem the rise of this toll, the WHO and the International Agency for Research on Cancer (IARC) argued that three issues in particular need to be tackled:

- tobacco consumption (still the most important immediate avoidable risk to health);
- healthy lifestyle and diet, in particular the need for frequent consumption of fruit and vegetables and the taking of physical activity;
- early detection and screening of diseases to allow prevention and cure.

In 2007, the World Cancer Research Fund published its latest global scientific review. This built on a path-breaking (and huge) report in the mid-1990s.[21] The 2007 report was equally weighty in pages and quality of evidence,[22] but applied more strict Cochrane standards of data quality, a method which takes note of robust evidence using double-blind randomised controlled trials. In other words, these new findings were even more robust. This too confirmed the picture from specialists in heart disease, diabetes and other non-communicable diseases about dietary impact on another category of disease. Significant numbers of cancer were diet-related and could be prevented by changed diets. The trends could be altered but ameliorative action was not happening. This was a situation of policy failure, the WCRF/AICR concluded, so two years later a report to policy-makers was produced making wide-ranging proposals, from the need to eat more plants to cutting back on processed meats.[23] The rise of childhood obesity globally was particularly shocking.[24] While some early analysts thought this was a sign of moral decline – loss of self-control, poor parenting, etc. – others connected it to changes in the food supply and economic signals.[25-28] In the 2000s, policy complacency was shaken by estimates of long-term healthcare costs from overweight and obesity, and popular books fuelled media interest.[29-33] In short, since the 2002 WHO report, all that changed was the amount of evidence and the moral concern.[34 35] The scientists of the 1930s–1950s would not have recognised this situation. They promoted Productionism as the resolution of hunger (to which we return below).[36-39] Yet here, just over half a century later, was a totally different world, one of continuing hunger, high in numbers still – 0.9 billion malnourished in 2010, although dropping as a proportion of the total world population[40] – but dwarfed by overweight and obesity at 1.4 billion, as was noted earlier.[7]

Table 3.1 lists some major reports since the 1980s from world bodies rather than academics. It documents the extent, severity and trends in diet-related diseases. The final one is by one of the world's most influential business consultancies, on the economic costs of obesity. Will this have impact where public health evidence has not?

Table 3.2 gives the top ten causes of all deaths globally for 2012. Even from this perspective, food-related ill health looms large, featuring significantly in five of the top ten. The final column in that table shows what the trends are. Table 3.3 then lists some major diet-related diseases, to 'populate' the food picture in more detail.

Could the proponents of Productionism have anticipated the scale of these most recent health concerns? To some extent, they could not. Even excessive intake of fats as causing ill health might have been something of a shock for the Productionist paradigm; it was almost heretical to argue that too much of a nutrient could be harmful to health.[49] The paradigm paternalistically assumed total knowledge of all variables needed to make good food policy; governments and companies could be trusted to look after the public health; the consumers' role was to select products to create their own (presumably balanced) diets. In fact, the scientific evidence has either been largely ignored, or it has been normalised and turned into a cultural discourse of blame. While, to some extent,

Table 3.1 Major reports giving evidence on diet's impact on health, 1980s–2010s

Date	Authors	Focus	Diet-related health problem
1982	National Academies of Science	USA/global	Diet and cancer
1983	Royal College of Physicians (UK)[26]	UK/global	Obesity
1988	Surgeon General (USA)[41]	USA	Non-communicable disease
1990	WHO [12]	Global	General diet and nutrition impact
1996	World Bank, WHO[4]	Global	Cost burden
1997	WCRF/AIRC[21]	Global	Diet and cancer
2000	International Diabetes Federation[42]	Global	Diabetes (Atlas 1st edition)
2000	UN SCN[43]	Global	Under-,over-, and mal-nutrition
2002	WHO[14]	Global	Diet and physical activity
2003	WHO, FAO[17]	Global	Diet and physical activity
2003	WHO, IARC[20]	Global	Cancer
2007	WCRF/AIRC[22]	Global	Cancer
2012	The Lancet (for WHO)[5]	Global	Total burden of disease (including diet and physical activity)
2011	WHO[44]	Global	Non-communicable diseases
2014	World Heart Federation[45]	Global	Heart disease
2014	International Diabetes Federation[46]	Global	Diabetes (Atlas 6th edition)
2014	McKinsey Global Institute[47]	Global	Global costs of obesity

a combination of commerce and consumerism has triumphed over public health, the situation is more complex. Consumers collude in their own downfall, argues industry; no one forces them to eat inappropriate diets, and if only they took more exercise, wouldn't it all be alright? The arguments continue at a cultural level (see Chapter 7) and a governance level (see Chapter 8).

Hunger

A constant in this modern picture of diet-related disease has been hunger. As a 1995 FAO review of the global picture starkly put it: '[H]unger . . . persists in developing countries at a time when global food production has evolved to a stage when sufficient food is produced to meet the needs of every person on the planet.'[50] The fact that there is hunger amidst plenty is not new. In the 1840s Irish Famine, ships exported food to Britain.[51] Over-consumption and under-consumption can and do coexist. The analysis that hunger is due to lack of food is simplistic. Others suggest that hunger exists because there is gross inequality of global distribution

Table 3.2 Leading causes of all death in the world, 2012

Rank	Cause of death	Numbers (millions)	%	Change 2000–12
1.	Ischaemic heart disease	7.4	13.2	Up
2.	Stroke	6.7	11.9	Up
3=	Chronic obstructive pulmonary disease (COPD)	3.1	5.6	same
3=	Lower respiratory infection	3.1	5.5	down
5.	Trachea and lung cancers	1.6	2.9	Up
6=	HIV/AIDS	1.5	2.7	down
6=	Diarrhoeal disease	1.5	2.7	down
6=	Diabetes mellitus	1.5	2.7	up
9.	Road injury	1.3	2.2	up
10.	Hypertensive heart disease	1.1	2.0	up
	Other causes		48.6	

Source: WHO 2014 [48]

Table 3.3 Some major diet-related diseases

Problem	Extent/comment
Low birthweights	30 million infants born in developing countries each year with low birthweight: by 2000, 11.9% of all newborns in developing countries (11.7 million infants)
Child under-nutrition	150 million underweight pre-school children: in 2000, 32.5% of children under 5 years in developing countries stunted, amounting to 182 million pre-school children. Problem linked to mental impairment. Vitamin A deficiency affects 140–250 million schoolchildren; in 1995 11.6 million deaths among children under 5 years old in developing countries.
Anaemia	Prevalent in schoolchildren; maternal anaemia pandemic in some countries.
Adult chronic diseases	These include adult-onset diabetes, heart disease and hypertension, all accentuated by early childhood under-nutrition.
Obesity	A risk factor for some chronic diseases (see above), especially adult-onset diabetes.[7] Overweight rising rapidly in all regions of the world.
Underweight	In 2000, an estimated 26.7% of preschool children in developing countries.
Infectious diseases	Still the world's major killers but incidence worsened by poor nutrition; particularly affects developing countries.
Vitamin A deficiency	Severe vitamin A deficiency on the decline in all regions, but sub-clinical vitamin A deficiency still affects between 140 and 250 million pre-school children in developing countries, and is associated with high rates of morbidity and mortality.

Source: adapted from ACC/SCN 2000[8]

and availability, which is itself due to lack of rights and organised demand – a view summarised brilliantly by Nobel Prize-winner Amartya Sen in his study of the Bengal famine of the 1940s.[52] According to the FAO there is sufficiency of calories worldwide. Western Europe, for example, has 3,500 kcalories available per person per day and North America 3,600, while sub-Saharan Africa has 2,100 and India 2,200.

In the mid-1990s, FAO calculated that by 2015, 6 per cent of the world's population (462 million people) would be living in countries with under 2,200 kcalories available per person, that by 2030, 15 per cent of sub-Saharan Africa would be under-nourished.[53] Table 3.4 gives the total picture in five snapshots: 1990–2, 2000–2, 2005–7, 2008–10, 2011–13. It shows that under-nourishment remains constant at less than 5 per cent in developed economies, while in the developing world it has declined from 23.6 per cent in 1990–2 to 14.3 per cent in 2011–13. By any count, this can be presented as – and is – a success story. But in absolute numbers, the number of undernourished people is stubbornly high, declining only slightly. Good but not great news. Table 3.4 also reveals how lower income regions can vary. South East Asia, for example, has lowered the percentage of under-nourishment from a third to a tenth, while Sub-Saharan Africa has managed to reduce its percentage from a third only to a quarter. Again, progress but not enough or fast enough. Within regions, too, there is an even more important story. Within Latin America, for example, which has lower under-nourishment to start with, the case of Brazil's *Fome Zero* anti-hunger policies showed what can happen if actions are championed by the President. Under President Jose 'Lula' da Silva that country made startling reductions in hunger.[54] Political will and social policies make the difference.[55]

Although there is enough food to feed the world, the FAO suggests that hunger is still correlated with supply. The more food there is in a region, the more likely its population is to be fed. Hunger tends to be widespread in countries with high poverty.[56] Figure 3.1 shows this clustering effect.

The nutrition transition

In a series of papers since the 1990s, Professor Barry Popkin and his colleagues at the University of North Carolina, USA, have shown that there is what they term a 'nutrition transition' occurring in the developing world, associated primarily with rising wealth.[57-59] Extensive country and regional studies show the process of dietary change and diet-related ill health previously associated with the more affluent West is now manifest in developing countries.[60] The nutrition transition is a shift of pattern from restricted diets to diets higher in saturated fat, sugar and refined foods, and lower in fibre. This transition is associated with two other historic processes of change: the demographic and epidemiological transitions. Demographically, world populations have shifted from patterns of high fertility and high mortality to patterns of low fertility and low mortality. In the epidemiological transition, there is a shift from a pattern of disease characterised by infections, malnutrition and episodic famine to a pattern of disease with a high

Table 3.4 Global under-nourishment, 1990/92–2011/13

	Number of under-nourished (millions) and prevalence (%) of under-nourishment				
	1990–92	*2000–02*	*2005–07*	*2008–10*	*2011–13*
World	1015.3	957.3	906.6	878.2	842.3
	18.9%	15.5%	13.8%	12.9%	12.0%
Developed Regions	19.8	18.4	13.6	15.2	15.7
	<5%	<5%	<5%	<5%	<5%
Developing Regions	995.5	938.9	892.9	863.0	826.6
	23.6%	18.8%	16.7%	15.5%	14.3%
Africa	177.6	214.3	217.6	226.0	226.4
	27.3%	25.9%	23.4%	22.7%	21.2%
Northern Africa	4.6	4.9	4.8	4.4	3.7
	<5%	<5%	<5%	<5%	<5%
Sub-Saharan Africa	173.1	209.5	212.8	221.6	222.7
	32.7%	30.6%	27.5%	26.6%	24.8%
Asia	751.3	662.3	619.6	585.5	552.0
	24.1%	18.3%	16.1%	14.7%	13.5%
Caucasus and Central Asia	9.7	11.6	7.3	7.0	5.5
	14.4%	16.2%	9.8%	9.2%	7.0%
Eastern Asia	278.7	193.5	184.8	169.1	166.6
	22.2%	14.0%	13.0%	11.7%	11.4%
South-Eastern Asia	140.3	113.6	94.2	80.5	64.5
	31.1%	21.5%	16.8%	13.8%	10.7%
Southern Asia	314.3	330.2	316.6	309.9	294.7
	25.7%	22.2%	19.7%	18.5%	16.8%
Western Asia	8.4	13.5	16.8	19.1	20.6
	6.6%	8.3%	9.2%	9.7%	9.8%
Latin America and the Caribbean	65.7	61.0	54.6	50.3	47.0
	14.7%	11.7%	9.8%	8.7%	7.9%
Caribbean	8.3	7.2	7.5	6.8	7.2
	27.6%	21.3%	21.0%	18.8%	19.3%
Latin America	57.4	53.8	47.2	43.5	39.8
	13.8%	11.0%	9.0%	8.0%	7.1%
Oceania	0.8	1.2	1.1	1.1	1.2
	13.5%	16.0%	12.8%	11.8%	12.1%

Source: FAO 2013[56]

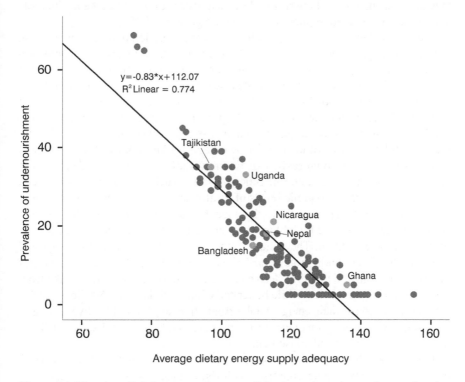

Figure 3.1 The close link between adequacy of food supply and prevalence of under-nourishment, FAO data

Source: FAO 2013 [56: p.25]

rate of the chronic and degenerative diseases. The nutrition transition amplifies and clarifies these. As populations become richer, they substitute cereal foods for higher-value protein foods such as milk, dairy products and meat, increased consumption of which is associated with Westernisation of ill health. Relatively better-off populations also consume a greater number of non-staple foods and have a more varied, if not healthier, diet. Thus there is a modern nutritional paradox: in the same low-income country there may be ill health caused by both mal-nutrition and over-nutrition; in the same rural area of a poor country both obesity and underweight can coexist.

This change of disease pattern is associated with a rise in urban and industrial lifestyles; this can be accompanied by income squeeze too.[61] Low-income countries undergo the nutrition transition but cannot afford to deal with its effects. Popkin argues that, while the nutrition transition brings greater variety of foods to people who previously had narrow diets, the resulting health problems from the shift in diet should not be traded off against the culinary and experiential gains. Consumers might enjoy the new variety of foods that greater wealth offers but they are often unaware of the risk of disease that can follow. Alas, the health costs tend to be downplayed in pursuit of economic growth.[32 62 63]

This would have been no surprise to an earlier generation of researchers whose writings had questioned the narrow Productionist view of progress. Nutrition and public health pioneers such as Professors Trowell and Burkitt had observed the lack of NCDs and good diet-related health of supposedly under-developed people in parts of Asia or African in the 1950s–1980s, which led them to question 'whether Western influence in Africa, Asia, Central and South America and the Far East is unnecessarily imposing our diseases on other populations who are presently relatively free of them'.[64] Trowell and Burkitt, both with long medical experience in Africa, could easily explain the variation in infectious diseases, but not the variation in rates of non-infectious diseases such as heart disease between countries at different economic levels of wealth and development.[65] In Africa in the post-World War II period, they witnessed the rise of key indicators for diseases such as heart disease and high blood pressure in peoples who had previously had little experience of them. The dietary transition is swift. In 2000, FAO reported how a study of very under-nourished Chinese people living on 1,480 kilocalories (kcal) per day showed that they derived three-quarters of their energy intake from starchy staples such as rice, while better-fed Chinese living on 2,500 kcal per day were able to reduce their energy intake from such staples and to diversify their food sources.[66] Rising incomes raises the range and nutritional value of foods that can be purchased and consumed.

Popkin has shown how this same process occurs in both urban and rural populations in developing countries with rising incomes. Figures 3.2 and 3.3, from a 1998 paper, show clearly the relationship between per capita income and what predominantly rural and predominantly urban populations eat as both get wealthier.[67] Both eat more meat and fats, and reduce carbohydrates, as a proportion of their overall diet. But there still remain differences between urban and rural populations, probably due to their different levels of activity, access to dietary ingredients and cultural mores.[59] Urbanised populations also consume more added sugars as they get wealthier, whereas the rural populations consume less. Rising sugar and use of refined carbohydrates are linked to rising consumption of processed foods, and particularly soft drinks.[68] The rise of sugar in diets has been a persistent theme in the analysis of critics of the nutrition transition from Professors Trowell and Burkitt or John Yudkin or Surgeon-General Cleave decades ago,[11 69 70] to Professor Robert Lustig's concerns with high fructose corn syrup today,[71] and the academic campaign Action on Sugar.[72] Popkin and his colleagues' point is that changing economic circumstances markedly shape the mix of nutrients in the diet and that lifestyle factors – such as the degree of urbanisation,[73] changing labour patterns and distribution of power between the genders – have a major effect on health.

Dietary and health changes can be rapid. In the Middle East, changing diets and lifestyles have changed patterns of both mortality and morbidity. In Saudi Arabia, for instance, meat consumption doubled and fat consumption tripled between the mid-1970s and the early 1990s; in Jordan, there has, in the same timescale, been

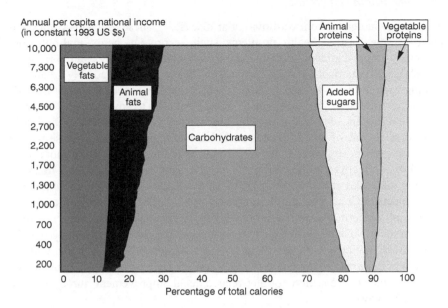

Figure 3.2 Relationship between the proportion of energy from each food source and Gross National Product per capita with the proportion of the urban population residing at 25%, 1990

Source: FAO/World Bank/Popkin (1998)[67]

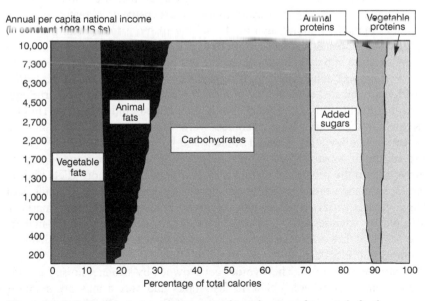

Figure 3.3 Relationship between the proportion of energy from each food source and Gross National Product per capita with the proportion of the urban population residing at 75%, 1990

Source: World Bank/FAO/Popkin (1998)[67]

a sharp rise in deaths from cardiovascular disease.[74] These problems compound older Middle Eastern health problems such as protein-energy mal-nutrition, especially among children. In China, the national health profile began to follow a more Western pattern of diet-related disease as the population gradually urbanised, leading to an increase in degenerative disease, that has accelerated with China's economic 'miracle'.[75 76] Consumption of legumes such as soyabean was replaced by animal protein, derived in part by feeding soy to meat animals. One early nutritional review of this problem concluded that exhorting the Chinese people to consume more soy when they were voting with their purses to eat more meat would be ineffective 'in the context of an increasingly free and global market'. [77] This view was reinforced in a 2013 study of the health and environmental food aspects of China's economic success story.[78] Such studies suggest the battle to prevent Western diseases in the developing world has already been lost on a massive scale. If China, one of the two most populous and economically powerful and growing nations, and with a famously low-fat plant-based diet, and with 22 per cent of the world's population living off only 7 per cent of its land, can change this fast, what chance is there for diet-related health improvements throughout the developing world?

But pessimism is not helpful. Neither China nor India, neither Latin America nor Africa nor the Middle East can afford the toll that follows from the nutrition transition. Even oil-dollar wealthy Middle Eastern countries are now troubled by the cost of rising diabetes type 2. Can India or China afford expensive western healthcare to tackle such ill health?[79] Indeed, can even rich economies afford rising demand for coronary by-pass operations or expensive drug treatments? Few expect that market solutions such as the selling of more foods with presumed health-enhancing benefits, or subscriptions to gyms and leisure centres rather than building exercise into daily life, will resolve this problem.

Do we simply conclude that the rising degenerative diseases and the Nutrition Transition are the cost of economic progress? Nobel Prize-winner Robert Fogel argued that health follows wealth; and the celebrated epidemiologist Thomas McKeown argued that rising incomes were more powerful than medical intervention in alleviating malnourishment and improving ill health.[80-83] Their views are much debated.[84] At a practical level, however, the issue is how to retain or regain the protective elements in traditional, indigenous diets high in legumes, fruit and vegetables while also accepting the nutritional advantages of eating a greater variety of foods and improved access to varied food markets, unaffected by vagaries of seasonality or market scarcity. Few policy-makers in the developing world have been prepared to fight to keep 'good' elements of national and local diets or to constrain the flow of Western-style foods and drinks into their countries. Their voices get drowned by stronger noise for trade liberalisation. Everywhere, US-style fast foods and soft drinks are symbols of modernity and choice; food culture has been 'burgerised'. The result is the emergence of three clear categories of malnutrition: the Underfed, Overfed and Badly Fed (see Table 3.5).

Table 3.5 Types and effects of mal-nutrition

Type of malnutrition	Nutritional effect	No. of people affected globally, in billions
Hunger	deficiency of calories and protein	at least 1.2
Micronutrient deficiency	deficiency of vitamins and minerals	2.0–3.5
Over-consumption	excess of calories, often accompanied by deficiency of vitamins and minerals	at least 1.2–1.7

Source: Gardner and Halweil (2000), based on WHO, IFPRI, ACC/SCN data[63]

The obesity epidemic

Obesity-related conditions include heart disease, stroke, type 2 diabetes and certain types of cancer; these are all preventable but are leading causes of death. The connection between overweight and health risk is alarmingly highlighted by the following list of the physical ailments that an overweight population (with a body mass index (BMI) higher than 25) is at risk of:[85]

- high blood pressure, hypertension;
- high blood cholesterol, dyslipidemia;
- type 2 (non-insulin-dependent) diabetes;
- insulin resistance, glucose intolerance;
- hyperinsulinemia;
- coronary heart disease;
- angina pectoris;
- congestive heart failure;
- stroke;
- gallstones;
- cholescystitis and cholelithiasis;
- gout;
- osteoarthritis;
- obstructive sleep apnea and respiratory problems;
- some types of cancer (such as endometrial, breast, prostate and colon);
- complications of pregnancy;
- poor female reproductive health (such as menstrual irregularities, infertility and irregular ovulation);
- bladder control problems (such as stress incontinence);
- uric acid nephrolithiasis;
- psychological disorders (such as depression, eating disorders, distorted body image, and low self-esteem).

The list is sobering. Rising obesity rates among children are particularly troubling, as they suggest a toll of degenerative disease in the future. In 2010 the WHO estimated there were globally 42 million overweight children under the age of 5 years. Its concern is that overweight and obese children are likely to stay obese into adulthood and thus are more likely to develop NCDs like diabetes and cardiovascular diseases at a younger age.[86] The rise in childhood obesity first noted in developed countries has now spread to developing countries. Ten per cent of the world's school-aged children (aged 5–17 years) and 6.7 per cent of pre-school-aged children (0–5 years) have been estimated as overweight or obese.[24 87]

In the USA, over a third of adults (34.9 per cent) are now obese, and the estimated annual medical cost of obesity in the USA was $147bn in 2008.[88] People who are obese cost healthcare $1,429 more than do people of normal weight. As early as 1948, medical researchers were beginning to research the incidence of obesity. There were official reports at country level by the early 1980s,[26] and there have been commercial and consumer concerns for a long time.[89-91] But it was not until the end of the 20th century that obesity was confirmed as an international medical problem by the WHO's Task Force on Obesity.[92] Within a decade, few in public health had to argue that overweight and obesity were a problem or that they are risk factors for chronic and non-communicable diseases. The issue was that they were actually being 'normalised', as the Chief Medical Officer of England noted in 2014,[93] accepted a bit like smoking in the 1980s, an evil but here and a matter of personal lifestyle choice.

In developing countries obesity is more common amongst people of higher socio-economic status and in those living in urban communities. In more affluent countries, it is associated with lower socio-economic status, especially amongst women and rural communities. Historically and biologically, weight gain and fat storage have been indicators of health and prosperity. Only the rich could afford to get fat. By 2000, the WHO was expressing alarm that more than 300 million people were defined as obese, with 750 million overweight, i.e. pre-obese: over a billion people deemed overweight or obese globally.[94] By 2013, as has been noted, this figure had been radically revised upwards to 1.4 billion people.[7]

Obesity is defined as an excessively high amount of body fat or adipose tissue in relation to lean body mass. Standards can be determined in several ways, notably by calculating population averages or by a mathematical formula known as 'body mass index' (BMI), a simple index of weight-for-height: a person's weight (in kilos) divided by the square of the height in metres (kg/m^2). BMI provides, in the words of the WHO, 'the most useful, albeit crude, population-level measure of obesity'. A personal BMI of between 25 and 29.9 is considered overweight; obesity is defined as a BMI of 30 and above; a personal BMI of less than 17 is considered underweight. BMI levels are a useful predictor of risk from degenerative diseases. Unutilised food energy is stored as fat. Currently, the US Centers for Disease Control and Prevention (CDC) considers that all adults (aged 18 years or older) who have a BMI of 25 or more are at risk from premature death and disability as a consequence of overweight and obesity.[95] There are debates about whether the BMI threshold for overweight should even be lowered to 23.

Men are at risk who have a waist measurement greater than 40 inches (102 cm); women are at risk who have a waist measurement greater than 35 inches (88

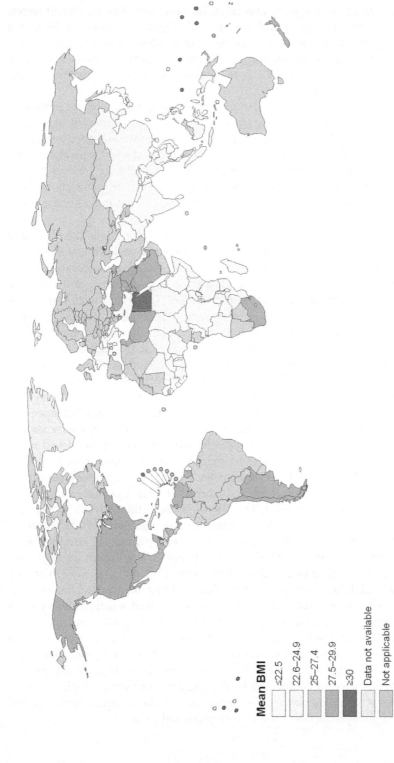

Mean BMI

≤22.5
22.6–24.9
25–27.4
27.5–29.9
≥30
Data not available
Not applicable

Figure 3.4 The spread of obesity worldwide, WHO

Source: WHO 2011[96]

cm). Whilst height is obviously also taken into consideration, we should regard these measurements as key health benchmarks. Figure 3.4 gives the 2008 data collected by WHO on obesity worldwide, with incidence worsening the darker the coloration. Sophisticated interactive maps are now available on the worldwide web, well worth a view.[96] The CDC, too, publishes a famous annual map of the growth across US states.[97]

Health education seems powerless before this rising tide of obesity. In the US, the homeland of fast food, President George W. Bush was so alarmed by the obesity crisis that back in 2002 he launched a national debate. He had good reason for concern; in 1986 the economic costs of illness associated with overweight in the US were estimated to be $39bn according to the CDC. And there are considerable differences between ethnic and age groups, with lower income groups more obese than the more affluent in rich countries and often the reverse in poorer countries, reflecting access to food, and its cost.

What can be done? One position – particularly voiced in the USA and associated with neo-liberal think-tanks – suggests that any critique of obesity is an infringement of personal liberty and is 'size-ist', an unnecessary cultural value statement.[98] If someone wants to be fat and is content and loved by others, goes this argument, what does it matter? The list of health problems given above is surely an answer to this position. Individual costs are actually borne by society as a whole. And how can children be responsible for their own weight or choice of foods from an early age? A pioneering analysis by antipodean researchers in the mid-1990s proposed that the obesity pandemic could only be explained in 'ecological' terms. Garry Egger and Boyd Swinburn set out environmental determinants such as transport, pricing and supply, arguing that environmental factors were so powerful in upsetting energy balances that obesity could be viewed as 'a normal response to an abnormal environment'. [28] So finely balanced are caloric intake and physical activity than even slight alterations in their levels can lead to weight gain. No amount of individual exhortation will reduce worldwide obesity. Transport, neighbourhood layout, home environments, fiscal policies and other alterations of supply chains must be tackled instead, they suggest. The UK Chief Scientific Adviser's Foresight programme explored such systemic thinking in a much cited 2007 report on obesity. This produced a celebrated systems map of the drivers of obesity. Figure 3.5 is one version of that in which the 'blackness' of the lines is an indication of each factor's significance in driving the central 'motor' of energy in/energy out (in the middle of the map). Multiple factors determine whether a person (or a population) puts on, reduces or maintains a balanced weight. This systems model captures the simple truth about weight gain; it is all about the match or mismatch of energy input and output. The web of competing drivers which operate on that central 'motor' can be clustered for policy action. Figure 3.6 gives the same basic systems map, but clusters the factors to show that policy-makers do not need to be daunted by the spaghetti-like (or ecological) nature of the interactions. Factors were grouped as individual psychology, societal influences, food production, food consumption and so on.

Map 27

Weighted
Causal Linkages

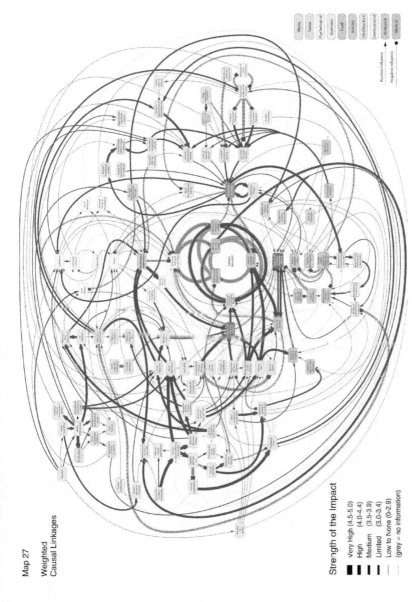

Strength of the Impact

■ Very High (4.5-5.0)
■ High (4.0-4.4)
▮ Medium (3.5-3.9)
— Limited (3.0-3.4)
— Low to None (0-2.9)
— (grey = no information)

Positive Influence
Negative Influence

Media
Social
Psychological
Economic
Food
Activity
Infrastructure
Developmental
Biological
Medical

Figure 3.5 The UK Foresight Obesity System Map, 2007

Source: Foresight 2007[99]

Map 0
Full Generic Map

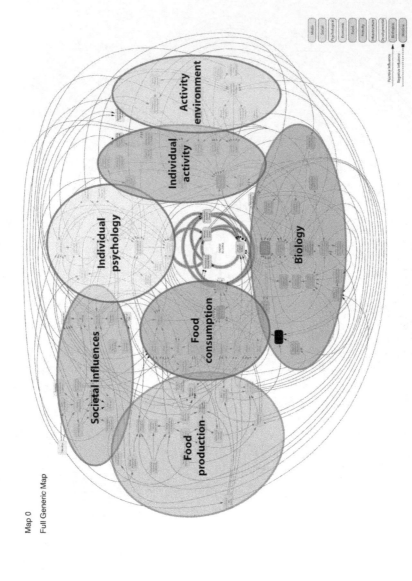

Figure 3.6 The UK Foresight Obesity System Map, with policy clusters, 2007

Source: Foresight 2007[99]

Coronary heart disease (CHD)

As was shown in Table 3.2, a quarter of all annual global deaths in 2012 – 14.1 million people – were attributed to cardiovascular diseases such as ischaemic heart disease and strokes. The main risk factors for heart disease are high blood pressure, smoking, fats, physical activity, obesity, diabetes, income and gender. Around those broad factors a more complex web is woven, in which diet has long been known to be key. In 2004, the WHO could state categorically that: '[w]hile genetic factors play a part, 80% to 90% of people dying from coronary heart disease have one or more major risk factors that are influenced by lifestyle'.[100] Nutritional risk factors include trans-fatty acids ('transfats'), high sodium intake (salt), overweight, and high alcohol intake. Recommendations for reducing CVD include: regular physical activity, consumption of diets based on plants, fish and fish oils, vegetables and fruits, including berries, lowering salt intake, and low to no alcohol intake.

Faced with high CHD incidence, public health policy has tended to take a number of approaches: health education as prevention, and improved medical treatment through drug, hospital and surgical care. It has also urged behavioural change, in particular a reduction of total fat intake and especially of saturated fats (mainly from animal meat and dairy fats). It might appear to have had some effect as rates of death from heart disease are declining in most affluent Western countries, after years of steady increase since the immediate post-World War II period. In fact the drop in deaths is mostly related to available healthcare. [101] And, barring miraculous healthcare magic bullets, the bill for that seems inexorably to rise.

CHD is rising in developing countries. There were steep rises in CHD in the newly independent countries of Eastern Europe when the Soviet Union collapsed. Even in countries considered to have a healthy diet, like Greece and Japan, social change is being accompanied by changing patterns of diet-related disease: Greece's CHD and obesity rates rose as it changed to a more Northern European diet high in animal fats, following entry to the European Union and increased tourism. In 1981 Trowell summarised the emergence of CHD amongst East Africans: autopsies in the 1930s had shown zero CHD in East Africa, and only one case among 2,994 autopsies conducted in Makere University Medical School over the period 1931–46. However, by the 1960s CHD in that region was becoming a major health problem.[69] In China, from the 1990s as it modernised and its economy expanded rapidly, the CHD incidence also grew. Its fat intake rose – from low levels compared to those found in Western populations – particularly among urban populations where a more affluent lifestyle was being adopted.[102] Daily intake of meat, eggs and cooking oil increased while proportional intake of legumes and cereals decreased. The official recommendation in China is to eat no more than 30 per cent of energy from fat,[78] but one study in 2012 found that the number of children eating a diet with more than that was now nearly 50 per cent. Fat intake growth was most advanced among low-income urban groups and high-income rural groups.[103]

Food-related cancers

In 1981, Richard Doll (who first produced firm epidemiological evidence on tobacco's role in premature death)[104] and Richard Peto suggested that dietary factors then accounted for around 30 per cent of cancers in Western countries, making diet second only to tobacco as a preventable cause of cancer.[105] The following year, the US National Academies of Science produced their landmark report stressing the diet-cancer connection.[106] Twenty years later, at the turn of the millennium, the 'Western' diet-cancer pattern had spread to developing countries, with diet accounting for an estimated 20 per cent of cancers.[107] In a few decades, the evidence and the process itself have both accelerated. The aetiology and the preventive messages to food policy-makers are remarkably clear and consistent. Diet and physical activity affect cell cycles, carcinogen metabolism, DNA repair, cell proliferation, hormonal regulation, inflammation and immunity, apoptosis (the process of programmed cell death), and more.[22] Much as with other nutritional impacts, cancer is affected by behaviour over the life-course. Figure 3.7 gives an illustration of how different factors work, taken from the major 2007 WCRF/AICR overview.[22]

Diet and physical activity are not the only reason cancers are increasing worldwide. They sit in a web of factors, of course; but they are highly significant. This is preventable. By virtue of steadily ageing populations, cancer could further increase by 50 per cent to 15 million new cases a year by 2020. In 2000, 6.2 million people died of cancer worldwide (12.5 per cent of all deaths), but 22.4 million were living with cancer. In 2012, 14.1 million new people were diagnosed with cancer and more than half the total incidence worldwide is now in developing countries. In 2012, an estimated 8.2 million people died from cancer worldwide, more than 60 per cent of these were in less developed regions of the world. Tobacco is the biggest cause, but diet, weight, alcohol and physical activity are causal factors in other common sites: breast, bowel and prostate. Red and processed meats are associated with bowel cancers.[108] Indeed, many cancers could be prevented by modifying dietary habits to include more fruits, vegetables, high-fibre cereals, fats and oils derived from vegetables, nuts, seeds and fish, and by limiting the intake of animal fats derived from meat, milk and dairy products. A number of published studies show that an increase in antioxidant nutrients such as beta-carotene, vitamins C and E, zinc and selenium could also decrease the risk of certain cancers and there seems to be strong evidence that eating a diet rich in fresh fruit and vegetables will reduce the risk of stomach cancer.[23] Yet the nutrition transition is being driven in a different direction – towards a diet actually higher in processed foods and animal fats, in short the output of key food industries within the Productionist paradigm.

Diabetes

The incidence of Type 2 diabetes is rising globally. This form of diabetes was formerly known as non-insulin dependent diabetes mellitus (NIDDM), occurring when the body is unable to respond to the insulin produced by the pancreas; it

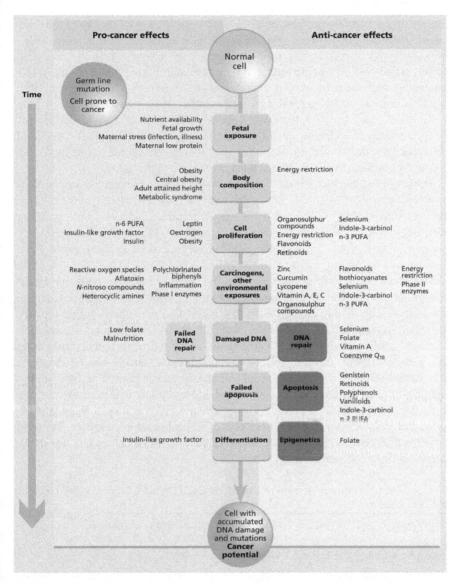

Figure 3.7 The influences of food, nutrition, physical activity and obesity on diet's impact on cancer processes

Source: WCRF/AICR 2007[22: p 45]

accounts for around 90 per cent of cases worldwide. In Type 1 diabetes (formerly known as insulin-dependent), the pancreas fails to produce the insulin which is essential for survival; this form develops most frequently in children and adolescents, but is now being increasingly noted later in life.[98] In 2000 the WHO anticipated that cases of Type 2 diabetes would rise from 150 million in 1997 to 300 million by 2025, with the greatest number of new cases being in China and India. In fact, by 2013 there were already 362 million, with 46 per cent of those estimated as undiagnosed. Table 3.6 gives the regional figures for 2013 and estimated growth in diabetes for each region, which is expected to rise by 55 per cent to 592 million by 2035. This is a major public health calamity.

The profile of diabetes has changed. The majority now live in low- and middle-income countries. China and India, the world's most populous countries, also have the most diabetics. The top ten by incidence are: China 98.4 million, India 65.1, USA 24.4, Brazil 11.9, Russian Federation 10.9, Mexico 8.7, Indonesia 8.5, Germany 7.6, Egypt 7.5 and Japan 7.2.[46] Every year 5.1 million deaths have diabetes as a factor.

Seventeen per cent of all live births in 2013 were to women with some kind of high blood glucose in pregnancy. Diabetes in adult life is associated with low birthweight,[109] and studies in India suggest that poor inter-uterine growth, combined with obesity later in life is associated with insulin resistance, diabetes and increased cardiovascular risk.[110] Once again, a single disease seems attributable to a pattern of poor nutrition related to the life cycle, and is one whose costs are externalised onto society as a whole and healthcare in particular. Devastating complications of diabetes, such as blindness, kidney failure and heart disease, are imposing a huge financial burden. The healthcare bill was now estimated to be $584bn in 2013, accounting for 11 per cent of all healthcare expenditure. Thus the drive for cheaper food has led to an externalised cost which is not paid for at the check-out or by the supply chain.

Table 3.6 Diabetes numbers worldwide, by International Diabetes Federation region, 2013 and 2035 estimates

IDF Region	People with diabetes, 2013	Growth anticipated by 2035, %
North America and Caribbean	37m	35.3
South and Central America	24m	59.8
Europe	56m	22.4
Middle East and North Africa	35m	96.2
Africa	20m	109.1
South East Asia	72m	70.6
Western Pacific	138m	46.0
TOTAL WORLD	**382m**	**592m**
Undiagnosed of the above	46%	

Source: IDF Atlas 2014[46]

Food safety and foodborne diseases

While the nutrition transition thesis has drawn attention on food's health impact towards non-communicable disease, foodborne diseases and other food safety concerns are a significant part of the overall health picture, too. The WHO reports that foodborne and waterborne diarrhoeal diseases kill an estimated 2.2 million people annually, mostly children. Diarrhoea is the most common foodborne illness caused by pathogens, much of this waterborne.[111]

Table 3.7 gives possible links for food and pathogens which threaten health. Food safety problems as a whole include risks from:

- veterinary drug and pesticide residues;
- food additives;
- pathogens (i.e. illness-causing bacteria, viruses, parasites, fungi and their toxins);
- environmental toxins such as heavy metals (e.g. lead and mercury);
- persistent organic pollutants such as dioxins;
- unconventional agents such as prions associated with bovine spongiform encephalopathy (BSE or 'mad cow disease').

Since the 1990s, scandals over food safety led to new regimes of traceability which put the onus on companies to ensure safety down the supply chain, and is a reminder that they have a big responsibility for public health.[112,113] The attention food safety receives is predictably higher in affluent countries when, on the evidence, the burden of ill health from poor food hygiene is actually far greater in the developing world, due to lack of investment and infrastructure, including drains, housing, water supplies and food control systems. In 2002, the WHO's World Health Report pointed out that, in developing countries, water supply and general sanitation remain the fourth highest health-risk factor, after underweight, unsafe sex and blood pressure.[14] In developing countries which are building their food export markets, there is too often a mixed structure, with higher standards for foods for export to affluent countries than for domestic markets. There ought to be a cascading down into internal markets of these higher standards.[114]

From the late 19th century, the early industrializing countries began to make dramatic public health improvements by vast investment in engineering and town planning – to secure water supplies, to create drains and to build safer street and covered markets.[84] Part of that investment included the introduction of effective monitoring and hygiene practice systems, such as the establishment of local authority laboratories and training, the packaging of foods and processes such as milk pasteurisation.[115] Today, public health proponents are actively trying to promote a 'second wave' of food safety intervention but this time using a risk-reduction management system known as Hazard Analysis Critical Control Points (HACCP), an approach first developed for the US space industry and designed to build safety awareness and control of potential points of hygiene breakdown into food handling and

Table 3.7 Pathogens which may be transmitted by an infected food handler

Astroviruses

Bacillus cereus

Campylobacter jejuni

Clostridium perfringens

Cryptosporidium species

Entamoeba histolytica

Shiga toxin-producing E. coli

Enterotoxigenic E. coli

Giardia intestinalis

Hepatitis A virus

Nontyphoidal Salmonella

Noroviruses

Rotaviruses

Salmonella Typhi*

Sapoviruses

Shigella species

Staphylococcus aureus

Streptococcus pyogenes

Taenia solium – cysticercosis

Vibrio cholera

Yersinia enterocolitica

Source: HHS/CDC 2014 [156]

management systems. HACCP also encourages the creation of a 'paper trail' of records to enable tracking along the production process, essential in order to obviate errors and enable learning (but adding bureaucratic burdens say the critics). Breakdowns in food safety have in the past led to major political and business crises, with governments under attack and new bodies responsible for food safety being set up in many countries. As food supply chains become more complex and as the scale of production, distribution and mass catering increases, so the chances for problems associated with food contamination rise; mass production breakdowns in food safety spread contamination and pathogens widely.

Antibiotics

Intensive agriculture, and particularly intensive meat production, are unlikely to have taken their current form without veterinary inputs and support. Key to this has been the widespread use of antibiotics or antimicrobials, but this use has had the effect of creating antibiotic or antimicrobial resistance (AMR). This is like a constant treadmill; bacteria which were constrained by an antibiotic drug build up resistance and make the drug ineffective. 'Wonder drugs' thus generate 'super-bugs'.[116] This effect follows Darwinian principles and was in fact anticipated as a danger even by developers. Alexander Fleming, the discoverer of penicillin, and René Dubos, the discoverer of gramicidin (an antibacterial agent that inhibits the growth of gram-positive bacteria) predicted AMR in the 1940s. When accepting his Nobel Prize in 1945 Fleming worried that the time might come when penicillin 'can be bought by anyone in the shops' and thus inevitably exposing microbes to non-lethal quantities of the drug and 'make them resistant'.[117] The fact of antimicrobial resistance had already been observed in the 1930s.[118]

Despite this, the new drugs received a light regulatory touch. After studies in the 1940s noted how animals grew faster if antibiotics were added as feed supplements, the USDA approved their use in 1951 without veterinary prescription.[119] In 1977, a few decades later, the US General Accounting Office raised concerns about the cost to human health of excessive use of antibiotics in intensive animal rearing.[120] Four decades later, the situation now alarms medical authorities.[121] The effectiveness of antibiotics which have saved tens of millions of lives in the last century is being radically undermined by:

*   excessive prescription by doctors;
*   poor use by patients (e.g. not completing a course of treatment);
*   routine use of antibiotics as growth promoters in intensive animal husbandry and in veterinary practice;
*   the capacity of bacteria to adapt and produce new antibiotic-resistant strains.

This is a depressing picture. By the turn of the millennium, there was ample scientific evidence for the gradual erosion of the effectiveness of antibiotics on farms, yet still their use continued.[122] In 1998, a UK House of Lords Committee on Science and Technology concluded that imprudent use of antibacterial drugs had made many new drugs worthless.[123] The only way to retain effectiveness was to restrict their use, their Lordships concluded. A 1997 WHO conference had also recommended the termination of the use of antibiotics as growth promoters in farm animals, if they were also used for human health.[124] In 1986, Sweden banned the use of growth promoters. The UK had banned the use of penicillin and tetracycline for growth promotion as early as the 1970s. Denmark banned virginiamycin in January 1998 and reduced its use of antimicrobials in food by 54 per cent by 2001. This caused a loss of €1.04 per pig; but there were no economic losses for poultry. The entire strategy was judged a success when the World Health Organisation reviewed it.[125] Other countries' health authorities called for voluntary reductions. This piecemeal policy situation rolled on, and on. Why? In the food system, a key reason is that antibiotics have been made central to the economics of intensive meat production. The US Food and Drug Administration (FDA) reported that 13.5 million kg of antimicrobials – about 80 per cent of the total – were sold for use in US agriculture in 2011.[126] This compared to less than

3.3 million kg of antibacterials sold for use in human medicine in the same year. [127] The US leads the world by intensity of use, but China is the largest antibiotics producer and consumer in the world. In a 2007 survey, the estimated annual antibiotics production in China was 210 million kg, of which 46.1 per cent were used in livestock industries.[128] There is a growing conviction among medical and veterinary science, and even amongst some food (alas, not all) producers, that this situation is unacceptable.

Antibiotics have been a 20th-century medical 'miracle', underpinning intensive animal husbandry and cheap meat. Researchers and official bodies have warned for decades about the growth of AMR. So alarmed by this situation was the English Chief Medical Officer, Professor Sally Davies,[121 129 130] that she persuaded the UK Prime Minister to ask a former Goldman Sachs chief economist Jim O'Neill to review the costs of inaction. His committee's 2014 report in effect restated what others had concluded but added a new twist by forecasting the penalties of not acting.[131] The prognosis was sober: 300 million preventable deaths by 2050, and a $100.2 *trillion* cost.[132] Will these figures stop routine animal use? Will doctors reduce prescriptions? Will pigs and poultry, if not fly, at least be managed in line with ecological public health principles? Time will tell. Certainly, the huge bill for inaction ought to encourage pharmaceutical giants who have not invested in new categories of drugs to begin to do so.

Inequalities, food poverty and food security

According to the World Bank in 2014, about 1 billion people live on a daily income of less than US $1, its definition of extreme poverty. And in 2010, 2.4 billion people lived on less than US $2, the average poverty line in developing countries and often used as an indicator of deep deprivation. This was only a slight decline from 2.59 billion in 1981.[133] Although most of the world's poor live in South and East Asia, sub-Saharan Africa has the fastest growing proportion of people who live in poverty. But poverty is not a matter of developing country economies. It's also something affected by gender and age; women, children and older people are at greatest risk. The health of the poor is, concomitantly, at risk from environmental hazards such as unsafe food and water, and from urban hazards such as air, and water pollution and accidents. In 1900 around 5 per cent of the world's people lived in cities with populations exceeding 100,000. Today, it is over half, and continues to rise.

It has been estimated that almost half of the world's wealth is now owned by just 1 per cent of the population, and the wealth of the 1 per cent richest people in the world amounts to $110 trillion, which is 65 times the total wealth of the bottom half of the world's population. That bottom half of the world's population owns the same as the richest 85 people in the world.[134] This is an unequal world.[135] This matters to food policy for two reasons. First, even the super-affluent people cannot eat as much as 3.5 million people, but the skewed wealth can hinder the poorest eating adequately for their health; lack of money stops them eating. And secondly, there is sound evidence that more equal societies are healthier.[136-138]

This is probably because they have a stronger sense of what Amartya Sen termed 'entitlement', the sense of having a right to food.[52] It also signifies a collective culture, a baseline of societal assumptions. Attempting to confront the startling rise in income inequality in 2000, the UN agreed some Millennium Development goals to be met by 2015.[133] Food featured in six out of eight of these goals.[139] These have been superseded by Sustainable Development Goals agreed in late 2015.[140] They include a commitment (yet again) to end rather than halve hunger, which was what the MDGs proposed back in 2000.

Worthy though such goals are, some critics argue they are unlikely to narrow inequalities unless there are firm policies to redistribute wealth from the rich to the poor,[138 141] but this is political anathema to right-wing politicians. Food sits in the crossfire. The widening disparity between social classes means that the rich, even in poor societies, have access to healthier dietary choices. The poor living in rural areas may have a more restricted diet based on a staple food usually grown on their own land, but the urban poor eat only what they can afford to buy, often suffering vitamin and mineral deficiencies as a result.

In public policy, the term 'food security' is often invoked in discussions about how to feed the under-nourished. The 1996 World Food Summit defined food security as the situation in which at the individual, household, national, regional and global levels 'all peoples, at all times, have physical and economic access to sufficient, safe and nutritious food to meet their dietary needs and food preferences for an active and healthy life'.[142] In the 1940s, when the WHO and the FAO were set up under the new United Nations, the policy priority was to increase food supply in every continent and nation; this macro-focus was still the dominant paradigm at the time of the 1974 World Food Conference, after which research on food security has mushroomed, and by the 1980s attention had shifted more towards household and individual access to food and towards improving what might be called 'micro-food' security. Four core foci emerged:

- *sufficiency* of food for an active healthy life;
- *access* to food and entitlement to produce, purchase or exchange food;
- *security* in the sense of the balance between vulnerability, risk and insurance;
- *time* and the variability in experiencing chronic, transitory and cyclical food insecurity.

Summing up thinking in this micro-focus period, one research team remarked that 'flexibility, adaptability, diversification and resilience are key words. Perceptions matter. Intra-household issues are central ... Food security must be seen as a multi-objective phenomenon, where the identification and weighting of objectives can only be decided by the food insecure themselves'.[143] Simon Maxwell, who coordinated the research after working on food security in the Sudan in the late 1980s, argued:

A country and people are food secure when their food system operates in such a way as to remove the fear that there will not be enough to eat. In particular, food security will be achieved when the poor and vulnerable, particularly women and children and those living in marginal areas, have secure access to the food they want. Food security will be achieved when equitable growth ensures that these people have sustainable livelihoods. In the meantime and in addition, however, food security requires the efficient and equitable operation of the food system.[141]

This debate about whether food security can be resolved by others or by the people themselves has long historical and political roots. The first director of the FAO, John Boyd Orr, mapped out what he thought the post-war vision to ensure food security should be. This was a vision largely framed by the experience of rich countries in the 'Hungry Thirties'.[144] The focus was on availability and to increase the food supply. Boyd Orr had mapped the policy package he thought was needed in 1942, before the FAO even was created, thus:[37]

1   Countries should set targets within a new global system and foster intergovernmental cooperation to help each other over good times and bad, to ease out booms and slumps in production.
2   Targets should be based on science, above all on nutrition and agricultural science.
3   Targets should be set to achieve health. Premature death from under-nutrition is inexcusable; investment in better food will yield health and economic gains and savings.
4   Agriculture should be financially and politically supported to produce more.
5   Industry should be geared to produce tools to enable agricultural productivity to rise: new buildings, tractors, equipment.
6   Trade should be encouraged to meet the new markets and to ease the over-productive capacity of some areas and match them with under-consumption in other areas.
7   International cooperation will have to follow the (proposed) UN Conference on Food and Agriculture.
8   New organisations will have to be created such as a new International Food and Agricultural Commission, National Food Boards to monitor supplies, Agricultural Marketing Boards, Commodity Boards.

Nearly a century later, this vision is still remarkable, but it lacks three elements central to the thesis of this book. First, there is no recognition that the model of agriculture envisaged could or would be a major driver of environmental damage, and that this would threaten food system viability in the 21st century – issues addressed in Chapter 4. Secondly, the Boyd Orr generation did not ever imagine that there could be the complicated health pattern associated with the Nutrition Transition and other health effects summarised in this chapter. And thirdly, believing in a benign, rational State, they hoped that policy-making would no longer centre on a narrow individualism, yet this in fact happened when the Keynesian economic thinking that dominated public policy from World War II was replaced in the 1970s by neo-liberal economic thinking. This demonised the state in general and state welfare in particular. Food poverty and inequality have grown once more as Chicago economics triumphed.[145]

The 1940s architects of Productionism would have recognised (but despaired to see) food poverty remain and even return. This, after all, was part of the motivation for Productionism: to provide more affordable food for all. From the 1980s, a new generation of researchers and campaigners began once more to identify pockets of food insecurity (as this was often now called) even in the richest economies. A series of studies in the UK, where inequalities in income and health widened under the Conservative government of 1979–97, as it introduced neo-liberal macro-economic policies, charted the gradual re-emergence of poor diet due to lack of disposable income.[146] Earlier studies had shown that welfare systems could trap people in low income, or subsidise low wages, but there was new evidence that relative poverty had an effect on diet and health. Welfare safety nets such school meals – championed as solutions by the 1930s thinkers – were not fulfilling their social intention to provide a decent nutritional meal.[147-149] Lower socio-economic groups were found to experience a greater incidence of premature and low birth-weight babies, and of heart disease, stroke and some cancers in adults. Risk factors such as bottle-feeding, smoking, physical inactivity, obesity, hypertension, and poor diet were clustered in the lower socio-economic groups whose diet traditionally derives from cheap energy forms such as meat products, full-cream milk, fats, sugars, preserves, potatoes and cereals with little reliance on vegetables, fruit and wholemeal bread.[150] Essential nutrients such as calcium, iron, magnesium, folate and vitamin C are more likely to be ingested by the higher socio-economic groups: their greater purchasing power creates a market for healthier foods such as skimmed milk, wholemeal bread, fruit and other low-fat options.[151]

A common picture of modern food poverty was being sketched in North America and the West.[152] New charitable welfare such as food banks, 'second harvest' (food from restaurants) sprouted, as welfare was cut back, putting pressure on health systems.[153] In the US, hunger rose during the 1990s, with the Census Bureau calculating that 11 million Americans lived in households which were 'food insecure', with a further 23 million living in households which were 'food insecure without hunger' (in other words at risk of hunger).[154] Other US surveys of the time estimated that at least 4 million children aged under 12 years were hungry and an additional 9.6 million were at risk of hunger during at least one month of the year. Despite political criticisms of these surveys, further research suggested that even self-reported hunger, at least by adults, is a valid indication of low intakes of required nutrients. It should be noted that, ironically, the US, spent over $25bn on federal and state programmes to provide extra food for its 25 million citizens in need of nutritional support.[154] By the 2010s, this situation had been normalised; the shock of the new had worn off. One US NGO summarised the situation, using official statistics, in 2012 thus:[155]

•   49.0 million Americans living in food insecure households, 33.1 million adults and 15.9 million children;

- 14.5 per cent of households (17.6 million households) were food insecure;
- 5.7 per cent of households (7.0 million households) experienced very low food security;
- households with children reported food insecurity at a significantly higher rate than those without children, 20.0 per cent compared to 11.9 per cent;
- households that had higher rates of food insecurity than the national average included households with children (20.0 per cent), especially households with children headed by single women (35.4 per cent) or single men (23.6 per cent), Black non-Hispanic households (24.6 per cent) and Hispanic households (23.3 per cent).

Implications

This chapter has sketched a complex global picture of diet-related health. For a century, epidemiologists and social researchers have generated many facts, figures and arguments about the role of food in the creation and prevention of ill health, linking what humans eat with patterns of disease. They raised and answered a number of important questions: how much of a risk does poor diet pose? A lot. What proportion of the known incidence of key diseases like cancer, heart disease, diabetes and microbiological poisoning can be attributed to the food supply? A lot. What levels of certainty can be applied to the many studies that have been produced? There is considerable scientific agreement about the core picture. Can people change their diet? Of course! The rapid nutrition transition shows people can and do, but the shape this takes carries major consequences.

For policy-makers, these questions and the uncomfortable evidence point to a policy failure. Exhorting the public to look after itself does not seem to stop the damage. In Chapter 5 we review the state of policy and whether the responses are adequate and in the rest of the book we consider why this is so. But first, in Chapter 4, we consider food's impact on the environment, asking: what is a sustainable food system? And how does human health fit eco-systems health?

References

1   Hill C. *The World Turned Upside Down: Radical Ideas during the English Revolution.* London: Temple Smith, 1972.
2   Popkin BM. An overview on the nutrition transition and its health implications: the Bellagio meeting. *Public Health Nutrition,* 5(1A) (2002): 93–103.
3   Popkin BM. The nutrition transition in the developing world. *Development Policy Review,* 21 (2003): 581–97.
4   Murray CJL, Lopez AD, eds. *The Global Burden of Disease: A Comprehensive Assessment of Mortality and Disability from Diseases, Injuries and Risk Factors in 1990 and Projected to 2020.* Cambridge, MA: Harvard School of Public Health on behalf of the World Health Organisation and the World Bank, 1996.
5   The Global Burden of Disease Study 2010. *The Lancet,* 380(9859) (2012): 2053–2260.
6   FAO. Hunger statistics and hunger map. <http://www.fao.org/hunger/en> Rome: Food and Agriculture Organisation, 2014.

7   WHO. *Obesity and Overweight*. Fact Sheet 311. Geneva: World Health Organisation, 2013.

8   Keys A. Coronary heart disease in seven countries. *Circulation*, 41(suppl. 1) (1970): 1–211.

9   McCarrison SR. *Nutrition and National Health: Three Lectures Delivered Before the Royal Society of Arts on February 10th, 17th and 24th, 1936*. London: Royal Society of Arts, 1936.

10 Burkitt D. Some diseases characteristic of modern Western civilisation. *British Medical Journal*, 1 (1973): 274–8.

11 Cleave TL. *The Saccharine Disease*. Bristol: John Wright & Sons, 1974.

12 WHO. *Diet, Nutrition and the Prevention of Chronic Diseases*. Technical Report Series, 797. Geneva: World Health Organisation, 1990.

13 Century CotNCots. *Ending Malnutrition by 2020: An Agenda for Change in the Millennium*. Final Report to the ACC/SCN. *Food and Nutrition Bulletin*, 21(2) suppl. (2000).

14 WHO. *World Health Report 2002: Reducing Risks, Promoting Healthy Life*. Geneva: World Health Organisation, 2002.

15 WHO. *World Health Report 2003: Shaping the Future*. Geneva: World Health Organisation, 2003.

16 WHO. Global strategy on diet, physical activity and health. 57th World Health Assembly. WHA 57.17, agenda item 12.6. Geneva: World Health Assembly, 2004.

17 WHO/FAO. *Diet, Nutrition and the Prevention of Chronic Diseases*. Report of the joint WHO/FAO expert consultation. WHO Technical Report Series, 916 (TRS 916). Geneva: World Health Organisation & Food and Agriculture Organisation, 2003.

18 Morris JN, Heady JA, Raffle PAD, *et al.* Coronary heart disease and physical activity of work. *Lancet*, 265(6796) (1953): 1111–20.

19 Morris JN, Kagan A, Pattison DC, *et al.* Incidence and prediction of ischaemic heart-disease in London busmen. *The Lancet*, 2(7463) (1966): 553–9.

20 WHO/IARC. *World Cancer Report World Health*. Geneva: World Health Organisation/International Agency for Research on Cancer, 2003.

21 WCRF/AICR. *Food, Nutrition and the Prevention of Cancer: A Global Perspective*. London and Washington, DC: World Cancer Research Fund/American Institute for Cancer Research, 1997.

22 WCRF/AICR. *Food, Nutrition, Physical Activity and the Prevention of Cancer: A Global Perspective*. Washington, DC/London: World Cancer Research Fund/American Institute for Cancer Research, 2007.

23 WCRF/AICR. *Policy and Action for Cancer Prevention – Food, Nutrition, and Physical Activity: A Global Perspective*. London/Washington, DC: World Cancer Research Fund/American Institute for Cancer Research, 2009.

24 De Onis M, Blössner M, Borghi E. Global prevalence and trends of overweight and obesity among preschool children. *American Journal of Clinical Nutrition*, 92(5) (2010): 1257–64.

25 Ritenbaugh C. Obesity as a culture-bound syndrome. *Culture, Medicine and Psychiatry*, 6(4) (1982): 347–61.

26 Royal College of Physicians. Obesity: report of the Royal College of Physicians. *Journal of the Royal College of Physicians of London*, 17(5) (1983): 5–65.

27 Tsomondo E, Jones J. Obesity: a disease of indolence and affluence. *Central African Journal of Medicine*, 20(1) (1974): 1–4.

28 Egger G, Swinburn B. An 'ecological' approach to the obesity pandemic. *British Medical Journal,* 315(7106) (1997): 477–80.

29 Critser G. *Fatland: How Americans Became the Fattest People in the World.* London: Penguin, 2003.

30 Spurlock M. *Don't Eat this Book: Fast Food and the Supersizing of America.* New York: Putnam/Penguin, 2006.

31 Schlosser E. *Fast Food Nation: What the All-American Meal is Doing to the World.* London: Allen Lane, 2001.

32 Patel R. *Stuffed and Starved: Markets, Power and the Hidden Battle for the World Food System.* London: Portobello, 2007.

33 Nestle M. *Food Politics.* Berkeley, CA: University of California Press, 2002.

34 Guthman J. *Weighing in: Obesity, Food Justice, and the Limits of Capitalism.* Berkeley, CA, and London: University of California Press, 2011.

35 Lustig R. *Fat Chance.* New York: Hudson Street Press, 2013.

36 Boyd Orr JL, Lubbock D. *The White Man's Dilemma.* London: George Allen & Unwin, 1953.

37 Boyd Orr SJ. *Food and the People: Target for Tomorrow No. 3.* London: Pilot Press, 1943.

38 Le Gros Clarke F, Pirie NW, eds. *Four Thousand Million Mouths: Scientific Humanism and the Shadow of World Hunger.* Oxford: Oxford University Press, 1951.

39 Fenelon K. *Britain's Food Supplies.* London: Methuen & Co., 1952.

40 FAO. *State of Food Insecurity 2012.* Rome: Food and Agriculture Organisation, 2012.

41 Surgeon General of the United States. *The Surgeon General's Report on Nutrition and Health.* Washington, DC: US Department of Health and Human Services, 1988.

42 International Diabetes Federation, World Diabetes Foundation. *Atlas of Diabetes.* 1st edn. Brussels: International Diabetes Federation, 2000.

43 Century. CotNCots. *Ending Malnutrition by 2020: An Agenda for Change in the Millennium.* Final Report to the ACC/SCN. *Food and Nutrition Bulletin,* 21(3 suppl.) (2000): whole issue.

44 Alwin DA, ed. *Global Status Report on Noncommunicable Diseases 2010.* <http://www.who.int/nmh/publications/ncd_report2010/en> Geneva: World Health Organisation, 2011.

45 World Heart Federation. *State of the Heart: Cardiovascular Report.* Geneva: World Heart Federation, 2014.

46 International Diabetes Federation. *Diabetes Atlas.* 6th edn. Brussels: International Diabetes Federation, 2014.

47 Dobbs R, Sawers C, Thompson F, *et al. Overcoming Obesity: An Initial Economic Analysis.* New York: McKinsey Global Institute, 2014.

48 WHO. The top 10 causes of death: The 10 leading causes of death in the world, 2000 and 2012. WHO Fact Sheets. Geneva: World Health Organisation, 2014.

49 Ewin J. *Fine Wines and Fish Oil: The Life of Hugh Macdonald Sinclair.* Oxford: Oxford University Press, 2001.

50 Alexandratos N, ed. *World Agriculture towards 2010: A FAO Study.* Chichester: J. Wiley & Son, 1995.

51 Woodham Smith CBFG. *The Great Hunger: Ireland, 1845–9.* London: Hamish Hamilton, 1962.

52 Sen A. *Poverty and Famines: An Essay on Entitlement and Deprivation.* Oxford: Oxford University Press, 1982.

53 Burninsma J, ed. *World Agriculture: Towards 2015/2030.* Rome and London: FAO and Earthscan, 2003.

54 Von Grebmer K, Ringler C, Rosegrant MW, *et al. Global Hunger Index 2012: The Challenge of Hunger: Ensuring Sustainable Food Security under Land, Water and Energy Stresses.* Washington, DC: IFPRI, Concern Worldwide and Welthungerhilfe and Green Scenery, 2012.

55 Von Braun J, Ruel M, Gulati A. *Accelerating Progress toward Reducing Child Malnutrition in India: A Concept for Action.* Washington, DC: International Food Policy Research Institute, 2008.

56 FAO. *State of Food Insecurity 2013.* Rome: Food and Agriculture Organisation, 2013.

57 Popkin BM. The nutrition transition in low-income countries: an emerging crisis. *Nutrition Reviews,* 52 (1994): 285–98.

58 Popkin BM. An overview on the nutrition transition and its health implication: the Bellagio Meeting. *Public Health Nutrition,* 5(1A) (2001): 93–103.

59 Caballero B, Popkin BM, eds. *The Nutrition Transition: Diet-Related Diseases in the Modern World.* New York: Elsevier, 2002.

60 Popkin B. *The World is Fat: The Fads, Trends, Policies and Products that are Fattening the Human Race.* New York: Avery/Penguin, 2009.

61 Pena M, Bacallao J. *Obesity and Poverty: A New Public Health Challenge.* Washington, DC: Pan American Health Organisation, 2000.

62 Lenfant C. Can we prevent cardiovascular diseases in low- and middle-income countries? *Bulletin of the World Health Organisation,* 79(10) (2001): 980–2.

63 Gardner G, Halweil B. *Underfed and Overfed: The Global Epidemic of Malnutrition.* Worldwatch Paper, 15. Washington, DC: Worldwatch Institute, 2000.

64 Robson I. Foreword, In: Trowell H, Burkitt D, eds. *Western Diseases: Their Emergence and Prevention.* London: Edward Arnold, 1981.

65 Trowell H, Burkitt D, eds. *Western Diseases: Their Emergence and Prevention.* London: Edward Arnold, 1981.

66 FAO. *State of Food Insecurity 2000.* Rome: Food and Agriculture Organisation, 2000.

67 Popkin BM. The nutrition transition and its health implications in lower income countries. *Public Health Nutrition,* 1(1) (1998): 5–21.

68 Popkin BM, Nielsen SJ. The sweetening of the world's diet. *Obesity Research,* 11(11) (2003): 1–8.

69 Trowell HC. Hypertension, obesity, diabetes mellitus and coronary heart disease. In: Burkitt DP, Trowell HC, eds. *Western Diseases: Their Emergence and Prevention.* London: Edward Arnold, 1981: 3–32.

70 Yudkin J. *Pure White and Deadly.* Harmondsworth: Penguin, 1972.

71 Lustig RH, Schmidt LA, Brindis CD. Public health: The toxic truth about sugar. *Nature,* 482 (2012): 27–9.

72 Action on Sugar. Launch of 'ACTION ON SUGAR'. <http://www.actiononsalt. org.uk/actiononsugar/Press%20Release%20/118440.html#sthash.EAaE1vx8.dpuf> London: CASH, 2014.

73 IFPRI. *Living in the City: Challenges and Options for the Urban Poor.* Washington, DC: International Food Policy Research Institute, 2002.

74 Verster A. Nutrition in transition: The case of the Eastern Mediterranean region. In: Pietinen P, Nishida C, Khaltaev N, eds. *Nutrition and Quality of Life: Health Issues for the 21st Century.* Geneva: World Health Organisation, 1996: 57–65.

75   Chen J, Campbell TC, Li J, *et al. Diet, Life-style and Mortality in China: A Study of the Characteristics of 65 Chinese Counties.* Oxford/Ithaca, NY/Beijing: Oxford University Press/Cornell University Press/People's Medical Publishing House, 1991.

76   Campbell TC, Campbell TM. *The China Study: The Most Comprehensive Study of Nutrition Ever Conducted and the Startling Implications for Diet, Weight Loss and Long-Term Health.* Dallas, TX.: Benbella, 2005.

77   Geissler C. China: the soybean-pork dilemma. *Proceedings of the Nutrition Society,* 58 (1999): 345–53.

78   Garnett T, Wilkes A. *Appetite for Change: Social, Economic and Environmental Transformations in China's Food System.* Oxford: Food Climate Research Network, University of Oxford, 2014.

79   Lang T. The public health impact of globalization of food trade. In: Shetty P, McPherson K, eds. *Nutrition and Chronic Disease: Lessons from Contrasting Worlds.* Chichester: John Wiley & Sons, 1997: 173–186.

80   Fogel RW. The global struggle to escape from chronic malnutrition since 1700. *Proceedings of the World Food Programme/United Nations University.* Rome: United Nations Food and Agriculture Organisation, 1997.

81   McKeown T. *The Modern Rise of Population.* London: Edward Arnold, 1976.

82   McKeown T. Fertility, mortality and causes of death: an examination of issues related to the modern rise of population. *Population Studies,* 32(3) (1978): 535–42.

83   McKeown T. *The Origins of Human Disease.* London: Basil Blackwell, 1988.

84   Rayner G, Lang T. *Ecological Public Health: Reshaping the Conditions for Good Health.* Abingdon: Routledge/Earthscan, 2012.

85   Stunkard AJ, Wadden TA, eds. *Obesity: Theory and Therapy.* New York: Raven Press, 1993.

86   WHO. Childhood overweight and obesity. <http://www.who.int/dietphysicalactivity/childhood/en> Geneva: World Health Organisation, 2014.

87   Finucane MM, Stevens GA, Cowan MJ, *et al.* National, regional, and global studies in body-mass index since 1980: systematic analysis of health examination surveys and epidemiological studies with 960 country-years and 9.1 million participants. *Lancet,* 377(9765) (2011): 557–67.

88   CDC. *Adult Obesity Facts.* <http://www.cdc.gov/obesity/data/adult.html> Atlanta, GA: Center for Disease Control and Prevention, 2014.

89   Stearns P. *Fat History: Bodies and Beauty in the Modern West.* New York: New York University Press, 2002.

90   Orbach S. *Fat is a Feminist Issue.* London: Hamlyn, 1978.

91   Dixon J, Broom D, eds. *Seven Deadly Sins of Obesity: How the Modern World is Making us Fat.* Sydney: University of New South Wales Press, 2007.

92   WHO. Obesity: *Preventing and Managing the Global Epidemic: Report of a Global Consultation on Obesity.* WHO/NUT/NCD/98.1. Geneva: World Health Organisation, 1998.

93   Davies SC. *Annual Report of the Chief Medical Officer: Surveillance Volume, 2012: On the State of the Public's Health.* London: Department of Health, 2014: 121.

94   WHO. *Obesity: Preventing and Managing the Global Epidemic. Report of a WHO Consultation.* WHO Technical Series, 894. Geneva: World Health Organisation, 2000.

95   CDC. *Defining Obesity and Overweight.* <http://www.cdc.gov/obesity/adult/defining.html> Atlanta, GA: Center for Disease Control and Prevention, 2014.

96   WHO. Mean Body Mass Index (BMI) 20+, age standardised: females 2008. <http://gamapserver.who.int/gho/interactive_charts/ncd/risk_factors/bmi/atlas.html> Geneva: World Health Organisation, 2011.

97   CDC. *State Adult Obesity Map.* <http://www.cdc.gov/obesity/data/adult.html> Atlanta, GA: Center for Disease Control and Prevention, 2014.

98   Tirosh Y. The right to be fat. *Yale Journal of Health Policy, Law, and Ethics*, 12(2) (2012): article 2 <http://digitalcommons.law.yale.edu/yjhple/vol12/iss2/2>

99   Foresight. *Tackling Obesities: Future Choices.* London: Government Office of Science, 2007. <https://www.nationalarchives.gov.uk/doc/open-government-licence/version/3>

100   MacKay J, Menshah GA. *The Atlas of Heart Disease and Stroke.* Geneva: World Health Organisation/CDC, 2004.

101   British Heart Foundation. *BHF 2012 European Cardiovascular Disease Statistics.* <http://www.bhf.org.uk/publications/view-publication.aspx?ps=1002098> Oxford and London: British Heart Foundation Health Promotion Research Group, University of Oxford, 2012.

102   Zhou B. Diet and cardiovascular disease in China. In: Shetty P, Gopalan C, eds. *Nutrition and Chronic Disease: An Asian Perspective.* London: Smith-Gordon, 1998: 47–49.

103   Cui Z, Dibley MJ. Trends in dietary energy, fat, carbohydrate and protein intake in Chinese children and adolescents from 1991 to 2009. *British Journal of Nutrition*, 108(7) (2012): 1292–9.

104   Doll R, Peto R, Borcham J, *et al.* Mortality in relation to smoking: 50 years' observations on male British doctors. *British Medical Journal*, 328 (2004): 1519.

105   Doll R, Peto R. The causes of cancer: quantitative estimates of avoidable risks of cancer in the United States today. *Journal of the National Cancer Institute*, 66(6) (1981): 1191–1308.

106   National Research Council. *Diet, Nutrition and Cancer.* Washington, DC: National Academy of Sciences, 1982.

107   Key TJ, Schatzkin A, Willett WC, *et al.* Diet, nutrition and the prevention of cancer. *Public Health Nutrition*, 7(1A) (2004): 187–200.

108   Cancer Research UK. *Worldwide Cancer: Cancer Statistics Key Facts.* London: Cancer Research UK, 2014.

109   Barker DJP, ed. *Fetal and Infant Origins of Adult Disease.* London: British Medical Journal, 1992.

110   Yajnik CS. Diabetes in Indians: small at birth or big as adults or both? In: Shetty P, Gopalan C, eds. *Nutrition and Chronic Disease: An Asian Perspective.* London: Smith-Gordon, 1998: 43–46.

111   WHO. *Strategic Plan for Food Safety Including Foodborne Zoonoses 2013–2022.* Geneva: World Health Organisation, 2013.

112   Nestle M. *Safe Food: Bacteria, Biotechnology and Bioterrorism.* Berkeley, CA: University of California Press, 2003.

113   Van Zwanenberg P, Millstone E. *BSE: Risk, Science, and Governance.* Oxford and New York: Oxford University Press, 2005.

114   Barling D, Lang T. Trading on health: cross-continental production and consumption tensions and the governance of international food standards. In: Fold N, Pritchard B, eds. *Cross-Continental Food Chains.* London: Routledge, 2005: 39–51.

115   Paulus I. *The Search for Pure Food.* Oxford: Martin Robertson, 1974.

116   Canon G. *Superbug: Nature's Revenge.* London: Virgin Publishing, 1995.

117   WHO. *World Health Day 2011: Combat Antimicrobial Resistance.* Geneva: World Health Organisation, 2011.

118 Davies J, Davies D. Origins and evolution of antibiotic resistance. *Microbiology and Molecular Biology Reviews*, 74(3) (2010): 417–33.

119 Castanon JIR. History of the use of antibiotic as growth promoters in European poultry feeds. *Poultry Science*, 86(11) (2007): 2466–71.

120 General Accounting Office. *Need to Establish Safety and Effectiveness of Antibiotics Use in Animal Feeds*. Washington, DC: General Accounting Office, 1977.

121 Davies SC, Grant J, Catchpole M. *The Drugs Don't Work: A Global Threat*. London: Penguin, 2013.

122 General Accounting Office. *The Agricultural Use of Antibiotics and its Implications for Human Health*. Washington, DC: General Accounting Office, 1999.

123 House of Lords. *Resistance to Antibiotics and Other Antimicrobial Agents*. Select Committee on Science and Technology. Seventh Report. London: HMSO, 1998.

124 WHO. *Medical Impact of the Use of Antimicrobials in Food Animals: Report of a WHO Meeting, Berlin, Germany, 12–17 October 1997*. Geneva: World Health Organisation, Division of Emerging and Other Communicable Diseases Surveillance and Control, 1997.

125 WHO. *Impacts of Antimicrobial Growth Promoter Termination in Denmark*. Geneva: World Health Organisation, 2002.

126 FDA. *2011 Summary Report on Antimicrobials Sold or Distributed for Use in Food-Producing Animals*. Washington, DC: Department of Health and Human Services, Food and Drug Administration, Center for Veterinary Medicine, 2013.

127 FDA. *Drug Use Review*. Washington DC: Department of Health and Human Services Public Health Service Food and Drug Administration Center for Drug Evaluation and Research Office of Surveillance and Epidemiology, 2012.

128 Hvistendahl M. China takes aim at rampant antibiotic resistance. *Science*, 336(6083) (2012): 795.

129 Davies SC. *Annual Report of the Chief Medical Officer*, vol. 1, *2011: On the State of the Public's Health*. London: Department of Health, 2012.

130 Davies SC. *Annual Report of the Chief Medical Officer*, vol. 2, *2011: Infections and the Rise of Antimicrobial Resistance*. London: Department of Health, 2013.

131 WHO. *Antimicrobial Resistance: Global Report on Surveillance 2014*. Geneva: World Health Organisation, 2014: 257.

132 Review on Antimicrobial Resistance. *Antimicrobial Resistance: Tackling a Crisis for the Health and Wealth of Nations*. London: Review on AMR; Wellcome Trust; HM Government, 2014.

133 World Bank. *Poverty Overview*. <http://www.worldbank.org/en/topic/poverty/overview> Washington, DC: World Bank, 2014.

134 Alvaredo F, Atkinson AB, Piketty T, *et al*. *The World Top Incomes Database*. <http://topincomes.g-mond.parisschoolofeconomics.eu> Paris: Paris School of Economics, 2013.

135 Oxfam. *Working for the Few: Political Capture and Economic Inequality*. Oxford: Oxfam, 2014.

136 Wilkinson RG. *Unhealthy Societies*. London: Routledge, 1996.

137 Wilkinson RG, Marmot M, eds. *Social Determinants of Health: The Solid Facts*. Copenhagen: World Health Organisation Regional Office for Europe, 2003.

138 Wilkinson RG, Pickett K. *The Spirit Level: Why More Equal Societies Almost Always Do Better*. London: Allen Lane, 2009.

139 United Nations Development Programme. *Millennium Development Goals*. <http://www.undp.org/mdg/basics.shtml> New York: United Nations, 2000.

140 United Nations. *Sustainable Development Goals*. New York: United Nations Department of Economic and Social Affairs, Division for Sustainable Development, 2014.

141 Piketty T. *Capital in the Twenty-First Century*. Cambridge, MA: Belknap Press of Harvard University Press, 2014.

142 FAO. *Rome Declaration on World Food Security and World Food Summit Plan of Action. World Food Summit 13–17 November 1996*. Rome: Food and Agriculture Organisation, 1996.

143 Maxwell S, Smith M. Household food security: a conceptual review. In: Maxwell S, Frankenberger T, eds. *Household Food Security: Concepts, Indicators, Measurements: A Technical Review*. New York/Rome: UNICEF and IFAD, 1995.

144 Boyd Orr J. *As I Recall: The 1880s to the 1960s*. London: MacGibbon & Kee, 1966.

145 Cockett R. *Thinking the Unthinkable: Think-Tanks and the Economic Counter-Revolution, 1931–1983*. London: HarperCollins, 1994.

146 Leather S. *The Making of Modern Malnutrition: An Overview of Food Poverty in the UK*. London: Caroline Walker Trust, 1996.

147 Walker C, Church M. Poverty by administration: a review of supplementary benefits, nutrition and scale rates. *Journal of Human Nutrition*, 32 (1978): 5–18.

148 Nelson M. Nutritional goals from COMA and NACNE: how can they be achieved? *Human Nutrition Applied Nutrition*, 39(6) (1984): 456–64.

149 Nelson M, Paul A. The nutritive contribution of school dinners and other mid-day meals to the diets of schoolchildren. *Human Nutrition: Applied Nutrition*, 37A (1983): 128–35.

150 James WPT, Nelson M, Ralph A, *et al.* Socio-economic determinants of health: the contribution of nutrition to inequalities in health. *British Medical Journal*, 314(7093) (1997): 1545–9.

151 Department of Health, *Low Income, Food, Nutrition and Health: Strategies for Improvement*. A report by the Low Income Project Team for the Nutrition Taskforce, London: Department of Health, 1996.

152 Riches G. *First World Hunger: Food Security and Welfare Politics*. London: Macmillan, 1997.

153 Seed B, Lang T, Caraher M. Exploring public health's roles and limitations in advancing food security in British Columbia, Canada. *Canadian Journal of Public Health*, 105(6) (2014): e324–329.

154 Eisinger PK. *Towards an End to Hunger in America*. Washington, DC: Brookings Institute Press, 1998.

155 Feeding America. *Hunger in America: Hunger and Poverty Statistics*. <http://feedingamerica.org/hunger-in-america/hunger-facts/hunger-and-poverty-statistics.aspx> [accessed July 2014]. Washington DC: Feeding America, 2014.

156 HHS/CDC. *Diseases Transmitted through the Food Supply*. Washington, DC/Atlanta, GA: Department of Health and Human Services/Centers for Disease Control and Prevention, 2014.

4   Food, environment and sustainability

In an unequal agricultural society, with primitive techniques, where men were at the mercy of nature and starved if the harvest failed; where plagues and warfare made life uncertain; it was easy to see famines and epidemics as punishments for human wickedness. As long as the level of technique was too low to liberate men from nature, so long were they prepared to accept their helplessness before a God who was as unpredictable as the weather. Sin, like poverty and social inferiority, was inherited.

Christopher Hill, historian, 1912–2003[1]

We basically have three choices: mitigation, adaptation and suffering. We're going to do some of each. The question is what the mix is going to be. The more mitigation we do, the less adaptation will be required and the less suffering there will be.
John Holdren, President, American Association for the Advancement of Science[2]

Core arguments

Food is a major source of environmental damage, yet food production and consumption are reliant on the existence and continuity of natural resources such as water, soil, land and the biodiversity of flora and fauna. Human health depends on good eco-systems, yet in pursuit of production, the food system is a major force in damaging what it depends upon. It is biting the hand that feeds us. The infrastructure of food systems is under strain, both by misuse of the land and by wasteful systems of farming, processing, distribution and consumption. Productionism was supposed to get rid of waste on and around the farm but it has serviced new forms of waste. Consumers in rich societies over-eat and under-value surplus food. This policy mess symbolises the Productionist era in which food production's environmental impact has become ever more intense, immense and extensive. Evidence has mounted that climate change, water stress, biodiversity loss, soil damage, pollutants and waste each pose major threats to food production. All effects are heightened by the historically rapid process of urbanisation and they pose a massive collective challenge to the 21st century aspiration to deliver good food for all everywhere. No wonder policy-makers are nervous about what to do. To eat within environmental limits, as implied by the Ecologically Integrated paradigm, suggests considerable change in what and how much Western societies eat, particularly in relation to meat and dairy consumption, since these have especially high environmental impacts. Supporters of Productionism, by contrast, see investment in the Life Sciences as a renewed commitment to maximise output, to keep and spread a model of food culture based on consumer choice and plenty. How society conceives of the environment – as something to be 'mined' and contained or as something to be nurtured and integrated into food systems – is a fundamental schism in the Food Wars.

The challenge posed by the food system's impact on the environment is incontrovertible. A UN review published in 2009 described it as a crisis for the food system,[3] echoing an earlier review of global eco-systems.[4] Food's role in shaping the environment is both real – in that the environment can be severely distorted or damaged by food production and consumption – and ideological – in that a food culture has emerged in which consumers are disconnected from the consequences of food supply chains. 21st-century policy-makers are thus in a double-bind. On the one hand, they ignore the environmental evidence at their peril. On the other hand, they want to keep consumers happy with plentiful, cheap food, which symbolises 20th century progress.

Stark choices loom: business-as-usual versus systems change. Ever more food versus low impact food for all. Or is a middle way possible: a kinder, softer food system not too different from what has emerged in the 20th century and now spreads worldwide? And is such a 'softer' approach even possible when food systems are being shaped by 'hard' consumer demands? Is run-away consumerism undermining its future capacity? These are general dilemmas but especially for food.[5] How humanity conceives of the environment and how seriously the environment is factored into food systems will be a key feature of 21st-century food systems thinking. New policies are almost certainly required which accept that human and environmental health are inextricably connected; they are two sides of the same coin. Human health depends on rich eco-systems, and eco-systems can be made or marred by how humans produce their food.

This chapter summarises the evidence of food's environmental impact in order to consider the track record of Productionism and to ask how food systems might better connect human health and the natural environment. By the end of the 20th century, it looked as though that attempt might be addressed, as dire warnings and evidence mounted. Climate change and water stress in particular captured growing political and sometimes popular attention, yet both issues became mired in political arguments about how they should be addressed: by regulations or informed consumer actions? By national governmental action or only through global negotiations? Intergovernmental talks about climate change, for example, had begun in the 1990s, and produced the 1997 Kyoto Protocol,[6] but were still unresolved two decades later, partly for ideological reasons.[7 8] Would reducing climate destruction mean a clamp on economic activity, particularly energy use and the economic growth versus planetary survival debate?[9] Meanwhile evidence of changing temperature became ever stronger.[10] The two emerging food paradigms began to offer different solutions for the future of food supply. The Life Sciences Integration paradigm suggested a new wave of technical solutions, while the Ecologically Integrated approach proposed a realignment of food systems with eco-systems. Policy-makers have found this split hard to grapple with, and tended to favour the technical route, not least since this seemingly offers opportunities for capital intensive investment. It does not threaten product or brand interests. But if ever more people consume ever more of less environmentally damaging food products, it is doubtful that this will realign human food eco-systems and it is likely to continue to threaten their resilience. No wonder scientists have talked of tipping points.[11] This chapter thus raises fundamental questions about how modern food systems can achieve long-term ecological sustainability. The alternative is that they continue to be major drivers of its destruction.

Rising environmental stakes

By the act of eating, we humans literally consume our environment. Food comes from the environment in one form or another – plants, animals, seeds. As food chains have become more complex, their environmental impacts become more diffuse and sometimes harder to appreciate. Food's environmental impact can be hidden from the consumer. We don't know what we are eating, nor necessarily understand or appreciate how food has this effect. Yet an increasingly important battle in the Food Wars is the conflict over which conception of the relationship of food and environment will triumph. Will the environment be seen as 'out there', something to be raided, used and mined to feed us, or will humans and food systems be understood and organised on the basis of a relationship of mutual dependency?

In the second half of the 20th century, soon after the Productionist paradigm was in place, evidence began to emerge about food's impact on the environment. [12] Questions began to be asked about food as a driver of environmental damage. Reports and campaigns began to point to new forms of food-related pollution, highlighting, for example, pesticides' impact on wildlife or the health and safety of poor farmers in developing countries,[13 14] nitrates leaking from man-made fertilisers into water systems,[15] weakened and polluted soil structures,[16] and many other issues associated with production methods. Towards the end of the century, the picture was fairly clear. The reliance of modern food systems on the environment was a major driver of its erosion and destruction. Viewing planet earth as fragile became more acceptable. New metaphors were produced: the earth as having boundaries, the earth as living, the earth as malleable; the earth as damaged by footprints left by production and consumption. These metaphors began to 'sell' a much better scientific understanding of food's impact on environment to the consuming public and to policy-makers.

The environment, however, is not just one or two single issues or even a handful. It's an 'ecological' web of many issues, all constantly changing and interacting. Soil is affected by climate change which affects water which affects plant growth which affects habitat for animals both tiny and large. Gradually what has emerged is that a complex eco-system has been altered for the worse by the success of increasing food output. The decline in biodiversity, the pollution of seas, the altering of earth's climate, all these are shaped by the pursuit of more food and by the methods pioneered and refined in the name of improving health and welfare, and of feeding more people as populations expand. Thus food policy emerged in the 2010s – as it had before – as an intersectoral challenge with the environment as a key focus. Only today, the intersectorality has expanded and become even more complex. The boundaries of food systems are being stretched ever wider. Food systems weave across human society, economy, politics, health, land use, demography, urbanisation, *and* environment, yet so often all these issues are presented as separate problems.

How could policy-makers respond to this environmental challenge? As the evidence summarised in this chapter emerged, the response was mostly to 'tweak' the format. Pesticide residues are causing some problems? Well, use them more discretely and minutely. Biodiversity loss is accelerating? Keep some cultivation-free strips round the edge of fields; or designate some national parks. Fish stocks are declining from over-fishing? Take more fishing boats out of action (but allow the others to get bigger). Meat is found to be a high impact farm product? Well, aim to lower its impact but sell more of it globally. Too often, attempts to address the environment have been simplistic, or contradictory, and fail to control the systemic drivers. But, to be fair, even powerful actors in food supply chains simply do not have the control over the whole food system they sometimes are said to have. Even mighty transnational corporations who want to do something about climate change are constrained by their need to remain profitable and competitive. To change towards sustainable production and consumption requires a change of framework. That's what only governments can do.

The good news is that there have been some real gains in prioritising the environment over thoughtless human food activity or commercial greed, and in awareness of the problems. The European Union's water controls are one example. Directives (laws) were put into place under which nation states accepted that water systems follow no boundaries.[17] They are public goods. Thus continental principles of conservation and high standards were put into place and huge investments were made to clean up sources and prevent future pollution. More prosaically, campaigns to reduce use of plastic bags in the 2000s also had some impact, but have not stopped massive use of plastics elsewhere. Waste is another issue on which shock at its 'rediscovery' sparked actions by citizens, companies and authorities.[18-20] The supposedly 'efficient' West was found to be not so efficient, after all. This suggested that Productionism had not resolved a key fault that its architects had decried in the mid-20th century.

A theme emerges from our analysis of food's environmental connection which goes to the heart of the paradigmatic choice explored in this book: can markets resolve food's problems? Or must there be new frameworks which put the environment at the heart of defining what a good food system is? Different approaches have emerged. Some point to technical solutions: better equipment; better filtration; better monitoring; better science; smarter management; more benign intensification of production. Others see solutions in the social sphere: recalibrating consumer demand; changed eating habits (less meat, more plants for example). Still others see a mixed approach: 'sustainable intensification'; 'climate-friendly agriculture'; 'nutrition-sensitive food systems'; and more. These can all be viewed through the language of the paradigms outlined in this book. Before we outline some of the key environmental impacts of food, we need to clarify the philosophical basis for thinking about the environment. What exactly is meant by environment? And what is ecological thinking?

Ecological and environmental thinking

The word 'ecological' was coined by the German writer Ernst Haeckel in 1866 to indicate the study of living contexts. In an intellectual world turned upside down by the theory of evolution by natural selection, first charted by Charles Darwin and Alfred Russel Wallace a few years earlier,[21 22] Haeckel wanted a word to capture the interconnections of everything in the natural world, including humans, and to show how life forms provide the context for each other.[23 24] Everything is connected. The role of science is to unravel those connections and to wonder at the complexity. This is post-Darwinian thinking, shaped by the realisation that slow powerful forces shape life.

In the last century and a half, the word 'ecology' and ecological thinking have both come to be associated more with the environment than with humans within the environment.[25] Seeing the environment as something apart from humans in part can be traced to Judaeo-Christian thought, which has tended to see nature as 'base', something at odds with the spiritual. 19th-century Darwinian evolutionary thinking, in contrast, had gradually realised that humans were animals who had evolved to achieve ecological dominance. This was truly shocking to traditional conservative hierarchical societies, a threat to the social order. Meanwhile yet another strand of thinking saw nature as something to be kept at bay or engineered away. Nature rather than society was the threat. Disease and ill health could be tamed. These different strands of thinking may be found in food policy, too, and indeed infuse the three paradigms explored in this book. Food is an intimate connection between humans and nature, that much is certain, but the more delicate matter is what form that connection takes.

Is the relationship between humans, food and the environment benign or exploitative? Mutually beneficial or destructive? What principles shape food systems – competition or cooperation? These are old philosophical questions. The Revd Dr Thomas Malthus's famous *Essay on the Principle of Population*, published in 1798 well before modern ecological let alone Darwinian evolutionary thinking, had offered a stark analysis.[26] Human population could grow faster than humans could increase their own food supply, he proposed. Disaster looms if population continues to grow as it was already beginning to with urbanisation and industrialisation. In fact, he was proven wrong; food supply did increase, shaped by science, technology and a new era of agricultural engineering, as we know.

But it is also now known that the environment is at risk,[12] and it is perfectly reasonable to conclude that a successful two-century period of food advance might be over. The UK Government's Chief Scientist mused that food is at the centre of a coming 'perfect storm'.[27] Huge Foresight reviews of the evidence on land use and future food concluded big changes are necessary to reduce environmental damage.[28 29] Australian scientific advisers wrote a report in similar vein,[30] as did France's.[31] The global figures are sobering, describing a food system which has stressed its eco-system's infrastructure to a point where the capacity of food supply to continue to grow is in question.

Many examples of food systems studied by scientists – climate change, water, fish, biodiversity, for example – illustrate why over recent decades scientists and official bodies have begun to ask political questions about the ecological relationship of humanity and the environment. How do effects noted in some areas relate to others? What is the impact of X on Y and can their mutual dependence survive stress? Terms such as risk and resilience have become more central in policy. A resilient system is one which can survive shock and stress. It bounces back; it can regenerate or recuperate; it can recover from difficulties.[32 33] With climate change, calculations of the likelihood of resilience becomes urgent. Could humans reduce or contain climate change sufficiently to allow eco-systems to adapt? How far can food caused destruction continue before it is irreversible? Far from being endlessly flexible and adaptive, is the planet now not showing signs of irresolvable stress?

Some have even gone so far as to propose that the planet earth is itself an organism or operates in myriad complex ways as though it is one. In the 1970s, James Lovelock, a British scientist, was working at NASA, the US space agency. Looking at the impact of chemicals on the atmosphere's ozone layer, he began to conclude that humans were altering how the planet actually functioned. Its organic processes of adaptation and change were being put out of kilter by chemical emissions. Lovelock was the first to chart the impact of chlorofluorocarbons (CFCs) on the ozone layer, and is credited with inspiring the worldwide ban on the use of CFCs in pressurised spray cans and refrigerators. Lovelock, like many scientists who emerged as defenders of the environment from the 1970s, began to realise that life forces may continue even if species do not. Darwin's slow evolutionary forces were being trumped by massive, crude human intervention. This new generation of scientists looking at the planetary level began to think that humans might be sowing the seeds of their own destruction, pushing planetary ecological processes into a new phase (one with species loss). Trying to make sense of multiple observations, Lovelock proposed an overarching theory – Gaia theory – in a 1979 book with that title.[34] He proposed that planet earth is a self-regulating organism. Actions such as man-made climate change will alter fundamental processes which the planet will incorporate but in ways that may show no respect for human survival. The planet's dynamics are more powerful than that of any species or subsidiary process. The name Gaia was chosen after the mythical Greek goddess of life, a suggestion by his friend the novelist William Golding, author of the bleak *Lord of the Flies*.[35]

Whether planetary interconnections work precisely in the macro-processes mooted by Lovelock continues to be debated, but the evidence that human progress has pushed resource use and various natural cycles or connections beyond their functional stasis or equilibriums now has considerable weight. Two linked and much-cited scientific papers in *Nature* and *Ecology and Society* in 2009 provide such a view. Interdisciplinary teams led by Professor Johan Rockström of the Stockholm Resilience Centre concluded that three out of ten ecological processes they reviewed at the planetary level had already exceeded their safe boundaries, with two more rapidly approaching that point.[36 37] They then revisited this analysis and the data in 2015, arguing that the evidence had become more certain on two of these – climate change and biosphere integrity – as having the capacity to 'drive the earth into a new state should they be persistently and substantially transgressed'.[38] The suggestion is that this is looking distinctly possible if human behaviour remains on its current trajectory. Figure 4.1 gives the 2009 summary diagram of the ten processes they reviewed: climate change, ocean acidification, stratospheric ozone depletion, the nitrogen cycle, the phosphorus cycle, global freshwater use, land systems change (sometimes 'land use change'

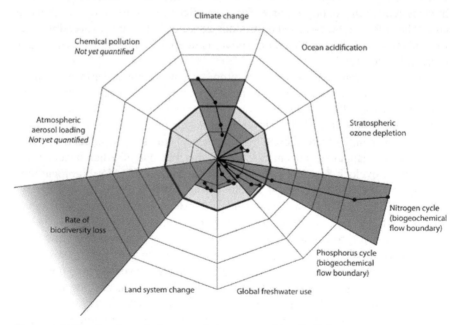

Figure 4.1 Planetary boundaries exceeded or approaching limits?

Source: Based on Azote Images/Stockholm Resilience Centre in Rockström *et al.* (2009).[36]

in the literature), biodiversity loss, atmospheric aerosol loading and chemical pollution. Food is a key factor in six of these ten processes: biodiversity loss, the nitrogen cycle, climate change, phosphorus, land use, chemical pollution.

Climate change

Perhaps the starkest illustration of how interconnected apparently single issues are comes from considering food and climate change. Although there is a persistent minority of opinion that, as Naomi Klein has shown,[39] skilfully denies that human activity causes climate change or that the consequences are of concern, the overwhelming majority of scientists now agree that the earth's temperature is rising, that this is man-made ('anthropogenic'), and that food is one of the major contributors to that upward trajectory.[10] Carbon dioxide (CO_2) and other greenhouse gases are altering the atmospheric layers around planet earth, and emissions are gradually heating the atmosphere up. As temperatures change, other natural cycles change, too, notably water precipitation, evaporation, plant growth, and a vast range of ecological interactions we know as life itself.

The Intergovernmental Panel on Climate Change, a huge collaboration of scientists which advises the UN, has documented for decades the evidence on climate change and has made it clear to policy-makers that changes in behaviour and economic activity need to happen to begin the twin processes of adaptation to further rising temperature and mitigation of its effects.[40] The consequences include rising sea levels and flooding, biodiversity change, weather volatility and storms, and the spread of infection and parasites. The implications are immense. It will hit already disadvantaged countries and social groups particularly hard because they cannot buy their way out of such crises. Wealth is no escape however. Even more affluent countries are affected by the spread of disease and crop failures. The world's commodity production – tea, coffee, grains – is highly likely to have to change as weather change occurs. The fear is that this will drive migration, geo-political and eco-systems pressures, particularly over land use. Farming and food production will be forced to change.

Climate change is not just something which is affecting and will affect food systems. Food systems are key factors in generating climate change. The IPCC estimates that agriculture, forests and land use account for 24 per cent of all greenhouse gas emissions.[40] Modern agriculture alone has been estimated to contribute around 14 per cent (perhaps even 18 per cent) of measured greenhouse gas emissions.[41] Of these, farm animals are responsible for 31 per cent, and fertilisers for 38 per cent.[42] In Europe, food is the European consumer's biggest source of greenhouse gas emissions.[43] Meat and dairy products account for 24 per cent of the average European's environmental impact.[43 44] 36 per cent of the calories produced by the world's crops are used for animal feed, and only 12 per cent of those feed calories ultimately contribute to the human diet as meat and other animal products.[45] This suggests that behind the much-vaunted efficiency of modern production sits a structural problem: as societies get richer, they eat more climate change-inducing foods. Food, of course, is not the only source of

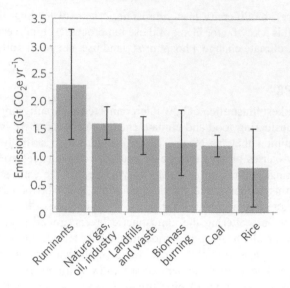

Figure 4.2 Some key sources of anthropogenic methane

Source: Ripple *et al.* (2013) [46]

anthropogenic (human-caused) emissions but Figure 4.2 shows the important impact of ruminants – cows, sheep – compared to other sources of methane, a potent greenhouse gas.[46] This highlights the importance of the debate about what a sustainable diet is – low impact and healthy – and of setting clear advice on how to achieve it.

Waste

Food waste began to capture public attention in the 2000s.[18 47] Some clarity about what is meant by waste is needed. Waste is food in the wrong place, rather as weeds are merely plants in the wrong place. All systems are wasteful. Part of the issue is what is done with it. In nature, there is no waste. Leaves and growth fall to the forest floor and microbes and insects turn it into humus, or soil. Darwin observed and wrote about this, how the earthworm, in particular, seemed to be crucial in turning vegetable litter into soil.[48] Today we know that bacteria, plants and other life forms coexist and feed off each other in the soil; they are part of but also create the conditions for life. Darwin used the memorable phrase 'the tangled bank' of life to capture this complexity and interaction. In an ideal world, this ecological interpretation of waste would imply that waste is impossible. Everything would serve a role in the tangled bank of life. In fact, we humans have created new forms of waste which break eco-systems. In the 1930s, one of the main criticisms of current systems was its waste: food rotting on or near farms or ports, for lack of infrastructure.[49] This was one of the great motivations for Productionism, the application of

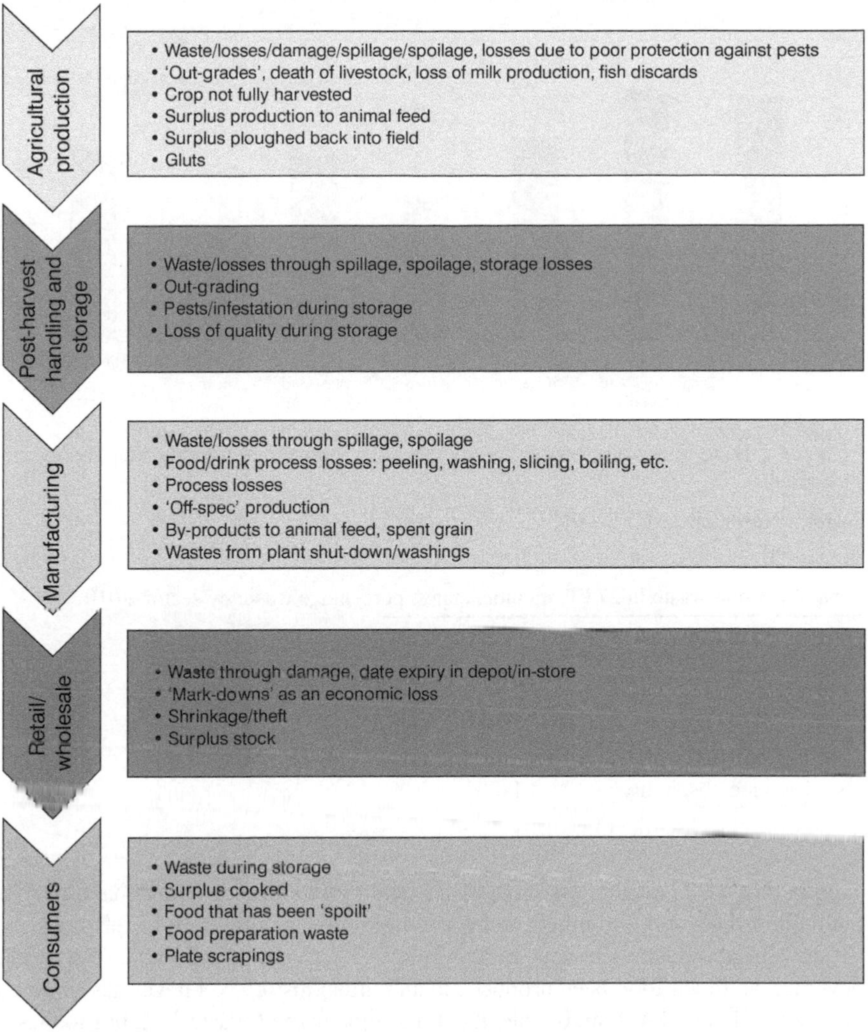

Agricultural production

- Waste/losses/damage/spillage/spoilage, losses due to poor protection against pests
- 'Out-grades', death of livestock, loss of milk production, fish discards
- Crop not fully harvested
- Surplus production to animal feed
- Surplus ploughed back into field
- Gluts

Post-harvest handling and storage

- Waste/losses through spillage, spoilage, storage losses
- Out-grading
- Pests/infestation during storage
- Loss of quality during storage

Manufacturing

- Waste/losses through spillage, spoilage
- Food/drink process losses: peeling, washing, slicing, boiling, etc.
- Process losses
- 'Off-spec' production
- By-products to animal feed, spent grain
- Wastes from plant shut-down/washings

Retail/ wholesale

- Waste through damage, date expiry in depot/in-store
- 'Mark-downs' as an economic loss
- Shrinkage/theft
- Surplus stock

Consumers

- Waste during storage
- Surplus cooked
- Food that has been 'spoilt'
- Food preparation waste
- Plate scrapings

Figure 4.3 The language of food waste along the supply chain

Source: House of Lords 2014[50]

science to old food systems to cut waste. Yet studies in the 2000s showed waste remaining at high levels. An FAO review and other studies showed continuing 'old' forms of waste in developing countries, now accompanied by 'new' forms.[20] Figure 4.3 provides a schema.

Globally, a new pattern of waste has emerged (see Figure 4.4). In developing countries, food mostly gets wasted on or near the farm or growing point due to lack of, or poor, storage, transport and preservation. When consumers in low-income societies buy the food, however, they waste almost nothing. Food is too expensive and too culturally precious. By contrast, in developed food economies, studies

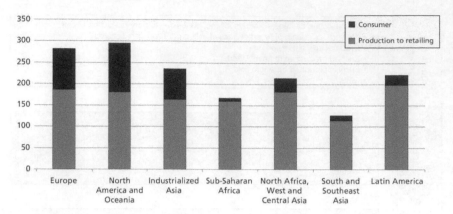

Figure 4.4 Food losses and waste (kg/year) at consumption and pre-consumption stages, by region

Source: FAO/Gustavsson *et al.* 2011[20]

Table 4.1 Food waste in 27 EU member states, percentage waste by sector, 2010

Source of food waste	% of total
Food drink manufacturing	39
Household	42
Retail/wholesale	5
Food service/hospitality	14

Source: UK House of Lords 2014[50]

suggest that waste occurs less on the farm, and more after consumers have bought the food.[20] They buy too much, throw it away, leave it at the back of refrigerators, get nervous about labels saying 'eat this before … xxx date', and so on. As a result, never has so much food been produced in all human history, yet never has so much been wasted. The FAO study calculated that global food waste is 220m tonnes a year, equivalent to the production of sub-Saharan Africa, whereas in high-income countries, consumers waste up to a third of what they buy.[20]

The European Union generates 89m tonnes of food waste each year. The UK wastes 15 of those. WRAP, the UK's food waste body, has calculated that UK food waste has a monetary value of about £950 per tonne.[50] In 2013, the EU calculated that the EU rate of waste was growing such that if not checked it would be 126m tonnes by 2020. EU food waste in 2010 is given in Table 4.1.

What can be done with waste? Historically, pigs have been used to 'recycle' edible food waste – Mao's China famously called the pig a 'fertiliser factory on four legs'[51 52] – but this can also be a route for spreading disease. 'Pigswill' has been blamed for major outbreaks of swine vesicular disease and foot and mouth disease, leading to devastating slaughter programmes.[53] And intensive rearing systems in the West, at least, have frowned on such use as a threat to biosecurity, preferring either closed loop feeding systems or heavily regulated and routinised feedstuff supply chains.

Policy-makers began to address the flow of waste food from urban households when pressures built up over landfill sites. In the 1990s, European households and commercial activities were pouring around 300kg per person per year into landfill. Targets to reduce this to 200kg were put in place.[54] They failed. Then a system of landfill taxes was introduced. That began to have effect. Then the European Commission introduced targets in 2011 to halve food waste by 2020, and once figures are given, performance can be monitored and the picture of systemic waste can be charted.

The 2007–8 commodity and financial crisis brought Western food waste to the fore, not least since food prices rose and became volatile with oil prices. Many countries introduced anti-waste campaigns, and NGOs began to campaign on the folly of surplus food failing to find needy consumers. In the UK, for example, consumers cut their food waste by 15 per cent in 2007–12, generating an average saving of £130 per household per year by doing so.[55]

Food and biodiversity

According to the UN's 2005 Millennium Ecosystems Assessment (MEA), a huge interdisciplinary scientific collaboration, modern food systems play a major part in destroying the earth's ecosystems; it documented major biodiversity loss, water depletion, land use stress and soil depletion.[3] Of 24 of the world's eco-system services studied by the MEA, five were being degraded or used unsustainably.

Food was a major source of this degradation.[56] Forests have been destroyed to create fields for cropping. With the trees the myriad insect and eco-systems above and below soil also disappear. This is not a new phenomenon. It began with settled agriculture around 10,000 years ago, and particularly with the development of field crops. To some extent this was the role of the farmer – to tame nature and to release land for crops desired for direct consumption. It epitomises the transition from hunter-gatherer societies to agriculturalist economies.[57] The slow creation of the major plant crops used by humans as staple foods – wheat, maize, rice, oats, barley, sorghum, etc. – initially spawned many hundreds of varieties. But in the 20th century, this legacy of successful agricultural experimentation and development was itself undermined by further intensification, specialisation and a tendency towards monoculture – using one variety on ever larger tracts of land.

The FAO has calculated that, in the 20th century, 75 per cent of the genetic diversity of domestic agricultural crops inherited by farming from earlier centuries was lost.[58] At the same time, farmers and growers had been encouraged to plant 'wonder' crops, thereby increasing genetic concentration.[59] Today an estimated 12 plant species account for 75 per cent of global food supply, and only 15 mammal and bird species account for 90 per cent of animal agriculture. [60 61] In the seas, where farming-type control has been hard to apply (because not even humans can tame the mighty oceans), the destruction of sealife has been on an awesome scale. The FAO has calculated that over half (52 per cent) of global wild fish stocks are now 'fully exploited'.[62] This raises a key health problem: is it right that governments continue to recommend the consumption of fish as part of a balanced diet?[63 64] Supporters argue that wild fish captured by the fishing industries has levelled off and the growth is now in aquaculture. Much of that, however, uses wild fish discards to feed the 'farmed' or captive fish. Fish farming still raids the environment – and indeed is judged in some studies as polluting it.[65 66]

Plant breeders have narrowed the strains of crops developed for intensive agricultural production, thus shrinking biodiversity in the field (as well as contributing to its diminution beyond the field).[67] In the US, by the 1990s, over 75 per cent of potato production came from four closely related varieties,[68] 76 per cent of snap beans from three strains, and 96 per cent of pea production from just two pea varieties;[69] 95 per cent of cabbage, 91 per cent of field maize, 94 per cent of pea, and 81 per cent of the tomato varieties have been lost; an estimated 80–90 per cent of vegetable and fruit varieties, strong in the 19th century, were lost by the end of the 20th century.[70] Of the 10,000 wheat varieties in use in China in 1949, only 1,000 remained by the 1970s. In India, farmers grew more than 30,000 traditional varieties of rice half a century ago; now ten modern varieties account for more than 75 per cent of rice grown in that country.[71] Often such 'modern' varieties of plants are grown for their weight, or volume, or predictability, or responsiveness to fertilisers, with not enough thought given to either the biodiversity or nutrition diversity.

Potatoes are another crop where the concentration of varieties has accelerated. In the Peruvian highlands, a single farm may grow 30 to 40 distinct varieties of potato, each having slightly different optimal soil, water, light and temperature regimes which, given time, a farmer can manage. Diversity like this requires skill. By contrast, in the Netherlands, a different skill has been applied to grow just one potato. One variety now covers 80 per cent of all potato land.[70]

Overall, according to the FAO, around three-quarters of the world's agricultural diversity was lost in the 20th century,[72] a direct result of the pursuit of uniformity, control and predictability, characteristics sought by businesses and processors. They want routine, regular supplies. Fast-food firms want the same potato worldwide. Together, agriculture and food culture are shrinking biodiversity in the field, and contributing to the sixth great period of extinction of the last half billion years,[73] and agriculture plays a significant part in this tragedy by losing and degrading habitats; it is also destroying pollinators whose presence enhances crop yields. Animal grazing is a significant threat to biodiversity too, as is drainage by destroying wetlands,[74] and there are concerns that precipitate use of biotechnology could release organisms into the environment that are impossible to control or retrieve and that will have their impact on biodiversity. In policy terms the choice is becoming clear: agriculture and food production can continue to degrade biodiversity or they can be redesigned to enhance it. The sustainable agriculture movement claims it offers the best approach.[75-77] Such protection measures are painfully slow to introduce as they threaten many interests, not least predators seeking to exert intellectual property rights. At the global level, the UN's Convention on Biological Diversity is now a framework under which, in theory, biodiversity is to be protected. This was first formulated in 1994 and by 2014 had held 11 further global conventions.[78] In 2010, the FAO and Bioversity International, one of the UN's research bodies, held a scientific meeting which agreed that diet and food systems must change if eco-systems are to prosper.[61]

Water

Water is critical not just for direct consumption and agriculture but for hygiene. Thirty-six per cent of the world's population – 2.5 billion people – lack improved sanitation facilities, and 768 million people still use unsafe drinking water sources, according to the 2013 UNICEF/WHO Joint Monitoring Programme for Water Supply and Sanitation.[79] At least 1.8 billion people drink water contaminated with faecal residues. Water covers 70 per cent of the planet, but 97.5 per cent is ocean, not usable in industry, agriculture or as drinking water. Of the 110,000 billion cubic metres of rainwater that falls on earth, only 12,500 billion are accessible and usable, yet this amount has been calculated to be sufficient for human use. Water for domestic use (i.e. drinking) accounts for only 8 per cent of the water available for human use, with agriculture using 70 per cent and industry 22 per cent.[80] Modern agriculture is, just like its consumer counterpart, a greedy consumer of water. The demand for water is expected to grow in all regions of the world over coming decades. A new water world order is emerging. Countries like the US and Canada have vast water resources; others such as Taiwan, Saudi Arabia and Germany are actually in water deficit. UNEP and all water monitoring agencies anticipate growing worldwide 'water stress' in coming years.

All food production relies on water, some crops massively so – rice particularly. Anyone who has looked at the breathtakingly beautiful contour rice paddies in Asia must marvel at the delicate but dogged way humans over many centuries have built small dams in line with hillside contours. Elaborate systems of water conservation, channels and sharing have been created. Rice requires heavy water use but it is only the most demanding of crops which all require water to grow. It is not surprising therefore that global agriculture consumes 70 per cent of all freshwater extracted for human use.[81] And some forms of farming then proceed to pollute good water. Intensive livestock production is probably the largest sector-specific source of water pollution.[41] Modern consumers, faced by multiple choices on supermarket shelves are mostly unaware of their dependency on water.

Water analysts make the distinction between Blue water (fresh surface or groundwater); Green water (rain which does not run off into rivers or replenish deep aquifers but remains in or near the soil); and Grey water (polluted or used). They also calculate how much embedded or 'virtual' water is in products.[82 83] Some processed foods – particularly meat and dairy – are very heavy water users. A Dutch study found it took 200 litres of water to produce a 200 millilitre glass of milk, and 2,400 litres of water to produce a 150 gram hamburger.[84] Table 4.2

Table 4.2 Indicative figures of water 'embedded' in some agri-food products

Product	Quantity	Embedded water (in litres)
Beef	1 kg	15,400
Cotton	1 kg	10,000
Milk	1 litre	1,000
Cane sugar	1 kg	1,800
Rice	1 kg	2,500
Wheat bread	1 kg	1,600
Beer	1 litre	300

Source: Hoekstra *et al.* (2011) and WWF *et al.* (2014) [85 86]

provides some indicative calculations of embedded water in some well-known food and agricultural products.[85 86]

Agriculture is both victim and perpetrator of water stress, and is acknowledged by the FAO as being mainly responsible for freshwater scarcity. FAO states that about 30 per cent of the food produced worldwide – about 1.3 billion tons – is lost or wasted every year, which means that the water used to produce it is also wasted. [87] Lack of water increases the chance of cropland being degraded. Agriculture also pollutes drinking-water quality, a problem less developed countries cannot afford to 'solve' by filtration of contaminants (but why, one wonders, pollute in the first place?). The water used by agriculture, drawn from lakes, rivers and underground sources, is taken mostly for irrigation, which helps provide 40 per cent of world food production. As a result, water tables are predicted to fall due to over-irrigation and intensive crop production, and salination (excess salt) to increase as a direct consequence. Large-scale irrigation is probably reaching its limits. Seventy per cent of water used is for irrigation. It would now be more sensible to invest in enhancing the capacity of soil to retain moisture, which is a feature of sustainable agriculture systems which focus on building healthy soil.

Pollution and pesticides

Over the half century since the publication Rachel Carson's *Silent Spring*, a study of the environmental impact of pesticides,[13] the arguments about pollution and pesticides have raged, with some arguing this is folly,[14 88-90] and others arguing that a global agrichemical market estimated to be worth $243bn by 2018 is what feeds humanity.[91 92] No one disagrees that food is a vector by which pollution can enter human bodies and create biological damage; the issue is whether levels are safe and management systems robust. Some of the food supply chain's environmental pollution is clearly related to intensification and poor application, affecting workers. Persistent organic pollutants (POPs) – toxic synthetic compounds – accumulate in the food chain, persist in the environment and travel by being bio-accumulated (as animals eat each other, so the POP is stored in fat and thus consumed and stored by humans). Most humans have around 500 POPs stored in their body fat that have been created since the chemical revolution of the 1920s.[93] Pesticides are a key route for POPs, notably through aldrin, chlordane, DDT, dieldrin, endrin and heptachlor. POPs have a malign impact on humans, wildlife, land and water.

Worldwide, 2m tonnes of pesticides are used annually: 45 per cent in Europe; 25 per cent in the USA; and 25 per cent in the rest of the world.[94] There is considerable variation in how much is used in different countries. Japan, for example, uses 12kg per hectare (kg/ha), Korea 6.6kg/ha and India only 0.5kg/ha. Agrichemicals have different functions in farming and the term is used to cover different purposes; insecticides to kill pests (insects), fungicides to control moulds and spores, for example. Although they are effective, concerns rose in the 20th century about the impact of agrichemicals on consumers, wildlife and workers.[95] One early study estimated that 25 million developing country agricultural workers are poisoned in one form or other by pesticides each year.[96]

A UK study found that 30 per cent of all foods contained detectable pesticide residues, with 1 per cent above maximum residue level, thus breaking the safety standard.[97] US diets were found to contain some food items contaminated by between three and seven POPs. The main POP-contaminated food items in the US have been found to be: butter, cantaloupe melons, cucumbers/pickles, meatloaf, peanuts, popcorn, radishes, spinach, summer squash and winter squash, all containing levels of POPs which may individually be within safety limits, but which collectively pose a risk, according to health standards set by the Center for Disease Control's Agency for Toxic Substances and Disease Registry and the Environmental Protection Agency.[90]

Although overall the quantity of pesticides applied in industrial agriculture has declined recently, the toxicity of what is used has increased by an estimated factor of 10- to 100-fold since 1975. The pesticides are packing a heavier punch. Despite this, resistance is spreading; POPs are becoming less effective: 1,000 species of insects, plant diseases and weeds are now resistant, an environmental impact known as the 'treadmill effect'. However, so extensive is the reach of POPs that even crops grown without them may contain them, while crops grown using pesticides contain much higher levels.[98] And despite strong evidence of the negative impact of POPs and their connection with pesticides, governments only recently agreed the Stockholm Convention to phase out their use. If governments want raised controls on pollutants, such as from pesticides, regulations are essential, although encouraging increased fruit and vegetable intake to reduce rates of CHD and cancers tacitly encourages increased intake of POPs through that route. However most epidemiological evidence suggests that the relative risk of POPs is offset by the gain from the nutrient intake of fruit and vegetables.[99]

From a policy point of view there need not be a trade-off of risks and benefits; why should this be an either/or when it is possible to aim for a win–win?

Soil and land use

It is said that all civilisation depends upon soil.[100] Arable land is estimated to be only 1/32 of the planet's surface.[101] Barring a vast future investment in hydroponics, human food production will always need healthy soil. Yet reviews of the global status of soil make sobering reading. Soil structure is being damaged by desertification, pollution, water damage, clear-cutting of forests and overgrazing, leading to loss of humus. Once lost, soil is irreplaceable in the short term. The average loss of soil humus in recent decades has been around 30 times more than the rate throughout the ten millennia of settled agriculture. [73] Demand for animal protein drives intensive cropping: if consumer demand for meat continues to rise, pressure to deforest will exacerbate climate change and land use.[102 103] Meat production and consumption help drive a vicious circle of energy and other fossil fuel inputs, chemicals, water and protein as feed, let alone labour.

In Africa, for instance, more and more population-pressured land has been taken into cultivation in response to commodity prices. The Montpellier Panel reported in 2014 that in sub-Saharan Africa, an estimated 65 per cent of soils are degraded, and unable to nourish the crops the chronically food insecure continent requires.[104] Poverty, climate change, population pressures and inadequate farming techniques are leading to a continuous decline in the health of African soils, whilst the economic loss is estimated at $68bn per year. Conversely, better land management practices could deliver up to $1.4 trillion globally in increased crop production, 35 times the losses.

UNEP has argued that much of the growth of production is simply due to the use of new land, yet hunger remains and land continues to be degraded as people search for new sources.[105] Soil erosion is exacerbated by increasing amounts of fertilisers and agrochemicals. In the Asia–Pacific region, an estimated 850m hectares or 13 per cent of cultivatable land is formally designated as 'degraded' due to a variety of reasons ranging from salinity and poor nutrient balance to contamination. 'Desertification' is spreading, despite there being a UN Convention to Combat Desertification since 1994: 12m hectares are lost each year, 23 each minute.[101] Loss of arable land is currently at about 30 times the historical rate of loss. An estimated 24bn tonnes of soil disappear each year – a scale that makes the US topsoil loss in the 1930s look slight.

According to UNEP globally nearly 2bn hectares of land are affected by human-induced soil degradation. Within Europe, assessments have identified problems such as sealing (under roads, house, concrete), erosion, contamination, acidification and degrading.[106] The European Agricultural Conservation Foundation has estimated that soil erosion and degradation caused by conventional agriculture affect c.157m ha (16 per cent of Europe, roughly three times the total surface of France).[107] Average soil erosion rates in Europe are judged to exceed the average rate of soil formation, with most EU countries affected. In the Mediterranean, soil erosion was judged as 'very severe'. Land is more valuable for housing than farming in some cases. This is a 'values' problem. With rising populations and urbanisation, land use is squeezed by competing demands for housing, fuel, food, water, wood and amenity. UNEP has estimated that, even if more land is brought into use, the world must prepare for only 0.2ha (1,970 m²) crop-able land per person by 2030.[108] This is highly sensitive politically because rich consumer societies are able to 'buy' the use of others' land by importing food (and water) as well as over-using resources.[109]

Urbanisation

For the first time in human history, more than half of humanity is urban (see Figure 4.5). In 2014, the UN calculated that 54 per cent of the world lived in urban areas, up from 30 per cent in 1950. And it estimates that 66 per cent will be urbanised by 2050. This has staggering implications for food systems. North America is 82 per cent urbanised; Latin America and the Caribbean 80 per cent; Europe 75 per cent; Asia 45 per cent; and Africa a mere 40 per cent. But Africa and Asia are urbanising rapidly now, with them expected to be 60 and 68 per cent urbanised by 2050.[110] With such rates of urbanisation, the questions arise: fed by whom, how and on what? Even allowing for urban food production – gardens, smallholdings, even window boxes – the majority of food in the city must be bought, and poor families often spend as much as two-thirds of their income on food, approximately a third more on food than their rural neighbours.

People all over the world migrate to the cities to look for employment and better economic prospects; usually they yearn for better access to amenities, services and food. This is not new,[111] but the result is increased strain on rural areas to feed growing cities. Cities and towns are magnets for access to education and healthcare. They are more prosperous. Caloric intake tends to rise with urbanisation, and children can usually attain a better dietary status. Set against these positive features is evidence of how urbanisation also can bring social marginalisation: increased poverty, inequalities, unemployment and dependency. Urbanisation is

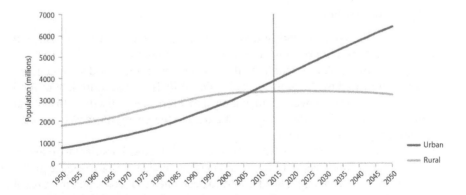

Figure 4.5 The changing urban rural split in world population, 1950–2050

Source: UNDESA 2014[110]

associated with changes in nutritional status. Intake of carbohydrates, meat, sugar, and edible oils increase. Towns expose their populations to marketing and junk foods high in salt and sugar and low in fibre; more traditional and restricted diets with lower intakes of unprocessed foods can fade.

Yet against these worrying features, the range of diet can actually improve in urban settings. Their populations have greater economic pulling power and usually offer more diversity of foods than is available in rural areas. In cities, the overwhelming preponderance of food is purchased rather than grown at home, although there is a surprisingly high amount of urban agriculture 'hidden' in urban areas. An estimated $500m worth of fruit and vegetables is produced by urban farmers worldwide.[112] Whilst urban areas have historically been fed by their hinterland, this is no longer the case in affluent societies. At the turn of the millennium, Londoners, for instance, consumed 2m tonnes of food annually, sourced from all over the world,[113] and the FAO has estimated that in a city of 10 million people, 6,000 tonnes of food needs to be imported on a daily basis.[114]

In the late 1990s, at its Habitat 2 conference in Turkey, the UN mapped out the urgency of the task, concluding that urban or peri-urban agriculture will have to make a comeback.[115] Back then in Kathmandu, 37 per cent of urban gardeners grew all the vegetables they consumed, while in Hong Kong, 45 per cent of demand for vegetables was supplied from 5 to 6 per cent of the land mass. But with land values rising due to population and housing pressures, the retention of land to grow food in cities is under pressure. Despite this, across the world, there is a burgeoning movement of local authorities, small farmers and ecology-conscious consumers arguing for this modern urban agriculture sector: 'growing our own', not least to retain a connection between urban and culinary culture and the land from which food comes. The Sustainable Food Cities movements in North America and Europe have urban agriculture as a key demand,[116] and have interesting exchanges with cities from the global south.

Energy and efficiency

The global food system is highly dependent on fossil fuels – a factor that will become even more critical with the prospect of producing more food to feed 9 billion people by 2050. It is estimated that 30 per cent of the world's total energy consumption is used by the food sector.[117] Fossil fuels are used throughout food supply chains – from powering irrigation and farm mechanisation, to natural gas to make chemical fertilisers and pesticides, to the energy needed to process foodstuffs and its transportation.

Reducing the food sector's fossil fuel use is also framed as important in reducing the sector's GHG emissions and energy saving is yet another reason to reduce global food waste. A major challenge ahead is how to decouple food production from fossil fuel use and to increase energy efficiencies and adapt food-related activities around renewable energy.[117] Energy use varies considerably within food sectors – not least between low-income producers and high-income production and consumption requirements. Even within countries, energy use can vary depending on agricultural methods. But conventional farming in the West tends to be fossil fuel dependent. Oil accounts for between 30 and 75 per cent of energy inputs of UK agriculture, for instance, depending on the cropping system.[118]

In agriculture energy use takes two forms: the direct use of fuels and electricity and indirect use of energy through energy-intensive inputs, most notably fertiliser and pesticides. Direct energy is used to run machinery for field operations such as planting, tilling and harvesting; to dry crops; for livestock use; and to transport goods. Figure 4.6 shows the relationship between direct and indirect energy use on US farms in 2001–11; it is notable just how significant fertilisers are in terms of on-farm energy inputs, second only to direct use as fuel.

Within the food system, energy use is by no means all on the farm. UK estimates put household energy use as the most significant sector, followed by net trade (see Figure 4.7). Net trade is the energy used in food exports less the energy used in food imports. These are followed by manufacturing, retail and transportation (see Figure 4.7).[120] As supply chains lengthen, they appear to have increased energy use. In the USA, a 2010 USDA study suggested that, as a share of the total US national energy budget, food-related energy use grew from 14.4 per cent in 2002 to an estimated 15.7 per cent in 2007.[121] All sectors except wholesale/retail saw a growth in energy use, 1997–2002 (see Figure 4.8). The sharpest rise was in the food processing sector. Comparing the sectors, household use of energy was the greatest energy user followed by processing, wholesale/retail and then agriculture.

The FAO has summed up the challenge of energy and fossil fuels in global food systems thus:

> The great challenge the world now faces is to develop global food systems that emit fewer GHG emissions, enjoy a secure energy supply and can respond to fluctuating energy prices while at the same time support food security and sustainable development.
>
> (FAO, 2011[117 p. iii])

A number of 'energy-smart' solutions or practices are being developed to reduce food sector energy reliance. One example is termed integrated food-energy systems (IFES), in which food and energy are produced hand-in-hand on the farm to achieve sustainable crop intensification. Figure 4.9 illustrates what such a food-energy based system might look like – it is proposed that such systems can be developed for both large- and small-scale farming systems.

Localisation and food miles

Most food is regionally or continentally produced and traded, but environmental analysts began to calculate whether the environmental 'footprint' of food varies according to where it comes from, not just how it is produced. The section just above showed how transportation is a significant energy user and that there

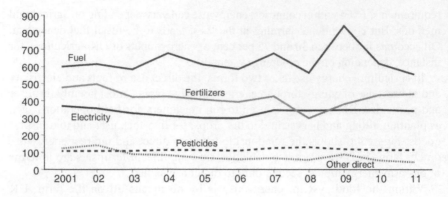

Figure 4.6 Energy inputs consumed on US farms, by component, 2001–11

Source: Beckman *et al.*, 2013[119]

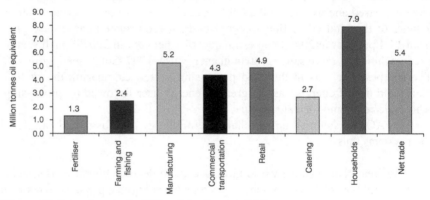

Figure 4.7 Energy use in the UK agri-food sector, 2011

Source: DEFRA 2014[120]

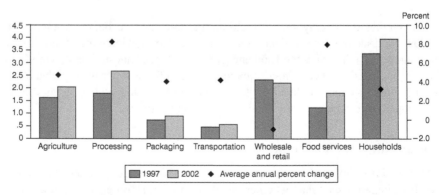

Figure 4.8 Change in US energy consumption by stage of production, 1997–2002

Source: Canning *et al.* 2010[121]

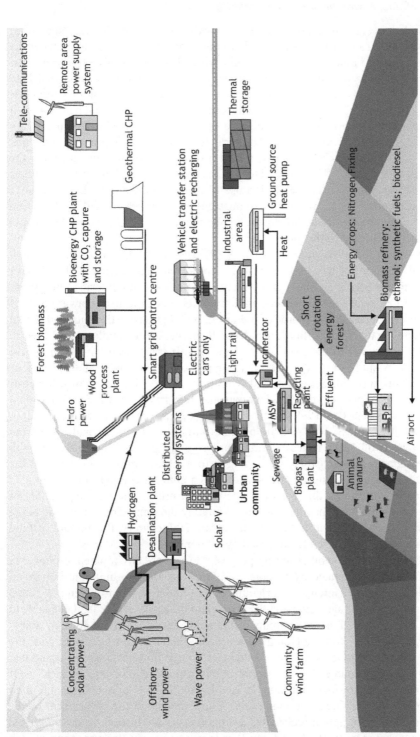

Figure 4.9 An example of an 'energy-smart' farming system using an integrated food-energy system

Source: IEA 2009[122]

is hidden energy in import/exports of foods. But can consumers spot this? A food may say on the label that it is the product of your country when its ingredients are assembled from far and wide. The notion of a 'footprint' was developed to try to measure environmental impacts as diverse as actual land use or greenhouse emissions or water use, and more.[109 123] Another term has been 'food miles', which proposed distance as a useful consumer indicator. It brought attention to the hidden mileage that food may travel between primary producer and end consumer.[124] The term led to a heated international debate. [125] Food exporting countries saw it as a protectionist notion, undermining the free trade in foods.[126] The originators of the concept argued that it may not be a perfect proxy for energy use but is a useful cultural indicator.[127] Long-distance food can certainly be lower in greenhouse gas emissions than local food.[128] And others from the fair trade movement argued that the notion of 'fair miles' would be a better cultural indicator.[129]

Despite (and even perhaps because of) this complexity and the debate it stirred, food miles has remained in the environmental discourse. The USDA for example uses it as one among many measures when exploring energy use in the food industry.[121] The food industry itself began to use it, aware of how transport can be a major source of pollution, both in transporting the food within the food system – from farmer to depot to processor to retailer – and then to get the consumer to the retailer.[130] Some of this transportation can be almost comical. A study of a West German yoghurt bought 'locally' showed that its ingredients were in fact assembled from hundreds or thousands of kilometres away.[131] Another US study of data on transport arriving at Chicago's terminal market from 1981 to 1999 found that produce arriving by truck from within the continental USA (i.e. excluding externally sourced food) had risen from an average of 1,245 miles to 1,518 miles, a 22 per cent increase,[132] with the national food-supply system using 4 to 17 times more fuel than did the localised system. Rising oil prices in 2007–8 and fierce competition encouraged businesses to rethink this creeping rise in food miles. The UK Department of Transport calculated that, despite approximately the same tonnage of food being consumed annually within the UK in the late 20th century, the amount of food being transported on roads increased by 30 per cent.[124]

'Ghost acres' is another notion proposed to capture environmental impact in products. It was created by George Borgstrom in 1973 to highlight bought-in feedstuffs used in intensive agriculture.[133] A Dutch study, for example, found that while the average human on the planet had 0.28ha of arable land available to feed him or herself, a citizen in the Netherlands actually used 0.45ha of arable land.[134] In other words, a Dutch citizen was reliant on 'ghost acres' outside the Netherlands for food. A study of London's resource flows in 2002 also found that food was a key source of the city's consumption and environmental impact.[135] Of the 6.9m tonnes of food consumed in London in 2000, 81 per cent was imported from outside the Greater London area. Food accounted for 14 per cent of the city's total consumption (calculated as imports plus production, less exports) and for 2 per cent of the waste. In addition, 94m litres of bottled water were consumed in one year, leaving an estimated 2,260 tonnes of plastic waste. Londoners each used 6.63ha of earth space, with food accounting for 41 per cent of them. For London to become sustainable, its overall consumption would need to reduce by 35 per cent by 2020 and by 80 per cent by 2050.

The creation of measures such as footprints, ghost acres, food miles or virtual water (see above) suggest the need for culturally as well as scientifically robust benchmarks by which the public could calculate resource flows and distribution. It needs help to recalibrate food choice away from mainly being made on the basis of the 'big three' – price, taste and convenience. The 21st-century challenge will partly be about making more efficient use of resources, and ensuring that some do not over-consume to the detriment of life chances for others. In no food matter is this more important than on meat and dairy.

Meat and dairy

Almost all studies of global trends assume rising global consumption of meat. New productionists cite this as a key reason that food production must increase by 70 per cent by 2050.[29 136 137] But we should be wary of assuming that because meat (or actually meat and dairy) is high impact its consumption must continue to rise. That said, meat and dairy do consistently emerge as highly significant food-related sources of greenhouse gas emissions, water use, land use and even biodiversity loss.[138] In 2013, the FAO calculated that livestock is responsible for 7.1 gigatonnes of CO_2e per year, which is 14.5 per cent of all human-caused greenhouse releases.[139] This was down from the calculated 18 per cent in a 2006 study.[109] Of all livestock impacts, beef and cattle milk account for the most emissions (beef 41 per cent, cattle milk 20 per cent). Greenhouse gas emissions per kilo of (meat) protein output is five times higher for beef than for pork or chicken, on average, although this varies by system of production. Feed is a huge factor, the largest source of emissions, followed by animals' enteric fermentation. Optimistically, the FAO 2013 study calculated that, if all livestock was managed as efficiently as the best, these emissions could come down by a further 30 per cent. But there was a sting in the tail on page 45: if meat consumption continues to rise globally, those efficiency gains would be neutralised.

Unlike food waste, now more accepted as a policy focus for and by business (see above), meat is highly contentious to many business interests. Meat and dairy sit in 'consumer choice' territory. Votes as well as food business profits depend on them. In the USA and Europe, and across the world, the meat and livestock industries are immensely powerful, and have long traditions of close engagement with public policy.[140 141] They receive or have received huge public subsidies, in the USA through the Farm Bill and in the EU through the Common Agricultural Policy. But meat is not the only subsidised industry; sugar is handsomely aided too.[142] Nevertheless, as Jeremy Rifkin argued twenty years ago, and Upton Sinclair over a century ago, 'cattle culture' is deeply entrenched in mainstream US politics.[143 144] Not without justification. According to the World Bank, livestock accounts for 30 per cent of agricultural gross domestic product in the developing world, and about 40 per cent of global agricultural GDP.[145] It is one of the fastest growing sub-sectors in agriculture. Players like the World Bank give it formidable policy support. To call for meat reduction means battling into strong policy headwinds.

In a food economy where half the grain grown is fed to livestock – a gross inefficiency – the grain and meat trades have a symbiotic relationship. One cannot contain meat production without tackling the feedstuffs sector. This is what is needed, and it gives a double win. Reducing meat production and consumption would cut CO_2 emissions *and* be a health gain.[146-149] Meat and dairy are prime sources of fat – as well as many other more beneficial nutrients – but the main health downside is the saturated fat. Processed meat products such as sausages, pies, salamis, where there are many additional ingredients, are also associated with some cancers.[99] Yet processed meats are popular and cheap foods in industrialised form. The growing consensus among scientists is that the rich world ought to reduce its meat consumption, and take it from the centre of the plate to its edge or, Chinese style, as flavouring and in slivers.

In 2007 McMichael, Powles and colleagues proposed that current global consumption of 100g of meat per person per day average should fall to an average 90g per person per day. That current average, they reported, disguised a ten-fold variation between high-consuming and low-consuming populations. They proposed 90g per day as an equitable 'working global target', with not more than 50g per day coming from red meat from ruminants (cattle, sheep, goats and other digastric grazers).[150] To meet this, big meat-eating cultures like the US would have to cut back considerably, while others might rise. (This is an example of the application of a 'contract and converge' policy approach, discussed later.[151]) McMichael and colleagues also recommended a shift from red to white meats as poultry has more efficient conversion ratios, even assuming some cereal use. A paper by Stehfest and colleagues echoed this general conclusion. This argued that a global transition to a low-meat diet would be good for human health, land use and CO_2 mitigation costs which are otherwise set to rise by 2050.[147] Another paper by Macdiarmid modelled a low GHG diet meeting health recommendations, based on foods most commonly consumed in the UK. This diet proposed 53g of meat a day, of which 27g could be red meat.[152] Another study on the effects of altering meat consumption and turning to alternative diets in the European Union showed that to halve the consumption of meat, dairy products and eggs would achieve a 40 per cent reduction in EU nitrogen emissions, 25–40 per cent reduction in greenhouse gas emissions and 23 per cent per capita less use of cropland for food production, and at the same time improve health risks.[153] These papers all imply significant not small dietary change ahead.

In any transition to more sustainable diets, messages will vary by country, level of economic development, consumer wealth, religion and cultural preference. For a largely Hindu culinary culture such as India, the main issue is dairy rather than meat. Each country needs to consider how to change its total diet, as Vaclav Smil recognised long ago.[154] A study by Baroni and colleagues confirmed that a process of dietary change first from meat-based conventional diets to lactovegetarian and then to ovo-vegetarian and ultimately to vegan diets would deliver a process of gradual eco-systems and resource use improvement, and public health gains.[155] This general effect – the improvement by *reducing meat while raising plant-based diets* – is the general picture given by most academic studies, whether they focus on climate change targets,[156 157] or the geographical region of Europe,[153] or individual countries such as Germany.[158]

The general verdict is that meat produced from grain is inefficient land use. It explains how meat production can have such a huge environmental footprint. But meat from land which otherwise cannot yield food – highlands, wet lands, marsh – may actually be ecologically efficient. This alters the debate from meat versus plants to grain-fed versus grass-fed meats. Grass-fed meat has lower use of cereals from land which could/should otherwise produce crops for direct human consumption; and feedlot-farmed cattle produce meat of higher fat content compared to grass-fed meat.[159] Grass-fed meat is generally lower impact and can have positive ecological factors – such as maintaining marshland habitats. In cultural terms, the emerging principle appears to be: 'consume meat and dairy infrequently if at all but pay more for better quality'. This eat better message troubles retailers, farmers and politicians but has advantages. Such messages are appropriate for meat-indulging, affluent cultures, but a curtailment of feeding animals in rich countries would not automatically improve diets for the poor in developing countries. IFPRI estimated two decades ago that a dramatic fall in meat consumption by 50 per cent, for example, would only deliver approximately 1 or 2 per cent decline in child mal-nourishment. [160] In short, messages about meat production and consumption may vary by place and income.

Sustainable diets: the answer to food's environmental impact?

In 2010, the FAO and Bioversity International (a UN affiliated body protecting biodiversity) hosted a scientific symposium on Sustainable Diets in Rome at the FAO headquarters.[61] This reflected growing interest in the need to have clearer guidelines on what would be a good diet for health, environment and social justice. The purpose of Dietary Guidelines is to set frameworks at population level by which dietary intake can be evaluated and to integrate health and environmental data into such advice. This goes to the heart of the Food Wars. How we define a 'good diet' symbolises which direction the food system might travel. In a world where cultural rules have been eroded or mixed up or made more flexible (or all of these), and in a world where food is ubiquitous and over-supplied in rich countries, where and what are the rules for eating? What contributes to a good or bad diet? What dietary signals are sent to supply chains? What are the principles for eating within environmental limits or with minimal negative environmental impacts? How can consumers know everything about their food? Does this matter? Or should change be left 'beneath the radar' to choice-editing orchestrated if at all by companies or governments?

Table 4.3 Food sustainability as a complex set of 'omni-standards' or 'poly-values'

Quality	Social values
Taste	Pleasure
Seasonality	Identity
Cosmetic	Animal welfare
Fresh (where appropriate)	Equality and justice
Authenticity	Cultural appropriateness
	Skills (food citizenship)

Environment	Health
Climate change	Safety
Energy use	Nutrition
Water	Equal access
Land use	Availability
Soil	Social determinants of health, e.g. affordability
Biodiversity	Information and education
Waste reduction and circularity	Protection from marketing

Economy	Governance
Food security and resilience	Science and technology evidence base
Affordability (price)	Transparency
Efficiency	Democratic accountability
True competition	Ethical values (fairness)
Fair return to primary producers	International aid and development
Jobs and decent working conditions	Trust
Fully internalised costs	
Circular economy (full recycling)	

Source: modified from SDC 2011[165]

The market model responds to such questions by saying that consumers should be left to decide, but are they really in control? There are no formal environmental labels giving information on water or carbon or impact on biodiversity. Some labels make a nod in those directions and appeal to those values but do not give neutral verifiable data. And could labels convey the complexity this chapter has summarised? There have been attempts to put carbon labels onto food,[161 162] but there are as yet no internationally agreed formats for doing so. There is also some commercial and consumer reluctance to add to information overload. That is where the notion of sustainable dietary guidelines becomes so attractive. They set broad population goals rather than individualised diktats. Such population goals might suit the fusion of the various environmental impacts considered in this chapter – water, biodiversity, quality, land use, soil impact, etc. – with other data and concerns considered in this book – health, quality, economic, governance.[152 163 164] One attempt to synthesise this was made by the UK's Sustainable Development Commission in its final 2011 report to the UK government before it was closed down. This presented food's sustainability challenge as requiring a framework of poly-values or 'omni-standards' (see Table 4.3). The next chapter considers how and whether policy-makers have faced up adequately to this challenge. Even if no new paradigm has yet triumphed, the criteria against which it will be judged are becoming more clear.

References

1   Hill C. *The World Turned Upside Down: Radical Ideas during the English Revolution.* London: Temple Smith, 1972.
2   Kanter J, Revkin AC. World scientists near consensus on warming. <http://www. nytimes.com/2007/01/30/world/30climate.html?_r=2&> *New York Times,* 30 Jan. 2007.
3   UNEP, Nellemann C, MacDevette M, *et al. The Environmental Food Crisis: The Environment's Role in Averting Future Food Crises.* A UNEP rapid response assessment. Arendal, Norway: United Nations Environment Programme/GRID-Arendal 2009.
4   Millennium Ecosystem Assessment. *Ecosystems and Human Well-Being: Synthesis.* Washington, DC: Island Press, 2005.
5   Gabriel Y, Lang T. *Unmanageable Consumer.* London: Sage, 2015.
6   UNFCC. Kyoto Protocol. <http://unfccc.int/kyoto_protocol/items/2830.php> New York: United Nations Framework Convention on Climate Change, 2015.
7   Grove VI, ed. *Global Warming and Climate Change: Ten Years After Kyoto and Still Counting.* Hamilton, Ontario, Canada: United Nations University, International Network on Water, Environmental and Health, 2008.
8   Marshall G. *Don't Even Think about it: Why our Brains are Wired to Ignore Climate Change.* New York: Bloomsbury, 2014.
9   Jackson T. *Prosperity without Growth: Economics for a Finite Planet.* London: Earthscan, 2009.
10 IPCC. *Climate Change 2014: Impacts, Adaptation, and Vulnerability.* Summary for Policymakers – IPCC WGII AR5 – released 30 Mar. 2014, Yokohama. Geneva: Intergovernmental Panel on Climate Change, 2014: 44.
11 O'Riordan T, Lenton T, eds. *Addressing Tipping Points for a Precarious Future.* Oxford: Oxford University Press/British Academy, 2013.
12 Sage C. *Environment and Food.* Abingdon: Routledge, 2012.
13 Carson R. *Silent Spring.* Boston/Cambridge, MA: Houghton Mifflin/Riverside Press, 1962.
14 Bull D. *A Growing Problem: Pesticides and the Third World Poor.* Oxford: OXFAM, 1982.
15 Lamb R. *Promising the Earth.* Abingdon: Routledge, 2012.
16 McGrath SP, Chaudri AM, Giller KE. Long-term effects of metals in sewage sludge on soils, microorganisms and plants. *Journal of Industrial Microbiology,* 14(2) (1995): 94–104.
17 European Commission. *The EU Water Framework Directive: Integrated River Basin Management for Europe.* <http://ec.europa.eu/environment/water/water-framework/index_en.html> Brussels: Commission of the European Communities, 2014.
18 Stuart T. *Waste: Uncovering the Global Food Scandal.* London: Penguin, 2009.
19 WRAP. Love food hate waste: UK household food waste campaign facts. <http://www.wrap.org.uk> Banbury (Oxon): Waste Resources Action Programme (WRAP), 2008.
20 Gustavsson J, Cederberg C, Sonnesson U, *et al. Global Food Losses and Food Waste: Extent, Causes and Prevention.* Rome: Food and Agriculture Organisation, 2011.
21 Wallace AR. On the tendency of varieties to depart indefinitely from the original type. Unpublished article sent to Charles Darwin 1858.

22   Darwin C. *On the Origin of Species by Means of Natural Selection, or The Preservation of Favoured Races in the Struggle for Life.* London: John Murray, 1859.

23   Haeckel E. *Generelle Morphologie der Organismen.* Berlin: Georg Reimer, 1866.

24   Haeckel E. *The History of Creation: Or the Development of the Earth and its Inhabitants by the Action of Natural Causes.* Trans. revised by E. Ray Lankester. 2 vols. London: Henry S. King & Co., 1876.

25   Rayner G, Lang T. *Ecological Public Health: Reshaping the Conditions for Good Health.* Abingdon: Routledge/Earthscan, 2012.

26   Malthus TR. *An essay on the principle of population, as it affects the future improvement of society with remarks on the speculations of Mr. Godwin, M. Condorcet and other writers.* London: Printed for J. Johnson, 1798.

27   Beddington J. Food, energy, water and the climate: a perfect storm of global events? Paper for 'Sustainable Development UK 09' conference, London, 19 Mar. 2009.

28   Foresight. Foresight Land Use Futures Project. <http://www.foresight.gov.uk/OurWork/ActiveProjects/LandUse/LandUse.asp> London: Foresight Programme of the Chief Scientist, 2008.

29   Foresight. *The Future of Food and Farming: Challenges and Choices for Global Sustainability. Final Report.* London: Government Office for Science, 2011: 211.

30   PMSEIC (Australia). *Australia and Food Security in a Changing World.* Canberra: Science, Engineering and Innovation Council of Australia, 2010.

31   Paillard S, Treyer S, Dorin B, eds. *Agrimonde: Scenarios and Challenges for Feeding the World in 2050.* Paris: Editions Quae, 2011.

32   Gunderson LH. Ecological resilience: in theory and application. *Annual Review of Ecology and Systematics,* 31 (2000): 425–39

33   Holling CS. Resilience and stability of ecological systems. *Annual Review of Ecology and Systematics,* 4 (1973): 1–23.

34   Lovelock J. *Gaia: A New Look at Life on Earth.* Oxford and New York: Oxford University Press, 1979.

35   Golding W. *Lord of the Flies: A Novel.* London: Faber, 1954.

36   Rockström J, Steffen W, Noone K, *et al.* Planetary boundaries: exploring the safe operating space for humanity. *Ecology and Society,* 14(2) (2009): 32 <http://www.ecologyandsociety.org/vol14/iss2/art32>

37   Rockström J, Steffen W, Noone K, *et al.* A safe operating space for humanity. *Nature,* 461(7263) (2009): 472–5.

38   Steffen W, Richardson K, Rockström J, *et al.* Planetary boundaries: guiding human development on a changing planet. *Science (Express),* 2015 <http://www.sciencemag.org/content/early/2015/01/14/science.1259855/suppl/DC1>

39   Klein N. *This Changes Everything: Capitalism vs the Climate.* London: Allen Lane, 2014.

40   IPCC. *Climate Change 2014: Synthesis Report.* Geneva: Intergovernmental Panel on Climate Change, 2014.

41   UN. *World Economic and Social Survey 2011: The Great Green Technological Transformation.* <http://www.un.org/en/development/desa/policy/wess/wess_current/2011wess.pdf> New York: United Nations Department of Economic and Social Affairs, 2011.

42   Stern N. *The Stern Review of the Economics of Climate Change: Final Report.* London: HM Treasury, 2006.

43   Tukker A, Huppes G, Guinée J, *et al. Environmental Impact of Products (EIPRO): Analysis of the Life Cycle Environmental Impacts Related to the Final Consumption of*

the *EU-25*. EUR 22284 EN. Brussels: European Commission Joint Research Centre, 2006.

44   Tukker A, Bausch-Goldbohm S, Verheijden M, *et al. Environmental Impacts of Diet Changes in the EU*. Seville: European Commission Joint Research Centre Institute for Prospective Technological Studies, 2009.

45   Cassidy ES, West PC, Gerber JS, *et al.* Redefining agricultural yields: from tonnes to people nourished per hectare. *Environmental Research Letters*, 8 (2013): 034015 (8pp).

46   Ripple WJ, Smith P, Haberl H, *et al.* Ruminants, climate change and climate policy. *Nature Climate Change*, 4 (2014): 2–5.

47   FAO. *Global Food Losses and Food Waste: Extent, Causes and Prevention*. Rome: Food and Agriculture Organisation, 2011.

48   Darwin C. *The Formation of Vegetable Mould, through the Action of Worms*. London: John Murray, 1882.

49   Brandt K. *The Reconstruction of World Agriculture*. London: George Allen & Unwin, 1945.

50   House of Lords EU Committee. *Counting the Cost of Food Waste: EU Food Waste Prevention*. 10th Report of Session 2013–14. London: The Stationery Office, 2014.

51   Schneider M. *Feeding China's Pigs: Implications for the Environment, China's Smallholder Farmers and Food Security*. Minneapolis, MN: Institute for Agriculture and Trade Policy, 2010.

52   Fertiliser use in China. Workshop on Environment, Resources, and Agricultural Policies in China, 19–21 June 2006, Beijing Session 3: Policy Options for China 2006; Beijing: OECD.

53   National Audit Office. *The 2001 Outbreak of Foot and Mouth Disease*. Report by the Comptroller and Auditor General. HC 939 Session 2001–2002: 21 June 2002. London: The Stationery Office, 2002.

54   Commision of the European Communities. *Directive on Packaging and Packaging Waste* (94/62/EC). Brussels: European Commission, 1994.

55   WRAP. Reports: food waste from all sectors. <http://www.wrap.org.uk/content/all-sectors> [accessed June 2014]. Swindon: WRAP, 2014.

56   Millennium Ecosystem Assessment (Program). *Ecosystems and Human Well-Being: Synthesis*. Washington, DC: Island Press, 2005.

57   Tudge C. *Neanderthals, Bandits and Farmers: How Agriculture Really Began*. London: Weidenfeld & Nicolson, 1998.

58   FAO. *Dimensions of Need: An Atlas of Food and Agriculture*. Rome: Food and Agriculture Organisation, 1995.

59   Khoury CK, Bjorkman AD, Dempewolf H, *et al.* Increasing homogeneity in global food supplies and the implications for food security. *Proceedings of the National Academies of Science*, 2014. <www.pnas.org/cgi/doi/10.1073/pnas.1313490111>

60   FAO. *Women: Users, Preservers, and Managers of Agro-Biodiversity*. Rome: Food and Agriculture Organisation, 1998.

61   FAO, Bioversity International. *Final Document: International Scientific Symposium: Biodiversity and Sustainable Diets – United against Hunger. 3–5 November 2010, Rome*. <http://www.eurofir.net/sites/default/files/9th%20IFDC/FAO_Symposium_final_121110.pdf> Rome: Food and Agriculture Organisation, 2010.

62   FAO. *The State of World Fisheries and Aquaculture 2006*. Rome: Food and Agriculture Organisation, 2007.

63 Royal Commission on Environmental Pollution. *Turning the Tide: Addressing the Impact of Fishing on the Marine Environment*, 25th report. London: Royal Commission on Environmental Pollution, 2004.

64 Pew Oceans Commission. *America's Living Oceans: Charting a Course for Sea Change*. Washington, DC: Pew Charitable Trusts, 2003.

65 Allsopp M, Johnston P, *Santillo D. Challenging the Aquaculture Industry on Sustainability: Technical Overview*. Greenpeace Research Laboratories Technical Note 01/2008. Amsterdam: Greenpeace International, 2008.

66 Hites RA, Foran JA, Carpenter DO, *et al.* Global assessment of organic contaminants in farmed salmon. *Science*, 303 (2004): 226–9.

67 Gardner G. *Shrinking Fields: Cropland Loss in a World of Eight Billion*. Washington, DC: Worldwatch Institue, 1996.

68 Halweil B. Where have all the farmers gone? *World-Watch*, 13(5) (2000):12–28.

69 Ausubel K. *Seeds of Change*. San Francisco, CA: HarperCollins, 1994.

70 Henry S. Sow few, so trouble. *Green Futures*, 31 (2001): 40–2.

71 McMichael P. The power of food. *Agriculture and Human Values*, 17(1) (2000): 21–33.

72 FAO. *State of the World's Plant Genetic Resources*. Rome: Food and Agriculture Organisation, 1996.

73 McMichael AJ. *Human Frontiers, Environment and Disease*. Cambridge: Cambridge University Press, 2001.

74 UNEP. *Global Environment Outlook 3*. London: Earthscan, 2002.

75 Pretty J. *Agri-Culture*. London: Earthscan, 2002.

76 Funes F, García L, Bourque M, *et al. Sustainable Agriculture and Resistance: Transforming Food Production in Cuba*. Oakland, CA: Food First Publications, 2002.

77 Altieri MA. *Agroecology: The Scientific Basis of Alternative Agriculture*. Boulder, CO: Westview, 1987.

78 Diversity CoB. Conference of the Parties (COP12) <http://www.cbd.int/cop> Rome: Convention on Biological Diversity, 2014.

79 UNICEF. *Water, Sanitation and Hygiene: WASH 2013 Annual Report*. Geneva: UNICEF, 2013.

80 Stockholm International Water Institute. *General Water Statistics*. Stockholm: SIWI, 2003.

81 WWF. *Thirsty Crops: Our Food and Clothes. Eating up Nature and Wearing out the Environment?* Zeist (NL): WWF, 2006.

82 Allan JA. Virtual water – the water, food and trade nexus: useful concept or misleading metaphor? *Water International,* 28 (2003): 4–11.

83 Allan JAT. *Virtual Water: Tackling the Threat to our Planet's Most Precious Resource*. 1st ed. London: I.B.Tauris, 2011.

84 Chapagain AK, Hoekstra AY. *Water Footprints of Nations*, vols 1 and 2. UNESCO-IHE Value of Water Research Report Series, 16. Paris: UNESCO, 2006.

85 Hoekstra AY, Chapagain AK, Aldaya MM, *et al. The Water Footprint Assessment Manual: Setting the Global Standard*. London: Earthscan, 2011.

86 WWF, Global Footprint Network, Institute of Zoology, *et al. Living Planet Report 2014: Species and Spaces, People and Places*. Gland: WWF International, 2014.

87 UNDESA. UN Decade of Water for Life: <http://www.un.org/waterforlifedecade/food_security.shtml> New York: UN Department of Economic and Social Affairs, 2014.

88   Dinham B. *The Pesticide Hazard: A Global Health and Environmental Audit*. London and Atlantic Highlands, NJ: Zed Books, 1993.

89   Jacobs MN, Covaci A, Gheorghe A, *et al*. Time trend investigation of PCBs, PBDEs, and organochlorine pesticides in selected n-3 polyunsaturated fatty acid rich dietary fish oil and vegetable oil supplements: nutritional relevance for human essential n-3 fatty acid requirements. *Journal of Agricultural and Food Chemistry*, 52 (2004): 1780–8.

90   Schafer KS, Kegley SE. Persistent toxic chemicals in the US food supply. *Journal of Epidemiology and Community Health*, 56 (2002): 813–17.

91   Avery DT. *Saving the Planet with Pesticides and Plastic: The Environmental Triumph of High-Yield Farming (1995)*. Washington, DC: Hudson Institute, 2000.

92   Markets and Markets. *Agrochemicals Market by Type (Fertilizers, Pesticides), by Fertilizer Type (Nitrogenous, Potassic, Phosphatic), by Pesticide Type (Organophosphates, Pyrethroids, Neonicotinoides, Bio-Pesticides), & Sub-types: Global Market Trends and Forecast to 2018*. Dallas, TX: Markets and Markets, 2014.

93   Lang T, Clutterbuck C. *P is for Pesticides*. London: Ebury, 1991.

94   De A, Bose R, Kumar A, *et al*. *Targeted Delivery of Pesticides Using Biodegradable Polymeric Nanoparticles*. New York: Springer, 2014.

95   Buffin DG. UK pesticides policy: a paradigm in transition? PhD thesis, City University London (Centre for Food Policy), 2009.

96   Jeyarraratnam J. Acute pesticide poisoning: a major problem. *World Health Statistics Quarterly*, 43 (1990): 139–44.

97   Pesticide Residue Committee (UK). *Annual Report of the Pesticide Residue Committee 2006*. York: Pesticide Safety Directorate, 2007.

98   Baker BP, Benbrook CM, Groth E, *et al*. Pesticide residues in conventional, integrated pest management (IPM)-grown and organic foods: insights from three US data sets. *Food Additives and Contaminants*, 19(5) (2002): 426–446.

99   WCRF/AICR. *Food, Nutrition, Physical Activity and the Prevention of Cancer: A Global Perspective*. Washington, DC/London: World Cancer Research Fund/ American Institute for Cancer Research, 2007.

100 Hyams E. *Soil and Civilization*. London: Thames & Hudson, 1952.

101 UNCCD. United Nations Convention to Combat Desertification. 17 June World Day. <http://www.un.org/en/events/desertificationday/background.shtml> Rome: UN Conventionl to Combat Desertification, 2015.

102 Smith P. Delivering food security without increasing pressure on land. *Global Food Security*, 2(1) (2013): 18–23.

103 Smith P, Haberl H, Popp A, *et al*. How much land based greenhouse gas mitigation can be achieved without compromising food security and environmental goals? *Global Change Biology*, 19(8) (2013): 2285–2302.

104 Glatzel K, Conway G, Alpert E, *et al*. *No Ordinary Matter: Conserving, Restoring and Enhancing Africa's Soils*. London: Montpellier Panel, c/o Agriculture for Impact, Imperial College London, 2014.

105 Oldeman LR, Hakkeling RTA, Sombroek EG. The extent of human-Induced soil degradation. In: Oldeman LR, ed. *World Map of the Status of Human-Induced Soil Degradation*. Nairobi: United Nations Environment Programme and International Soil Reference and Information Centre, Wageningen, the Netherlands, 1991: 1–18.

106 European Environment Agency, UN Environment Programme. *Down to Earth: Soil Degradation and Sustainable Development in Europe. A Challenge for the 21st Century*. Copenhagen: European Environment Agency, 2000.

107 ECAF. *Conservation Agriculture in Europa*. ECAF First Report Brussels: European Conservation Agriculture Federation, 1999.

108 UNEP. *Assessing Global Land Use: Balancing Consumption With Sustainable Supply*. A Report of the Working Group on Land and Soils of the International Resource Panel. Nairobi: UN Environment Programme, 2014.

109 Global Footprint Network. *Living Planet Report 2010: Biodiversity, Biocapacity and Development*. Gland: WWF, Institute of Zoology, Global Footprint Network, 2010.

110 UNDESA. *World Urbanisation Prospects: The 2014 Revision, Highlights*. ST/ESA/ SER.A/352. New York: United Nations Department of Economic and Social Affairs, Population Division, 2014.

111 Harrison P. *The Third Revolution: Environment, Population and a Sustainable World*. London: I.B.Taurus & Co. (with World Wide Fund for Nature), 1992.

112 Millstone E, Lang T. *The Atlas of Food*. London: Earthscan, 2003.

113 Garnett T. *City Harvest: The Feasibility of Growing More Food in London*. London: Sustain, 1999.

114 FAO. *State of Food and Agriculture*. Rome: Food and Agriculture Organisation, 1998.

115 Smit J, Ratta A, Nasr J. *Urban Agriculture: Food, Jobs and Sustainable Cities*. New York: UN Development Programme Habitat II Series, 1996.

116 Sustainable Food Cities. Sustainable Food Cities network. <http://sustainablefoodcities. org> Bristol, London, Brighton, 2014.

117 FAO. *Energy-Smart Food for People and Climate*. Rome: Food and Agriculture Organisation, 2011.

118 Woods J, Williams A, Hughes JK, *et al*. Energy and the food system. *Philosophical Transactions of the Royal Society*, 365 (2010): 2991–3006.

119 Beckman J, Borchers A, Jones CA. *Agriculture's Supply and Demand for Energy and Energy Products*. Washington, DC: USDA Economic Research Service, 2013.

120 DEFRA. *Food Statistics Pocketbook 2013 (May 29 Update)*. London: Department for Environment, Food and Rural Affairs, 2014.

121 Canning P, Charles A, Huang S, *et al*. *Energy Use in the U.S. Food System*. Washington, DC: USDA Economic Research Service, 2010: 39.

122 International Energy Agency. *Cities, Towns and Renewable Energy – YIMFY – Yes In My Front Yard*, I. Paris: International Energy Agency, 2009.

123 Wackernagel M, Rees WE, Testemale P. *Our Ecological Footprint: Reducing Human Impact on the Earth*. Gabriola Island, BC: New Society Publishers, 1996.

124 Paxton A. *The Food Miles Report*. London: Sustainable Agriculture, Food and Environment (SAFE) Alliance, 1994.

125 Desrochers P, Shimizu H. *Yes, we have No Bananas: A Critique of the 'Food Miles' Perspective*. Washington, DC: George Mason University Mercatus Centre, 2008.

126 Stancu C, Smith A. *Food Miles: The International Debate and Implications for New Zealand Exporters*. Auckland: Landcare Research, 2006.

127 Lang T. Food miles. *Slow Food* (2006): 94–7.

128 Smith A, Watkiss P, Tweddle G, *et al*. *The Validity of Food Miles as an Indicator of Sustainable Development*. Report to DEFRA by AEA Technology. London: Department for the Environment, Food and Rural Affairs, 2005.

129 Rai Chi K, MacGregor J, King R. *Fair Miles: Recharting the Food Miles Map*. Oxford: International Institute for Environment and Development & Oxfam GB, 2009.

130 IGD. *Sustainable Distribution in 2008*. Radlett Herts: IGD, 2008.

131 Boege S. *Road Transport of Goods and the Effects on the Spatial Environment.* Wuppertal: Wuppertal Institute, 1993.

132 Pirog R, Benjamin A. *Calculating Food Miles for a Multiple Ingredient Food Product.* Ames, IA: Leopold Center for Sustainable Agriculture, Iowa State University Ames, 2005.

133 Borgström G. *The Food and People Dilemma.* North Scituate, MA: Duxbury, 1973.

134 Van Brakel M, Zagema B. *Sustainable Netherlands.* Amsterdam: Friends of the Earth, 1994.

135 BFF Ltd, GLA. City Limits: a resource flow and ecological footprint analysis of Greater London.   <http://www.citylimitslondon.com/downloads/Complete%20report.pdf> London: Greater London Authority & Best Foot Forward Ltd., 2002.

136 Ministry of Defence. *The DCDC Global Strategic Trends Programme: 2007–2035.* Shrivenden: Ministry of Defence Development Concepts and Doctrine Centre (DCDC), 2007.

137 Commission for Future Generations. *Now for the Long Term.* Oxford: Oxford Martin School, University of Oxford, 2014.

138 Steinfeld H, Gerber P, Wassenaar T, *et al. Livestock's Long Shadow: Environmental Issues and Options.* Rome: Food and Agriculture Organisation, 2006.

139 Gerber PJ, Steinfeld H, Henderson B, *et al. Tackling Climate Change through Livestock: A Global Assessment of Emissions and Mitigation Opportunities.* Rome: Food and Agriculture Organisation of the United Nations, 2013.

140 Nestle M. *Food Politics.* Berkeley, CA: University of California Press, 2002.

141 Cannon G. *The Politics of Food.* London: Century, 1987.

142 Farmsubsidy.org. EU farm subsidies for all countries, all years. <http://farmsubsidy. openspending.org/EU> [accessed June 2014]. London: Farmsubsidy.org 2014.

143 Rifkin J. *Beyond Beef: The Rise and Fall of the Cattle Culture.* New York: Dutton, 1992.

144 Sinclair U. *The Jungle.* Harmondsworth: Penguin, 1906/1985.

145 World Bank. *Minding the Stock: Bringing Public Policy to Bear on Livestock Sector Development.* Washington, DC: World Bank, 2009.

146 Walker P, Rhubart-Berg P, McKenzie S, *et al.* Public health implications of meat production and consumption. *Public Health Nutrition,* 8(4) (2005): 348–56.

147 Stehfest E, Bouwman L, van Vuuren DP, *et al.* Climate benefits of changing diet. *Climatic Change,* 95 (2009): 83–102.

148 Sinha R, Cross A J, Graubard B I, *et al.* Meat intake and mortality. *Archives of Internal Medicine,* 169(6) (2009): 562–71.

149 Lang T, Caraher M, Wu M. Meat and policy: charting a course through the complexity. In: d'Silva J, Webster J, eds. *The Meat Crisis: Developing More Sustainable Production and Consumption.* London: Earthscan, 2010: 254–274.

150 McMichael A, Woodruff R, Hales S. Climate change and human health: present and future risks. *The Lancet,* 367(9513) (2006): 859–69.

151 Royal Society. *People and the Planet.* London: Royal Society, 2012.

152 Macdiarmid J. Is a healthy diet an environmentally sustainable diet? Proceedings of the Nutrition Society 2012.

153 Westhoek H, Lesschen JP, Rood T, *et al.* Food choices, health and environment: effects of cutting Europe's meat and dairy intake. *Global Environmental Change* (2014) <http://dx.doi.org/10.1016/j.gloenvcha.2014.02.004>

154 Smil V. *Should we Eat Meat? Evolution and Consequences of Modern Carnivory*: Wiley-Blackwell, 2013.

155 Baroni L, Cenci L, Tettamanti M, *et al*. Evaluating the environmental impact of various dietary patterns combined with different food production systems. *European Journal of Clinical Nutrition*, 61 (2007): 279–86.

156 Hedenus F, Wirsenius S, Johansson DJA. The importance of reduced meat and dairy consumption for meeting stringent climate change targets. *Climatic Change*, 124 (2014): 79–91.

157 Carlsson-Kanyama A, Gonzalez AD. Potential contributions of food consumption patterns to climate change. *American Journal of Clinical Nutrition*, 89(suppl.) (2009): 1S–6S.

158 Meier T, Christen O. Environmental impacts of dietary recommendations and dietary styles: Germany as an example. *Environmental Science and Technology*, 47(2) (2013): 877–88.

159 Cordain L, Eaton SB, Sebastian A, *et al*. Origins and evolution of the Western diet: health implications for the 21st century. *American Journal of Clinical Nutrition*, 81(2) (2005): 341–54.

160 Rosegrant MW, Leacha N, Gerpacio RV. Meat or wheat for the next millennium? *Proceedings of the Nutrition Society*, 58 (1999): 219–34.

161 Carbon Trust. Tesco and Carbon Trust join forces to put carbon label on 20 products. <http://www.carbontrust.co.uk/News/presscentre/29_04_08_Carbon_Label_Launch. htm> London: Carbon Trust, 3 June 2008.

162 Harvey F. Snack-maker aims for green consumers with carbon labels. *Financial Times*, 16 Mar. 2007: 4.

163 Garnett T. *What is a Sustainable Diet? A Discussion Paper*. Oxford: Food & Climate Research Network, 2014: 31.

164 Van Dooren C, Marinussen M, Blonk H, *et al*. Exploring dietary guidelines based on ecological and nutritional values: a comparison of six dietary patterns. *Food Policy*, 44 (2014): 36–46.

165 Sustainable Development Commission. *Looking Forward, Looking Back: Sustainability and UK Food Policy 2000–2011*. <http://www.sd-commission.org.uk/ publications.php?id=1187> London: Susainable Development Commission, 2011.

5 Policy responses to food's role in human and environmental health

War is probably the single most powerful instrument of dietary change in human experience. In time of war, both civilians and soldiers are regimented – in modern times, more even than before. There can occur at the same time terrible disorganization and (some would say) terrible organization. Food resources are mobilized, along with other sorts of resources. Large numbers of persons are assembled to do things together – ultimately, to kill together. While learning how, they must eat together. Armies travel on their stomachs; generals – and now economists and nutritionists – decide what to put in them. They must do so while depending upon the national economy and those who run it to supply them with what they prescribe or, rather, they prescribe what they are told they can rely upon having.

Sidney Mintz, anthropologist of food, USA, b. 1922[1]

Core arguments

In the 20th century there was a growth of policy awareness of the evidence of food's impact on both health and the environment. This led to battles over the shape of policies, with different vested interests vying for influence over actions or whether to do anything. Positions ranged from radical change to maintaining the status quo. In nutrition, these battles ran throughout the century, whereas for the environment battles over policy really began later. In the 20th century, understanding of food's role in meeting public health objectives fluctuated considerably driven by a mix of scientific advances, new evidence together with social upheavals, from wars to lifestyle changes. During this time nutrition split into two major strands: one focused on social objectives such as poverty reduction, the other on bio-chemical mechanisms. In the 21st century a third, newer, strand emerged as increasingly important: nutrition's connections to the environment which re-emerged with growing evidence about food's heavy impact on eco-systems and use of resources. For the Productionist paradigm, dietary guidelines have been the main battleground for nutrition and environmental food policy. There is now a renewed interest in creating an integrated approach to food, diet, health and the environment. This objective has been articulated through the concept of sustainable diets from sustainable food supply chains. This sits comfortably with the Ecologically Integrated paradigm and, in part, with the Life Sciences Integrated paradigm, too. There are differences in how such policies are being framed, such as between developing and developed countries, but also between the food rich and the food poor within the same country.

Introduction

In this chapter we outline the beginnings of policy response to the enormous issues raised in Chapters 3 and 4 which document how evidence has grown on food's impact on health and the environment. The story we tell now is of a patchy response. There have been bigger actions in some areas than others. Overall, however, the picture is of slow, laboured and disjointed policy engagement. This criticism is not given lightly; nor is it a denunciation. The evidence and arguments summarised in the book so far are immensely complicated. And food policy-making has tended to be dominated for decades by economic imperatives.

The best brains in the world have not yet managed to persuade politicians, business leaders or the public to act coherently on major challenges such as obesity or climate change. So we are not surprised that the complex challenges raised in this book so far elude policy-makers. There are health policies; there are environmental policies; there are industry or farming policies. What there is not, however, is a framework for addressing all these simultaneously. This is the 21st-century challenge. We present this chapter, half way through the book, to capture the big picture of policy, and to give some key examples of where innovative or different policy developments are beginning to emerge, building up pressure for coherence.

The scientific evidence for diet-related disease has not gone unnoticed within governments or the business sector. The same is true for food's impact on the environment. In fact, government ministries have helped create the evidence through statistical surveys and funding academic and other scientific studies. That said, the case we make in this chapter is that what policy actions there are have in many instances not done what is necessary. The evidence on food's impact on health and environment is often very strong, for instance on climate change or heart disease. Yet actions to shift population diet or to change food production systems to address climate change or to drastically reduce obesity have too often been blunted. Good intentions and recommendations are left to gather dust on shelves.

The reflex is too often to maintain the status quo or to deal with problems exposed by the evidence and science in small and 'soft' policy ways. But it is now clear that food policies are needed as a key mechanism for achieving better integration between health and the environment. The health and environmental crises cannot be resolved unless food is part of the solutions. One difficulty is that the state apparatus of the Productionist paradigm is controlled by the ministries most associated with production: usually agriculture and not health or environment. Until recently, any connection between diet, disease, environment and food supply has been subverted or resisted. In the case of health, by the 1990s, there were more than 100 authoritative scientific reports produced between 1961 and 1991 recommending dietary change in relation to disease and health published throughout the world.[2] Equally there were many environmental reports suggesting the urgent need to reduce modern food system's impact on the environment. While a new mantra emerged in the 2000s to 'produce more from less' – summarised in the new phrase of 'sustainable intensification'[3] – policy frameworks to deliver this seemed elusive and unclear.

In this chapter we describe the 100-year food battle to bring nutrition policy to the forefront in public policy, and the half century battle to address food and the environment. We suggest that these agendas are beginning to be seen as two sides of the same coin: a matter to be resolved in different ways by the respective Food Wars paradigms. We see this discussion as crucial since it will demonstrate the deep roots of the tensions between the Life Sciences Integrated paradigm and the Ecologically Integrated paradigm and the urgent need for an ecological approach to nutrition, health and environmental policies as they affect food.

Changing conceptions of public health

A much-cited 1998 definition of public health proposed that it is 'the science and art of preventing disease, prolonging life and promoting health through the organised efforts of society'.[4] This 'science and art' phrase was actually first coined in 1920 by Charles Edward Winslow, a Yale University bacteriologist.[5] At the core of many definitions of public health is the notion that health is not an individual but a societal phenomenon and that the social and natural environments shape the chances of people contracting or preventing disease. There are many traditions of what is meant by public health.

In 1998 the WHO proposed a 'new public health' focused upon 'lifestyles and living conditions [which] determine health status', and whose challenge is to 'mobilise resources and make sound investments in policies, programmes and services which create, maintain and protect health by supporting healthy lifestyles and creating supportive environments for health'.[6]

This new public health tried to locate public health policies as:[7]

> an approach which brings together environmental change and person preventative measures with appropriate therapeutic interventions, especially for the elderly and disabled. [H]owever, the New Public Health goes beyond an understanding of human biology and recognises the importance of those social aspects of health problems which are caused by lifestyles.
>
> In this way it seeks to avoid the trap of blaming the victim. Many contemporary health problems are therefore seen as being social rather than solely individual problems; underlying them are concrete issues of local and national public policy, and what are needed to address these problems

are 'Healthy Public Policies' – policies in many fields which support the promotion of health. In the New Public Health, the environment is social and psychological as well as physical.

This interpretation is for health as the outcome of all public policy, not just a healthcare policy. It implies that all policies need to be health-proofed, asking what is the impact on health if this or that policy is followed. In policy terms, this conceives of public health as a 'cross-cutting' issue which cannot be separated from other policy spheres. The WHO has defined this in 2010 as follows:[8]

> Health in All Policies is an approach to public policies across sectors that systematically takes into account the health implications of decisions, seeks synergies, and avoids harmful health impacts, in order to improve population health and health equity.

Rayner and Lang have outlined five major traditions or models of public health which have emerged since industrialisation over the last two and half centuries:[9]

1   Sanitary-Environmental: this suggests that health is determined by the external environment. This is the model of health as maintained by building sewers, providing clean water and regulation to prevent the adulteration of foodstuffs.
2   Social Behavioural: this focuses on the social and cognitive dimensions of existence which affect health, human behaviour and habits. In this model providing 'information' is regarded and critical to healthy behaviour.
3   Bio-Medical: this is the dominant model at present. It sees health as dependent on medical interventions, pharmaceuticals and expert knowledge. Theories of disease explain how people become ill or not.
4   Techno-Economic: this suggests that health is the outcome of economic and technical progress. Better houses and infrastructure improve health. People get healthier when they become wealthier. Public health is thus dependent on wealth creation.
5   Ecological Public Health: this stresses the coexistence of humans and eco-systems and the natural world. It proposes that public health is ultimately a matter of shaping the conditions for good health.

These different traditions are useful for analysing the various approaches to diet's impact on health and the environment.

The nutrition pioneers: a 100-years war

Nutrition policy is a distinct part of public health policy which, like public health, has fought hard to gain a place at the political table. The 18th century doctor James Lind, although not the first to note the connection between diet and ill health, is often credited with putting modern nutrition onto a scientific basis. With trade routes dependent upon the health of ships' crews, the problem of scurvy was a major threat: it could devastate entire ships' crews.[10] In 1753 Lind published the results of the first controlled study and established conclusively that scurvy could be prevented and cured by introducing citrus fruit into the diet.[11] This was an early indication of how the science of nutrition could contribute to economic and even military well-being. Napoleon Bonaparte is famously stated to have said that an army marches on its stomach and to have initiated in the late 18th century the technology search that resulted in canning as a safe means to perfect, portable and long-lasting food. (He also encouraged the development of the French sugar beet industry!)

Two and a half centuries on, nutrition covers a vast field ranging from social nutrition (e.g. studying 'at risk' social groups), nutritional epidemiology (plotting the contribution of diet to diseases), biochemistry (the study of the bio-chemical interaction of nutrients and the body), sports and animal nutrition (optimising physiological performance) and psychophysiology (including the study of food choice).[12] Partly fuelled by huge pharmaceutical and food-industry research funds, it is biochemistry that dominates nutrition today, with its researchers seeking profitable health benefits from within the diet. This is very much within the Life Sciences Integration paradigm. This direction is sometimes associated with Sir Frederick Gowland Hopkins's discovery in 1901 that the human body could not make the amino-acid *triptophan*, an essential part of protein, and that it could only be derived from the diet, demonstrating the principle that, without a proper diet, bodily function could be impaired or deficient. Hopkins proved the existence of what he called food hormones or 'vitamines', and by the end of the 1930s most 'vitamins' are subsequently 'discovered',

Nutrition, like any study concerning humans, is inevitably framed by social assumptions.[13-15] Some see the pursuit of better nutrition as an individual responsibility, while others view nutritional science as a tool of greater social efficiency or as an end in itself. Throughout the 20th century, nutrition was a battleground with some forces using it as an opportunity for social control and others arguing that it could liberate human potential. This tension between social control and democracy – 'top-down' science versus people-oriented science – still characterises the policy discourse about food and health.

W.O. Atwater, an influential late 19th century American nutritionist, was an early critic of the US national diet, but he also pursued a mechanistic approach to understanding food as fuel in physical labour: he calculated how much or little nutrient intake was required by different grades of manual workers, according to whether they were engaged in moderate or heavy work;[16 17] he produced estimates of the protein, fat and carbohydrate required of workers performing light, heavy or moderate work. His work was taken up east of the Atlantic by B. Seebohm Rowntree, scion of a giant UK chocolate dynasty (now owned by Nestlé) and a founding father of UK food and welfare policy.

Throughout the first half of the 20th century, Rowntree conducted both domestic and industrial surveys in his home town of York, England, based on Atwater's calculations of nutritional need.[18-21] Rowntree used Atwater's minimalist approach to nutrition in order to assess whether the poor were meeting their needs. Rowntree's minimalist approach to poverty made his UK findings all the more shocking. He found that 27.8 per cent of the population of York was living in total poverty, unable to afford the means for even a minimal level of healthy existence. In such findings, Rowntree was using a 'scientific' lens to produce conclusions that other investigators of the Victorian working-class existence before him such as Booth, founder of the Salvation Army, had also come to.[22] Deficiency was the hallmark of the diet of the poor. Thus early nutrition science was inevitably entering a highly charged political and policy sphere. What to do about food poverty? Thereafter, the paradigms and approaches to nutrition diverged; some seeing the solutions through social change such as by raising wages, others as resolvable by technical improvement.

In 1915 at the start of World War I, Thomas Wood and Frederick Gowland Hopkins made a plea that nutrition should be taken seriously (in the war effort). They summarised the value of applying nutrition to public policy goals:[23]

> The human body, though doubtless in many of its aspects something more than a mere machine, resembles the steam-engine in two respects. It calls for a constant supply of fuel, and as a result of doing work, it suffers wear and tear. The body must burn fuel in order that the heat it is always giving off may be continuously replaced; and it must burn still more fuel whenever it does work. From this necessity there is no escape . . . It is, of course, the food eaten which provides these fundamental needs of the body; and if we are to understand properly the nutrition of mankind, we must bear in mind the two distinct functions of food – its function as fuel and its function as repair material.

Arguments about whether diets should be calculated at a minimal or adequate level exposed how, not for the first or last time, the purpose of food can be tinged with morality. Nutrition's focus was on diet rather than the supply chains. During the 1930s recession, as British wages and already thin welfare weakened further, British physicians courageously argued that state welfare should not be based on an Atwater-type subsistence or bare minimum diet, but on one conducive to maintaining both 'health and working capacity',[24] optimum rather than minimum nutrition. Aware that the nutrition evidence suggested that poor health was the outcome of poor diet, in 1939, the British Medical Association organised a conference of many such thinkers to review the food policy situation. Lord Horder, the President of the BMA, could unashamedly comment in radical terms thus:[25]

> A short time ago I was so bold – even so impertinent – as to express the wish that the Ministers of Health, of Agriculture and of Transport, with the Governor of the Bank of England, might be locked in a room together and kept there until they had solved the problem of food production and of food distribution in this country. This was only another way of saying that I believed the problems of malnutrition, of food, and of poverty in the midst of

plenty – that is surely not an overstatement – could never be solved if dealt with compartmentally, but that they could be solved if taken together and dealt with by a long-term policy.

Such radical language from an establishment figure was not uncommon in the 1930s. Major-General Sir Robert McCarrison was another such voice. He would today be seen as an early exponent of an integrated approach to food, nutrition and the environment. While he was Director of the Army's Indian Medical Service, McCarrison questioned the view of food as mere fuel, having been alerted to the impact of poor nutrition by the lamentable health of British army recruits.[26 27] This had first surfaced as a major issue following the second Boer War (October 1899–May 1902) when the British government set up an interdepartmental Committee on Physical Deterioration to address the eugenic argument that, unless more attention was paid to ensuring that the 'fitter' members of society produced more children and fed them well, the national 'breeding stock' would be weakened. In its 1904 report, the Committee adopted a line of national self-interest and promoted optimal feeding, especially in the light of the parlous state of children's nutrition.[28]

McCarrison, only a few years later, was arguing something different: although a degree of self-interest was in order, nutrition does not act as an on–off switch, with the consumer having only enough or not enough. McCarrison argued that the quality of the food mattered. His view was that optimum nutrition was essential for a sound society, being the lubricant between good agriculture and good health. Education was the key, he argued: science should inform citizens and not control them. He did not favour technical fixes to solve national food problems. These required integrated policy action from agriculture to the school.

In this, he was supported by the work of Sir John (later Lord) Boyd Orr, the first Director-General of the FAO at the end of World War II. Boyd Orr had founded the Rowett Research Institute at Aberdeen, which became Europe's largest nutrition institute, in order to explore and promote better scientific links between farming and health. Inclined to favour market solutions, he was gradually convinced of the need also for state action to tackle societal problems. He had conducted a highly influential study of poverty, *Food, Health and Income*, published in 1936, which was widely cited and made headline news around the world.[29] Its findings made news because it exposed food poverty at the heart of the British Empire. The study concluded that the key solution was adequacy of income: without income above a certain threshold, people could not purchase a nutritionally appropriate diet. Boyd Orr's study calculated that 50 per cent of the UK population was unable to afford a diet deemed adequate – up on Rowntree's estimate of food poverty in York 30 years earlier. These data were taken up with vigour by campaigners for women and families, who argued that, because it was ultimately mothers who controlled food within homes, it was they who should be provided with food-oriented aid and education.[30]

Post-World War II developments

By the end of World War II, the view that income was the key to health had triumphed in Western public policy. Yet the FAO, established in 1945, operated primarily as a production-oriented world body while the WHO, set up in the following year, building on the remnants of the 1930s League of Nations' International Health Organisation,[9] remained locked in the bio-medical model of health. Boyd Orr had been pioneering the case for better integration of health and agriculture for many years, and by 1945 the idea of joining health and agriculture had already been considered by scientists from all over the world.[31] Such a food policy was reinforced at the Hot Springs Conference on Food and Agriculture in 1943. This conference, planning for a food system after World War II, accepted the need to have the agriculture and health sectors collaborate.

The USA had joined the War but also was struggling with the legacy of 1930s 'Dust Bowl' crisis and recession. An assignment given to a journalist James Agee and photographer Walker Evans in 1936 by *Fortune* magazine produced the now classic account of the dire circumstances of Southern US sharecroppers – their book was published in 1941.[32] The plight of small farmers is today associated more with developing countries but is also an issue that has confronted the world's richest economy in its recent history. In the 1940s, therefore, the USA was prepared to consider a new policy architecture for food, health and environment at a global level, but only up to a point. Boyd Orr's vision was global: feed the under-consuming parts of the globe by unleashing the productive capacity of Western and particularly US agriculture.[33] This is when the Productionist Paradigm takes centre stage in public policy, focused on quantity of output. But within a decade of it becoming the dominant view, a new nutrition critique emerged, centered on the quality of diets. This is still being fought over.

Professor Ancel Keys's pioneering research in the 1950s showed that diet was a crucial factor in degenerative disease patterns. In his famous 'Seven Countries' study, he compared the diets and health profiles of seven different countries: Finland, Greece, Italy, Japan, Netherlands, USA and Yugoslavia. He found that the inhabitants of the island of Crete (a cohort of the Greece study) suffered least from the circulatory diseases.[34 35] These data are the beginnings of the modern policy reverence for the Mediterranean diet, viewed as significantly healthier than the US diet. Why was there heart disease in New York and not in poor regions like Crete, Keys was asking. This heralded a significant shift from nutrition's policy concern with micronutrient deficiencies to the prospect that macro-nutrients such as fats and sugar could have a significant role in diet-related disease.

Dr Hugh Sinclair, today scantly remembered in the world of public health nutrition for his work on essential fatty acids, until his death in 1990 promoted a view that is central to this book: namely, that the relationship between food and health requires total food supply chain thinking.[36] The pursuit of quality should not be sacrificed on the altar of quantity. In 1961, Sinclair argued:[37]

[W]e can now see clearly that the nutritional problems confronting the world are more urgent and serious than any others. They can be divided into two broad classes: the provision of adequate food for a rapidly increasing world population, and the disasters caused by the processing and sophistication of food in more privileged countries.

More than 55 years later, Sinclair's insight is as pertinent as ever.

The emergence of modern food and environment policy

Just at the time when the new nutrition focusing on non-communicable diseases was emerging in the 1960s and 1970s, another body of evidence began to emerge questioning the direction of post-World War II food systems and food policy. This concerned the rise of environmental problems associated with food, particularly those which are the results of human actions. External environmental health factors with physical, chemical or biological determinants, actually have human-made and societally-framed determinants. According to the WHO, major environmental health problems include pollution, indoor and outdoor air quality, biodiversity depletion, waste production, toxic chemicals and food supply and production.[38]

Arguably, food's environmental impact entered modern policy discourse with the publication of Rachel Carson's *Silent Spring* in 1962.[39] Carson, a biologist employed by the US government, had come to the conclusion that the application of agrichemicals to control pests was disturbing ecological relationships in nature. She argued that what had been unleashed for understandable intentions had spawned unintended consequences, a chain of destruction across species. *Silent Spring* was met by a furious counter-attack from the main US agricultural establishment which accused Carson of scientific illiteracy and 'neo-Luddism'. Alas, Carson died of cancer in 1964 only two years after the book came out, but her thesis and case gradually won support, as evidence began to emerge of agrichemicals' impact on diverse eco-systems.[40] A spray may kill a pest or a fungus on a crop but the processes of Darwinian evolution mean that this is not a fixed war. Bugs change. Pests and plants develop resistance. Human health can be damaged by misuse and overuse. Carson's argument was that residues could emerge all round the world, and that some of the chemicals were persistent. They decayed only very slowly. Water systems are polluted; the web of life is frayed.

The arguments over pesticides were just one of many issues to emerge from the 1960s on food's impact on the environment (summarised in Chapter 4). These became campaigning issues for the new generation of NGOs and civil society organisations. This is the era when Greenpeace, Friends of the Earth and a welter of new environmental organisations were founded. They disagreed with the earlier conservation movements from the 19th century which they saw as seeking 'islands' of conservation in a 'sea' of thoughtless pollution. They began campaigns to change public policy, specifically to prevent new pollution, new adulteration and the system-wide environmental impact of modern food and farming. They were questioning, in effect, the direction of progress, and were proposing a model of food production and consumption which aimed to work with nature rather than tame or destroy it. Oxfam, the large development NGO, for example, picked up the profligate use of agrichemicals in the developing world,[41] and joined others in what was in effect a new global civil society alliance to rein in unnecessary and dangerous use of pesticides and fungicides.[42 43] The WHO came under pressure to take this seriously and to use its co-chairing of the Codex Alimentarius Commission – the global food standards body – to reduce pesticide use in the name of human health, if not necessarily environmental health.[44]

Within 30 years of Carson, at national, regional and global levels of public policy-making, there was a tightening up of regulations on the agrichemical industry, and increasing scrutiny of its workings.[45] Some industry advocates, of course, saw this as unnecessary bureaucracy adding costs to the development of new chemicals for use on food. Activists countered that nature was too precious not to defend. A recent example of this continuing argument is the case of whether use of neonicitinoids – themselves introduced in the 1990s to replace more dangerous agrichemicals – have reduced populations of bees and other pollinators.[46] The EU introduced a ban on them in 2012 and the USA began a major review in 2014.

Many specific fights have occurred and continue within wider food and environment wars. Following Carson's lead, by the 1970s, the environmental critique of Productionism was becoming scientifically more respectable and, perhaps more importantly, resonating with the public. In 1971, for example, Francis Moore Lappé and Joe Collins produced a huge-selling *Diet for a Small Planet*,[47] and created an NGO around it. Environmental activism began to articulate public health arguments when charting different lifestyles.

The rise of sustainable development

As such activism emerged – over plastics, toxic chemicals, nitrate and nitrite residues in water from fertilisers, species loss, deforestation and waste – governments began to create or ramp up their policy frameworks on the environment. Regional and global bodies began to meet to address the scientific and NGOs' case that environmental hazards were too high risk and to manage them more effectively. The growth of public policy was accelerated by the shock of the 1970s oil crisis, when the Middle East producers cartel unilaterally raised oil prices. This exposed how reliant the West's model of food development and economic growth itself were on oil. Food was oil. Fertilisers were oil. Mechanisation and the replacement of human and animal labour were oil. Packaging was oil. The case for exploiting environmental resources in modern supposedly efficient food systems looked vulnerable.

New analyses emerged with highlighted food system dependency on non-renewable resources.[48-50] The Club of Rome, a meeting of global industrialists, captured this threat in their much-cited 1972 report *The Limits to Growth* which used computer modelling of resource use to suggest constraints upon economic development were not far off.[51] In fact, economic growth and intensive food and farming systems continued, albeit modifying some excesses. Forty years later, however, when oil prices rocketed, in 2007–8, these questions again surfaced.

Table 5.1 Examples of key global reports and policy statements on environment, sustainable development of relevance to food systems

Date	Parent body	Output/report	Comment
1972	UN Conference on the Human Environment	Known as the Stockholm Conference	First major modern environmental strategy emerges
1972	Club of Rome	Limits to Growth	Industry-funded forecast of future resource squeeze
1987	UN World Commission on Environment and Development	Brundtland Report 'Our Common Future'	Defined sustainable development as inter-generational legacy
1992	UN Conference on Environment and Development	The Rio Declaration	Tries to bridge environmentalism, sustainable development and business. 'Think globally, act locally'. Identifies food as a key area for policy action.
1992	Convention on Biodiversity	Framework to conserve biological diversity	This came out of the Rio conference (above). Came into force in 1993.
1994	UN	International Conference on Population and Development (ICPD)	Cairo conference represented a shift from previous thinking about controlling population growth to seeing economic development as the key factor
1998	Kyoto Conference	World agreement on measures to address climate change through reducing GHG emissions	Came into force in 2005, but needed governments to turn into nationally binding targets
2000	UN	Millennium Development Goals (MDGs) set by UN Millennium Declaration	Aimed to address the gap between rich and poor countries, including 'halving hunger'
2002	UN	World Summit on Sustainable Development	Johannesburg meeting that reaffirms sustainable development
2006	UK government	Stern Report on Climate Change	Major economic review funded by UK for global consideration. Estimated costs from climate change and saw agriculture as key threat, accounting for c.14% of greenhouse gas emissions.
2012	UN Conference on Sustainable Development	Rio + 20 conference	Reconvenes 1992 process
2015	UN	Sustainable Development Goals (SDGs)	Aim to take over and extend the MDGs

Supporters of the status quo prophesied that this would be another 1970s and that farmers would produce more food and prices of both oil and food would decline. In fact, this did not happen and prices edged up again, after an initial fall; this time the concern became price volatility.[52]

Over the last half century, there have been many global statements about the environmental challenge, suggesting the importance for food policy in the pursuit of sustainable development. Table 5.1 gives some major reports, conventions and conferences at which governments met to address problems of the environment and economic development, at which significant concerns about food were voiced. In effect, this shows the development of international thinking on environment, and a transition from seeing the environment as something to be almost cordoned off; instead putting it at the heart of development. Policy concern about food and agriculture have featured significantly in this shift. While food and agriculture had received policy attention in and immediately after World War II, high-level interest declined in the 1950s and 1960s (partly because of Productionism's success) until the shock of the 1970s oil crisis reminded policy-makers that food could not be taken for granted. The response was to invest in another round of technical investment, the so-called Green Revolution, which focused on developing world agriculture, and promulgated a policy mix that included F1 hybrid seeds dependent on agrichemicals, irrigation, monocropping, machinery and other labour-shedding managerial approaches.[53-55] Research and Development investment then slowly declined until the late 2000s.[56] In the 1970s, however, a concern about the impact of the growing population (re)emerged, with environmentalists arguing that the earth was approaching environmental limits, and that there would be Malthusian crises in the future unless either population was controlled or food production was ramped up.[57 58] This is a theme that stalks policy circles to this day, with population estimates of 9 billion people by 2050 and arguments still rage about whether this can be met if the world eats a more meat-based and energy-heavy diet.[59]

Health policy responses

Where does health fit into sustainable development thinking? Health, in the WHO's 1946 founding charter, is defined as: 'a state of complete physical, mental and social well-being, and not merely the absence of disease or infirmity'.[60] While everyone knows when they are not feeling well, many people do not actually think about their health until they are not well or they are reminded of the fragility of life, such as when a relative gets ill or dies. Well-being tends to be a normative issue: we say we are well when we can continue to do what we normally do. The issue with much diet-related disease, however, is that one is not necessarily aware of the growth of disease. One only learns about arteries being furred up with the heart attack, which is too late. Thus public health strategists have tended to argue that health promotion is ill-health prevention. This was summarised in the WHO's Ottawa Charter in 1986.[61] From a public policy perspective, environment and health have to be seen as linked.

The challenge is how to deliver a state of population health which is optimum and permanent and which meets ecological and economic criteria. As was suggested at the start of this chapter, there are five major traditions of what is meant by public health. Of these, until recently, ecological public health has been at the margins. Most strategies have either focused on technical approaches exemplified by the huge success of antibiotics (now at risk from over-use) or medical interventions (which are expensive and draining national health budgets or insurance schemes). With life expectancy associated with income, the argument that health improvement follows from economic advance has considerable support in Ministries of Finance; health follows wealth. This is the techno-economic approach, championed by the great British epidemiologist Thomas McKeown (1912–88),[62 63] and Nobel Prize-winner Robert Fogel.[64] In reality, health advance requires investment in infrastructure, too. The reason the Victorians invested in drains, roads, water and food adulteration prevention was because even the rich could not avoid the diseases spread by poor sanitation. A mix of policies is almost certainly required.

Health policies, however, cannot work unless there is a 'public' focus in the public health strategy. Neo-liberal market economic thinking argues that health can be left to personal responsibility. Geoffrey Rose, an epidemiologist, countered this position arguing that public policy should aim to deliver social structures that allow individuals to remain well. If policy-makers target only 'at risk' groups, they lose the population effect. All populations, he said, will have a range of behaviours. The key point of policy is to shift the range of population behaviour in the desired direction, not to separate off the 'worst' sections of the population. If high blood pressure patients are to get well, it is best not to see them as a

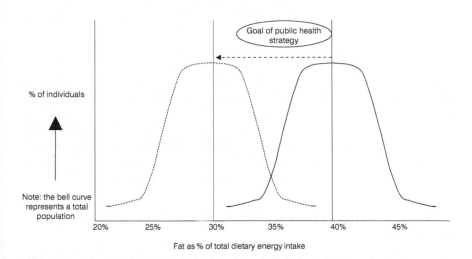

Figure 5.1 Shifting a population in a healthier direction: a hypothetical example of fat intake

different category of people but as the extension of the median; they are at one end of the normal statistical distribution curve. Figure 5.1 illustrates this view of public policy, with a hypothetical distribution of fat intake as a percentage of total calorific intake. If the median of the population is currently 40 per cent, and the desirable is 30 per cent, the goal of public policy is to work out how to frame conditions so as to shift the total population in the desired direction.

This argument has become critical in the case of obesity. As was shown in Chapter 3, obesity rates have risen dramatically across the world in recent decades. Appealing to individual behaviour change is not working. The Chief Medical Officer of England in 2014 voiced an echo of the Geoffrey Rose analysis when she said that rising obesity was now being accepted as 'normal' – i.e. people assuming that a fatter body shape was in fact acceptable.[65] Targeting whole populations provides governments with better chances of public health success, whereas targeting 'at risk' individuals can be socially divisive.

How does this population health thinking get translated into public policy? Table 5.2 lists some global reports and declarations on health which had particular resonance for food. There are in fact many such conventions, reports, declarations and conference resolutions. Not all are legally binding; they require governments to ratify them. But collectively they indicate the global dialogue and the slow incremental manner in which – just at the global level – there has been some engagement with the evidence about food and health summarised in Chapter 3.

Table 5.2 begins with the 1948 UN Declaration on Human Rights. Fresh from World War II, this landmark UN binding agreement is the foundation of global policy about food rights. It suggested only in very general terms that all humans have a right to food, but this abstract principle has been much fought over ever since and a slow policy growth can be charted, as it has been by academics.[15 66] Table 5.2 illustrates some key moments in the development of food rights thinking, including: the 1974 World Food Conference, the 1989 UN study *The Right to Adequate Food as a Human Right*, and the important 2004 *Voluntary Guidelines to support the progressive realization of the Right to Adequate Food in the context of national food security*.[67] The latter provided a template by which member states can be held to account on whether they deliver food security to their population. This was the culmination of efforts by mainly Norwegian lawyers and global food activists to create a new public policy framework about minimum standards for food availability and access.[68 69] The creation of a UN Special Rapporteur on the Right to Food, reporting to the UN Secretary General through the Committee on Economic, Social and Cultural Rights, has strengthened this approach, with an avalanche of reports and country audits conducted by the Rapporteur.[70 71]

Some specific policy responses: safety, antibiotics, waste, food security, obesity

As we said at the start of this chapter, the problems raised by contemporary food policy analysis are immensely complex. Policy-makers work within structures inherited from the past to address problems ahead. There is always a time-lag in capacity. Most governments set up and still have some kind of Ministry of Agriculture and Food, the same for Health, and Finance. Mostly these were created in the 19th century. Ministries of Environment are more recent, and tended to emerge with the growth of modern environmentalism; they in a sense illustrate the effective campaigning of the new breed of 'green warriors'.[72]

Political scientists have shown that the institutions and architecture of governance have undergone a major shift in the last half century. No longer is policy the preserve of governments – whether democratically elected or authoritarian in style – today, food governance has become a looser and messier process. The rise of transnational corporations means that they can be more powerful over food systems than a government, however well resourced. Whose opinion on food safety matters more: Walmart, Carrefour and Tesco (the top three global food retailers) or the government of a small developing country? Policy is about power. Food policy illustrates this rule, and has long been recognised as such.[73-75] At the same time, the rise of noisy civil society and campaigning NGOs means that consumers have organised voices also shaping public debate and daring to act and speak on their behalf. Some NGOs have bigger memberships than political parties.[76] The importance of the media is also a contributing factor to this new policy milieu, with media being keen to report on food-related public debates, especially if contentious. This process has accelerated with the rise of 'new' media (emails, worldwide web, twitter, Facebook, etc.). In sum, the old 'top-down' model of state-dominated governance has been eroded and replaced by a more diverse set of processes.

The world of food policy has been totally changed in this transition. [77] It can even be argued that governments no longer govern food systems. A new architecture has emerged of agencies at arms-length to government, of collaborations such as public–private partnerships, of consultation processes, of stakeholder engagement, of 'independent' inquiries (which can be ignored or not at the whim of government), of think-tanks (who lobby from particular perspectives), and of lobbyists. The rise of lobbies is immensely important. Huge sums are expended by food companies to ensure that their interests are protected and promoted when new regulations or responses to health or environmental matters are being considered. The world of food is enmeshed in the world of corporate responsibility.[78 79]

Public–private partnerships (PPPs) – to take one of the new features of modern food policy making approaches – have emerged as a major means by which government and companies can present themselves as acting on health or environmental problems. The mechanism of PPPs is often used by policy-makers to develop supposedly consensual actions or even to manage projects. [80] Some see PPPs as contentious; others as the way forward. For example, an International Food and Beverage Alliance was set up by a coalition of ten big companies in 2008,[81] in response to the WHO and FAO's major 2004 Global Strategy on Diet, Physical Activity and Health (DPAS),[82] which momentarily looked as though it might threaten industry interests by calling them to account. An audit of the top 25 food companies in the world conducted in 2005–6 had

Table 5.2 Some key global reports and policy statements on health relevant to food policy

Date	Parent body	Output/report	Comment
1948	UN Declaration of Human Rights	Sets legal basis for human right to food	First major international statement on food rights; the basis for work today by UN Rapporteur on Right to Food
1974	UN	World Food Conference	Makes Universal Declaration on the Eradication of Hunger and Malnutrition
1986	WHO	Ottawa Charter on Health Promotion 'Health for All'	Sets goals for health promotion as health and well-being
1989	UN	Convention on the Rights of Child	Sets framework for food protection for children, which opens up work on marketing targeting children
1990	WHO and Unicef	Innocenti Declaration on Breastfeeding	Introduces international framework to support breastfeeding and protect it for health, in opposition to commercial interests
1990	WHO	Diet, Nutrition and the Prevention of Chronic Diseases	Major technical report giving overview of diet's impact on NCDs by UN. Sets the tone for subsequent policy development
1992	WHO and FAO	International conference on Nutrition	Set out the goals for nutrition improvement in the developing world
1996	FAO and WHO	World Food Summit	Rome Declaration Rome Declaration on World Food Security and World Food Summit Plan of Action
1997	World Cancer Research Fund/ AICR	'Food, Nutrition and the Prevention of Cancer' report	First major overview of diet's impact on cancers. (this was updated in 2007 using the 'evidence-based' Cochrane approach)
2004	WHO and FAO	Global Strategy on Diet, Physical Activity and Health	Sets out strategic framework to tackle Non-Communicable Diseases

Year	Organisation	Title	Description
2004	UN	Voluntary Guidelines to support the progressive realisation of the Right to Adequate Food in the context of national food security	This set out guidelines to be used to audit whether countries were meeting their population's right to food
2010	UN	UN Resolution on Right to Safe and Clean Water and Sanitation	Sets clean, safe water as a human right, as part of UN Decade of Water for Life 2005–15
2010	OECD	'Obesity and the Economics of Prevention: Fit Not Fat' report	This report costed the impact of rising obesity in rich countries
2011	UN	High Level Meeting on Prevention and Control of Non-Communicable Diseases	Sets out to inject urgency into the growth of NCDs and obesity, particularly in relation to developing countries
2014	WHO and FAO	2nd International Conference on Nutrition (ICN2)	Revisited the 1992 ICN and to set a new framework for public health nutrition globally
2015	WHO	Draft Global Action Plan on Antimicrobial Resistance	After consultation, WHO tries to create a new framework to stop misuse of antibiotics

shown that, despite a plea by the WHO to take NCDs seriously, few giant food companies were doing so.[83] The creation of the International Food and Beverage Alliance two years later was a belated response to DPAS. Part of it expressly set out actively to partner with NGOs to support healthy lifestyle programmes geared to preventing obesity and NCDs.

That this Alliance was actually set up can thus be perhaps judged as an attempt to stave off more interventionist actions by governments, such as putting in place new legislation.[84] PPPs themselves are not without risk or complexity but they become a brake on (and substitute for) active, speedy, interventionist government. That is their purpose. They are voluntary not mandatory. The development of official or quasi-official PPPs has been much used in food, health and the environment.[85] The EU, for example, created a Platform for Action on Diet, Physical Activity and Health, launched in 2005 to encourage food industries to work with NGOs.[86] And a Platform was created on Resource-Efficiency, including food in 2012.[86] For 'Platform' in the EU read PPPs!

Food safety

In food safety, a different approach has been the more recent hallmark of public policy. In the 1990s, as food safety scandals rocked the developed world, governments came under pressure to reform their own institutions which were clearly failing to prevent safety problems. The mechanisms adopted were not PPPs or other 'soft' mechanisms but stand-alone, arms-length agencies with statutory remits. The model here was the USA's Centers for Disease Control (CDC) and the Scandinavian experience – particularly Sweden – of creating new bodies to eradicate the source of unsafe food. The stakes were high. The rise in food poisoning, and debates about the systematic use of additives, had undermined public confidence. How could consumers protect their own health?

A clear divide had emerged, with proponents of health arguing that prevention was the key, while market theorists still believed in consumers making responsible choices in response to market 'signals' or through companies 'self-correcting' market failures.[87] If the choice is over fatty or sugary foods, this might be plausible, but with food pathogens which are invisible and not declared on a label, how could consumers protect themselves? The answer is that they could not and cannot. There has to be a principle of prevention all down the supply chain. Regulation and law were inevitable. Indeed, from the 19th century that had been the legal tradition across the world. Yet, here, policy and practice were once more being weakened. Risks were being put onto the consumer without prior informed consent. Was institutional reform therefore necessary? The arguments became stark. Food safety is a matter of public trust. If the consumer loses faith in food, the viability of food supply chains, specific products and categories may be threatened. This was the crisis of food governance in the 1990s. Bovine Spongiform Encephalopathy (BSE, known as Mad Cow Disease) brought it to a head in Europe, with the exposé that cows had been turned into cannibals by consuming recycled meat residues in industrially produced feedstuffs.[88] Studies suggested that governments and companies were trusted to act independently for public health.[89] A new food agency could, as one UK government minister told one of the present authors, 'put blue water between us (the government) and political damage'. The argument for reform centred on who might conduct the risk analysis, risk management and risk communication. Some EU governments separated these functions. The EU itself did so.[90] The EU White Paper in 2000 set up the European Food Safety Authority (EFSA) while risk management remained with the Commission itself.[91]

At the global level, a similar tension had emerged when, in 1987, a new round of intergovernmental negotiations to set world trade rules started. The General Agreement on Tariffs and Trade (GATT) had been created in the immediate post-World War II economic reconstruction. Economists at the time had wanted food and agriculture to be part of this agreement, to prevent a repetition of what they saw as ruinous protectionism following the 1930s recession and the Wall Street Crash. In the negotiations leading to the 1948 GATT, however, the USA had blocked food and agriculture in the deal. By the 1980s, the tune had changed and the USA, like most Western economies, was in thrall to the neo-liberal market emphasis for trade policy. The Round of negotiations that started in 1987 took seven years to complete and in 1994 at Marrakesh a new GATT was signed, including agreements to reduce tariffs (taxes at borders) on food and farm commodities. NGOs and some developing countries had grave reservations about the implications. They worried that this would open up developing countries to subsidised exports and to an avalanche of processed foods and drinks. (In fact this happened; it was the moment when the nutrition transition accelerated.)

Part of the GATT agreement was a sub-agreement knows as the Sanitary and Phytosanitary Standards (SPS) agreement. This said that countries could not use food safety or other food standards to block imports from other countries. So who could decide what were good standards? The GATT negotiators chose to involve a hitherto advisory body of the UN, the Codex Alimentarius Commission (known as Codex) to adjudicate standards. Codex thus became a critical body in food policy. A study at the time exposed, however, that Codex was a governmental body in name more than reality, since national delegations were full of vested industrial interests.[92] Some 'cleaning up' of delegation membership then followed, not least since consumer bodies were expressing nervousness.[93] Ever since 1994, therefore, a new architecture of processes and institutions has been defining food safety policy.

Antibiotics: policy failure

The development of antibiotics is frequently cited as one of bio-medicine's success stories. Millions of lives have been saved. These are remarkable aids for public health. Their use in food production, however, is part of the crisis facing their continued viability and effectiveness. Major reports have for decades expressed alarm about profligate use of antibiotics generating antimicrobial resistance.[94] Antibiotics have been routinely put into animal feeds, partly to accelerate meat growth and partly to counter-act the squalor of industrial intensive livestock production methods.[95] Around 75 per cent of antibiotics fed to animals passes directly through the gut into faecal waste and then is applied back onto the land, becoming a means of environmental entry of both antibiotic residues and gene-based determinants of antibiotic resistance (genes, plasmids, etc.).[96] Agriculture is a major user of antibiotics everywhere. In a 2007 survey, the estimated annual antibiotics production in China was 210m kg, of which 46.1 per cent were used in livestock industries.[97] Norway, however, has a strong policy to restrict use. Norway uses 20mg to produce 1kg of meat, compared to the Netherlands using 180mg to produce 1kg of meat. The USA uses an estimated 300mg to produce 1kg of meat.[98: Fig 4.3 pg 55 99] According to the FDA, 13.5m kg of US antimicrobials – some 80 percent of the total – were sold for use in agriculture in 2011,[100] compared to less than 3.3 million kilograms sold for human medical use.[101]

The danger of overuse and misuse is not a new worry. In 1956, the US National Academies of Science hosted a major conference expressing concern.[102] In 1969, the UK Swann Report highlighted the misuse of antibiotics on farms.[103] In 1977 the US Food and Drug Administration (FDA) tried to regulate or ban routine use of antimicrobials in animal feeds.[104 105] This met concerted industry resistance. It has taken decades for public campaigns to emerge arguing for restricted use, yet half a century after scientific evidence and concern began to be published, systemic overuse continues.

Food waste: slow engagement

Never has so much food been produced in all human history yet, as was outlined in the previous chapter, world food waste remains at unacceptable levels, globally at 220 million tonnes (mt) a year, equivalent to the total food production of sub-Saharan Africa.[106] In low-income countries, food waste occurs near the farm while consumers waste very little; whereas in high-income countries, consumers waste up to a third of what they buy.[106] For example, in SE Asia losses of rice can range between countries from 37 per cent to 80 per cent of total production, the equivalent of 180mt per year. Each year in the European Union it has been estimated 89mt of food waste is generated, of which 15mt are in the UK. This has a monetary value of about £950 per tonne.[107] The EU rate of waste is growing such that if not checked it is estimated it will be 126mt by 2020.

The UK policy response is interesting, mainly having a consumer to industrial focus. Modern UK food waste action began when the Waste Resources Action Programme (WRAP) was created in 2000.[108] After the 2008 financial crisis, WRAP funding was cut and became more dependent on consultancy work. WRAP's overarching role was set in the Courtauld agreements, a partnership of government and industry to reduce food waste. Courtauld Agreement 1 ran in 2005–9, Courtauld 2 in 2010–12 and Courtauld 3 in 2012–15.[109] In the mid-2000s, WRAP conducted a well-financed and high-profile public *Love Food Hate Waste* campaign to highlight consumer behaviour while reinforcing the case for industry change. WRAP evidence suggests this has been effective. Whereas in 2009, 8.3mt of food and drink in the UK were wasted, this halved to 4.4mt by 2014. That still represented 15 per cent of total edible food and drink purchases, valued at £480 per year per household. Waste remained associated with particular foods. Banana waste declined from 1.7million to 1.4million wasted a day; tomatoes from 2m to 1.5m; yogurts from 1.7m to 1.2m; and bread from 37m slices to 24m slices.[110] As part of this success story, the food industry cut its waste too.[111] Food which went to landfill goes to charities or is recycled; 3.3mt of CO_2e were estimated to be saved in 2005–9.[109]

WRAP first had an industrial, then a consumer, focus. In the mid 2010s, it began to promote a more systemic policy model of the 'circular economy' also being championed by the EU and others such as the Ellen MacArthur Foundation, McKinsey and the World Economic Forum.[112 113] This focuses on the flow of materials (see Figure 5.2). Whereas a linear economy is characterised by 'dig, make, use, dispose', the circular economy places value on what was seen as 'waste' and prioritises recapture and reuse. This certainly is an advance on

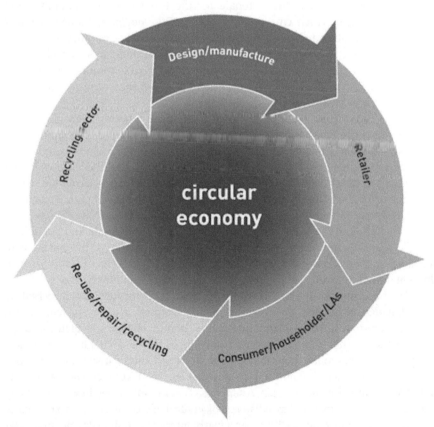

Figure 5.2 Waste and the circular economy

Source: WRAP 2014[112]

previous thinking, and can be applied to food, but can imply that the way forward is mainly the task of re-engineering. This approach is in effect a modernisation of the Victorian sanitary engineering approach, which was to disperse everything or dump it into land holes or the sea, and replaces that with a modern bio- or carbon flow engineering approach. This is an improvement but food waste is not just a matter of material flows, however cleverly conceived. Waste is also shaped by the meaning of food, social norms, habits, definitions of acceptability, and subject to economic pressures not always captured by the circular economy analysis: e.g. if food is cheap, where are the incentives to recycle or consume differently? Engineering alone is not enough. Culture change is needed too.

Food security: rationale for productionist modernisation?

If any one topic has grabbed the attention of global policy-makers since the turn of the millennium it is the issue of food security. Since World War II, the focus of food and farming policy was on producing more food to counter hunger.[66] This is a central plank in the rationale for the Productionist paradigm. Food Security is a term which emerged from the 1970s, but now summarises this continuing aspiration to feed all adequately and to enable production to meet need. In practice, food security has taken on multiple meanings. 'Need' can be defined as feeding people and be framed in social justice terms or as meeting the dynamics of the market. One study estimated that food security has over a hundred different meanings![114]

Although the figures for hunger have remained stubbornly high in absolute numbers, as a proportion of the rising population of the world, they have fallen since the 1980s but with fluctuations. Still, in 2015, around one in nine people were estimated to be hungry. This is concentrated in a limited number of developing countries. There is no shortage of food if it is shared equitably across the world. Food Security is undergoing renewed national and international policy attention, particularly since the 2007–8 price spike which scared even the rich world's policy-makers into the reality that markets were now vulnerable and potentially volatile after decades of relative stability and predictability. This linked with recognition of environmental and resource threats to food supply from prolonged drought, material shortages and the growing impact of climate change.

It has been argued that this is an era of 'new fundamentals' characterised by a combination of old and new features: population pressures, eco-systems threat, climate change, water stress, land use squeeze, market change, changed consumer eating patterns (nutrition transition), competition between countries for scarce resources and geo-political instability. This shock spawned a new round of policy engagement. The West's complacency was shaken. The FAO met. G8 met. OECD met. World Bank became engaged. A supposed 'land-grab' ensued in which wealthy countries and 'investors' and speculators took control of vast land masses particularly in Africa and South America.[115] A particularly interesting policy development was the reform of the Committee on World Food Security, which saw the UN's cross-agency committee include more stakeholders from civil society.[116]

A policy divide emerged in the 2010s. On one side, there were those who argued that the situation could be resolved by food justice, a more equitable distribution of power and food and aiming for more 'people fed per hectare'.[117] On the other side were those arguing that this was a wake-up call for another bout of intensification of production, and the application of new technical fixes to agriculture such as robots, nanotechnology, biotechnology and precision farming – another modernisation of the Productionist paradigm. This tension highlights the importance of health: agriculture for whom, and is it about the quantity of food or its quality (the latter being especially important for nutrition well-being)? The future could be producing factory meat or dairy-based products or it could reducing meat and dairy consumption and making a shift to more plant-based diets.

Obesity: a policy 'time-bomb'

Warnings from public health analysts about the rising problem of obesity began decades ago and have continued to grow in numbers, almost in line with the growth of obesity itself. In 1950, a single physician set up one of the first clinics up for the study and treatment of obesity in children, in London.[118] By 1983 the Royal College of Physicians was expressing grave alarm at the prospect of rising obesity in the population.[119] In the USA, where the epidemic emerged rapidly and early,[120] the American Medical Association defined obesity as a disease only as recently as 2013. The WHO recognised obesity as an epidemic in the 2000s. In 2000 the US Surgeon-General produced his hard-hitting 'Call to Action'.[121] In countries across the globe, similar processes were underway, with inquiries, calls for evidence and new action plans being created. In Europe, the long-established Nutrition Council of Norway was given a new emphasis to look at diet (input) and physical activity (output),[122] and in Sweden a national working party was set up, charged to produce a new national policy and action plan. The WHO European Region made obesity its core theme with the Istanbul Declaration in 2007.[123] Latin America, meanwhile, was also noting a worrying rise, as processed foods spread.[124] Even in poor sub-Saharan Africa obesity was emerging. Everywhere obesity sparked and still sparks a moral debate: is this self-inflicted ill health? As a 1974 journal article about African obesity put it: is it a matter of 'indolence and affluence'?[125] Three problems arise. What is causing this rise? What can be done about it? Who can do what in the policy world to prevent as well as treat the consequences of obesity?

In 2001, the UK's National Audit Office – charged with assessing financial matters in government – produced a report suggesting a huge externalised cost to the Treasury if obesity was allowed to continue. In 2003–4 the UK Parliamentary Health Committee ran a major inquiry and also recommended urgent action.[126] Government did what it often does – and set up another inquiry! This one was different, however. The 2007 Foresight report, under the UK's Chief Scientific Adviser, was a two-year study watched by many countries around the world. It produced a now celebrated 'systems map' which charted how multiple drivers collectively shaped population obesity.[127] At the centre of this system map was the simple 'engine' of energy in and energy out. Around it, shaping how fast and in what direction it revolved, were multiple other factors. Figure 3.6 is one of the Foresight systems maps with the spaghetti like lines of factors shaping the central obesity engine. These were what the vast literature on obesity suggested and included issues such as genetic and psychological factors, food supply, prices, the built environment, family norms and styles, marketing, and more. The Foresight report put these into clusters to show that obesity could not be resolved by the pulling of mythical single levers of change or interventions. It concluded that, if governments want to address obesity, in effect they have to aim for systemic change.

Most interventions to date, however, have pretty systematically downplayed if not ignored this advice and have tended still to focus on individual change. This approach has had little or no impact on rising obesity worldwide. Policy based on health education alone has been largely ineffective.[128] The conventional policy reflex to offer single solutions for a single problem clearly does not fit the etiology of obesity which requires multi-level multi-faceted solutions.[129] Key tensions in how policy might address obesity have centred on:

- environmental versus genetic causes;
- targeting 'at-risk' people versus a population approach;
- the relationship and roles between corporate and public spheres;
- the relative importance of diet versus physical activity;
- fostering 'real' change versus public relations approaches to health education;
- government's role: the 'hands-on' versus the 'hands-off' state;
- people's responsibility: are they simply consumers or 'food citizens'?;
- agriculture and the food industry producing and marketing inappropriate foods and beverages.

As a result of tensions over these issues, and the inevitable sectoral in-fighting for policy position and influence, there is often no clear-cut or uncontested set of policy options and interventions. The result is systems drift.

Much has emerged from the short summaries of the various food issues summarised in previous pages. There has been no standard response or even mechanism of response. PPPs suggest an aspiration of partnership and 'getting together' but how can that resolve obesity? The food safety example was a reminder of how 'hard' policy and new institutions, such as food agencies, are needed. Governments had to govern and lead. The food waste example, however, showed how governments can put responsibility onto consumers but, even so, the case for systemic change resurfaced in the shift from linear to circular economic thinking. The food security example illustrated the continuing battles over paradigms, where the dominant policy reflex is still to analyse the problem as insufficient supply, within a 'top-down' approach. In all of these examples, the major tension remains between population and individualised approaches, and by whom and how decisions are implemented. Table 5.3 summarises some differences in these approaches that have emerged in the discussion so far.

Table 5.3 Individualist and population approaches to food and health

Policy focus	Individualist public health approach	Population public health approach
Relationship to general economy	Trickle-down theory; primacy of market solutions; inequality is inevitable	Health as economic determinant; public-private partnerships; inequalities require societal action
Economic direction for health policy	Individual risk; personal insurance; reliance on charity	Social insurance including primary care, welfare and public health services
Morality	Individual responsibility; self-protection; consumerism	Societal responsibility based on a citizenship model
Health accountancy/costs	Costs of ill health not included in price of goods	Costs internalised where possible
Role of the State	Minimal involvement; avoid 'nanny state' action; resources are best left to market forces	Sets common framework; provider of resources; corrective lever on the imbalance between individual and social forces
Consultation with the end user	As consumer; dependent on willingness to pay	Citizenship rights; authentic stakeholder
Approach to food and health	The right to be unhealthy; a medical problem; individual choice is key driver; demand will affect supply; niche markets	The right to be well; entire food supply geared to deliver health

Table 5.4 then summarises the range of policy measures available to policy-makers. These vary from 'soft' policy measures (at the top of the table) such as health advice or labelling to 'harder' interventions (at the bottom of the table) such as taxation or legal bans and rationing. Each has consequences and implications. The scope and range of policy options is often wider than is actually used.

Dietary guidelines: from nutrition to a 'sustainable diet'?

A central element of international public health and nutrition policy response to the food and health data summarised in Chapter 3 was the creation of national dietary guidelines and goals for populations. These are often translated into advice to the public on what constitutes a healthy diet and even suggesting individual changes, alongside product labelling. Dietary guidelines have been controversial and often occasions of battle in the Food Wars, having, in theory, formed the basis of much government and nutrition policy since the 1980s.

Although there were some nutrition guidelines prior to World War II, the modern era of guidelines began in Europe with Norway, Sweden and Finland producing the first recorded governmental dietary guidelines in 1968, with the US following in 1970 and New Zealand in 1971. Initially, with the Nordic exception, these were

Table 5.4 The range of public policy measures to shape supply and consumption

Measure	Main sources	Comment
Advice	National or local government; official bodies; company nutritionists; popular media	Tends to be weak and with low impact
Labelling	State or company	This emerges often after long 'consultation' and debate. It puts the onus on consumers who can suffer from information over-load
Education	Historically the remit of the state but increasingly is used by companies, e.g. in 'educational' materials	Long time to be effective; works best when coupled with other measures. With the rise of social media, people self-educate
Advertising and marketing	Corporate. Sometimes funded by states or levies on trade	Ranges from advertising and marketing to virtual and web-based media
Endorsement and sponsorship	Corporate	Increasing use of celebrity, sporting events and product placement on TV and film. This blurs the lines between media content and advertising
Welfare support	State and philanthropic/charity	Used in food security, e.g. by the World Food Programme in emergency feeding crises; and Food Banks
Product/ compositional standards	Historically, this was set by the state but now there are diverse sources: industry, NGO and state	This is increasingly used by 'partnerships', e.g. in Fairtrade, Marine Stewardship Council; supply chains increasingly use common standards set by retailers
Subsidies	States at national and intergovernmental levels (e.g. EU)	Deeply opposed by neo-liberals as market distorting, but remain widespread.
Competition rules	State	Many rich societies have competition bodies which conduct inquiries and have leverage, e.g. through fines
Taxes & fiscal measures	State	Critics see them as distortions; supporters see them as effective compensation for market failures
Bans	State, companies and civil society	Consumer boycotts are a form of ban; corporate standards using 'choice-editing' are a version; governments use them rarely but when they do they are effective (e.g. tobacco, transfats, some agrichemicals and other drugs)
Rationing	State	Tends to be used in times of war. Markets of course 'ration' by creating equilibrium between supply and demand.

guidelines produced by expert societies with government approval, rather than by government itself; but gradually most developed economy governments took responsibility for their production. Once created, guidelines inevitably have to be kept under review, given the constant shifts of both consumer eating patterns and scientific knowledge. Today, a general consensus exists, with relatively minor country-by-country variations, on what national dietary guidelines should be.[2] Generally, they promote a variety of foods; maintaining weight within an ideal range; eating foods with adequate starch and fibre; avoidance of too much sugar, sodium, fat (especially saturated fat), transfats and cholesterol; and drinking alcohol only in moderation. People are also strongly advised not to smoke tobacco and to take regular physical exercise. Breast-feeding and pre-conceptual care are also usually recommended.

The early editions of dietary goals were received with due attention, while nutritionists remained alarmed by the problems generated by deficiency diseases. Dietary guidelines began to be met by resistance from vested interests, however, once scientific bodies started criticising excess consumption, particularly as high-fat diets came under suspicion,[130] and also sugar.[131 132] Food producers feared loss of sales, particularly industries dependent on purveying fat, salt and sugars. Nevertheless, Recommended Daily Allowances and reference values became the benchmark for dietary guidelines.[133] The first edition of 'Dietary Goals for the United States' was published in 1977 and was described as a 'revolutionary document'. The EU got to this stage 30 years later with the production of Eurodiet guidance in 2000.[134] Population-based dietary guidelines have often been converted into consumer-oriented campaigns such as '5 a-day', pioneered in California and replicated in many countries: the consumption of at least five portions of fruit and vegetables daily to deliver the right mix of positive nutrients to protect health. In fact, countries vary in their advice. Danes proposed '6-a-day', the Greeks, parents of the Mediterranean diet, recommended '9-a-day' and the USA raised its public advice to 7-a-day.

While dietary guidelines have been a policy response to public health concerns, evidence on food's impact on the environment did not get included into diet advice. Why not? Should dietary guidelines, therefore, be revised to include an environmental element? Do environmental and health concerns align? What is the relationship between cultural, environmental and health food advice? These questions now have fuelled the debate known by the phrase 'sustainable diets'.[135]

A number of proposals for public advice on how to fuse environmental, health and cultural considerations in diet have emerged in recent decades at national and local levels, increasingly as concerns about climate change rose. The first forays into this terrain came from environmental activists decades ago. In 1971, a small US think-tank issued a book which became a bestseller: *Diet for a Small Planet*.[47] Forty years later, activists in British Columbia, Canada, proposed that people eat food only from within 100 miles of Vancouver.[136] And in Fife, north Scotland, a group of households committed in 2007 to eat 80 per cent of their diet from food grown in Fife, their county (i.e. locally). Given this northern

Table 5.5 Five examples of sustainable dietary advice: Some principles on sustainable eating compared

Source/country	Environmentally effective food choices (Sweden) [142]	Sustainable Shopping Basket (Germany) [143]	Guidelines for a healthy diet: the ecological perspective (Netherlands) [144]	UK Green Food Project, 8 principles [145]	Brazilian Food Based Dietary Guidelines [146]
Date	2009	1990s→2013 (4th edition)	2011	2013	2014
Lead Body	National Food Administration & Environmental Protection Agency	German Council for Sustainable Development	Health Council of the Netherlands	UK Government working party	Ministry of Health. Brazil
Prime concerns	Pro health and environment to reduce climate change and promote non-toxic environment	To integrate advice from many sources for daily food shopping	Linking gains in public health nutrition to lower ecological impact	To combine health and environmental advice	To promote public health; and to realign health and food culture
Actual Advice	Eat less meat. Replace it with vegetarian meals; choose local meats or organic if available	Follow the food pyramid	Move to a less animal-based, more plant-based diet – this is the key advice	Eat a varied balanced diet to maintain a healthy body weight	1. Prepare meals from staple and fresh foods
	Eat fish 2–3 times a week from sustainable sources	Eat less meat and fish but savour them	Lower energy intake, and eat fewer snacks	Eat more plant based foods, including at least five portions of fruit and vegetables per day. Value your food. Ask about where it comes from and how it is produced. Don't waste it	2. Use oils, fats, sugar and salt in moderation
	Eat fruit, vegetables, berries: a good rule of thumb is to choose seasonal, local and preferably organic products	Follow 5-a-day on fruit and vegetables	Eat two portions of fish a week but from sustainable sources	Moderate your meat consumption, and enjoy more peas, beans, nuts, and other sources of protein	3. Limit consumption of ready-to-consume food and drink products
	Choose locally grown potatoes and cereals rather than rice	Eat seasonally and regionally as your first choice	Reduce food waste		4. Eat regular meals, paying attention, and in appropriate environments

Choose pesticide-free or organic when possible	Eat organic products	Choose fish sourced from sustainable stocks. Seasonality and capture methods are important here too	5. Eat in company whenever possible
Choose rapeseed oil rather than palm oil fats	Choose fair trade products	Include milk and dairy products in your diet or seek out plant based alternatives, including those that are fortified with additional vitamins and minerals	6. Buy food at places that offer varieties of fresh foods. Avoid those that mainly sell products ready for consumption
Eat fish 2–3 times a week from sustainable sources	Choose drinks in recyclable packaging	Drink tap water	7. Develop, practice, share and enjoy your skills in food preparation and cooking
Eat fruit, vegetables, berries: a good rule of thumb is to choose seasonal, local and preferably organic products	Use designated certification schemes (many are cited in the document)	Eat fewer foods high in fat, sugar and salt	8. Plan your time to give meals and eating proper time and space
Choose locally grown potatoes and cereals rather than rice			9. When you eat out, choose restaurants that serve freshly made dishes and meals. Avoid fast food chains
			10. Be critical of the commercial advertisement of food products

climate, one might have feared this choice would consign adherents to the Fife Diet to slow hunger, but in fact the movement grew and they became fitter.[137] Their motivation was to do something – to take control – and not wait for others higher up to act. Such public experiments are often localist or 'locavore', and put a premium on plant-based locally sourced food. This may be termed food bio-regionalism. It has been influential in reasserting traditional and seasonal foods, as well as local sourcing. But how does this cultural emphasis fit health and environmental concerns?

In 2010, the FAO and Bioversity International (the UN-affiliated CGIAR body working on biological diversity) held a scientific conference to address the issue of sustainable food systems and of diet in particular.[138] This generated a now widely cited broad definition of what is meant by sustainable diets:[139]

> Sustainable Diets are those diets with low environmental impacts which contribute to food and nutrition security and to healthy life for present and future generations. Sustainable diets are protective and respectful of biodiversity and ecosystems, culturally acceptable, accessible, economically fair and affordable; nutritionally adequate, safe and healthy; while optimizing natural and human resources.

Various attempts to create national sustainable dietary guidelines have been made at governmental level, and even more by unofficial or independent academics.[140] Table 5.5 gives five such efforts. The German Council on Sustainable Development perhaps made the earliest such public advice and continues to revise it on an ongoing basis. The most comprehensive is that jointly produced by the Swedish National Food Administration and the Environmental Protection Agency in 2008. This was presented to the European Food Safety Authority (EFSA) for EU-wide approval but withdrawn a year later, reputedly under pressure from the meat lobby. The 2011 advice from the Netherlands was created by its nutrition expert panel, and the 2102 UK advice was from a project set up by the Department for Environment, Food and Rural Affairs. That, too, appeared to be pushed 'into the long policy grass', with Ministers choosing not to respond. The 2014 Brazilian advice is the latest official nutrition guidelines. While not overtly addressing the environment, the guide was based on research showing how rising obesity and NCDs reflect Brazil's nutrition transition to consuming more 'ultra-processed' foods.[141] After extensive national consultation, the advice has been summed up in three 'golden rules' for the public:

- make fresh and minimally processed foods the basis of your diet;
- use oils, fats, sugar and salt in moderation when preparing dishes and meals;
- limit consumption of ready-to-consume food and drink products.

As the evidence continues to mount about food's impact on health and environment, it surely becomes harder for national policy makers not to integrate health, environment and cultural dietary advice, using best scientific evidence. Surely, the proper policy way forward is to aim for sustainable diets from sustainable food systems everywhere? This cannot be undertaken without engaging with how food is produced, manufactured and retailed. Thus in the next chapter, we set out what is meant by food business and consider examples of how the food industry is (and is not) addressing the challenge of sustainability and public health nutrition.

References

1   Mintz S. *Tasting Food, Tasting Freedom: Excursions into Eating, Culture and the Past*. Boston, MA: Beacon Press, 1996.

2   Cannon G. *Food and Health: The Experts Agree*. London: Consumers' Association, 1992: 230.

3   Garnett T, Godfray C. *Sustainable Intensification in Agriculture: Navigating a Course through Competing Food System Priorities*. Oxford: Food Climate Research Network and the Oxford Martin Programme on the Future of Food, University of Oxford, 2012.

4   Acheson SD. *Committee of Inquiry into the Future Development of the Public Health Function* Public Health in England. London: HMSO, 1988.

5   Winslow C-EA. The untilled fields of public health. *Science*, 51(1306) (1920): 23–33.

6   Nutbeam D. *Health Promotion Glossary*. Geneva: World Health Organisation, 1998.

7   Ashton J, Seymour H. *The New Public Health: The Liverpool Experience*. Milton Keynes: Open University Press, 1988.

8   WHO. *Health in All Policies: The Adelaide Statement*. Report from the International Meeting on Health in All Policies, Adelaide 2010. Geneva: World Health Organisation, 2010.

9   Rayner G, Lang T. *Ecological Public Health: Reshaping the Conditions for Good Health*. Abingdon: Routledge/Earthscan, 2012.

10   Harvie D. *Limeys: The Conquest of scurvey*. Stroud: Sutton, 2002.

11   Lind J. *A treatise of the scurvy: Containing an inquiry into the nature, causes and cure, of that disease*. Edinburgh: printed by Sands, Murray & Cochran for A. Kincaid & A. Donaldson, 1753.

12   Mann J, Truswell S, eds. *Essentials of Human Nutrition*. Oxford: Oxford University Press, 2012.

13   Fieldhouse P. *Food and Nutrition: Customs and Culture*. 2nd edn. London: Chapman & Hall, 1995.

14   Scrinis G. *Nutritionism: The Science and Politics of Dietary Advice*. New York: Columbia University Press, 2013.

15   Lang T, Barling D, Caraher M. *Food Policy: Integrating Health, Environment and Society*. Oxford: Oxford University Press, 2009.

16   Atwater WO. *Investigations on the Chemistry and Economy of Food*. US Department of Agriculture Bulletin, 21. Washington, DC: Department of Agriculture, 1891.

17   Atwater WO. *Foods, Nutritive Value and Cost*. US Department of Agriculture, Farmers Bulletin, 23. Washington, DC: US Department of Agriculture, 1894.

18   Rowntree BS. *Poverty: A Study of Town Life*. London: Macmillan, 1902.

19 Rowntree BS. *How the Labourer Lives*. London: Thomas Nelson & Sons, 1913.

20 Rowntree BS. *The Human Needs of Labour*. London: Longmans, 1921.

21 Rowntree BS. *Poverty and Progress*. London: Longmans, 1941.

22 Booth W. *In Darkest England, and the Way Out*. London: International Headquarters of the Salvation Army, 1890.

23 Wood TB, Gowland Hopkins F. *Food Economy in War Time*. Cambridge: Cambridge University Press, 1915.

24 Burnet E, Aykroyd WR. Nutrition and public health. *Quarterly Bulletin of the Health Organisation*, 4(2) (June 1935): 2–145.

25 British Medical Association. Nutrition and the public health: proceedings of a national conference on the wider aspects of nutrition, Apr. 1939. London: British Medical Association, 1939.

26 Burnett J. *Plenty and Want: A Social History of Diet in England from 1815 to the Present Day*. London: Nelson, 1966/1979.

27 McCarrison SR. *Nutrition and National Health: Three Lectures Delivered Before the Royal Society of Arts on February 10th, 17th and 24th, 1936*. London: Royal Society of Arts, 1936.

28 Inter-departmental Committee on Physical Deterioration (chaired by Sir Almeric W. Fitzroy). *Report of the Inter-Departmental Committee on Physical Deterioration*, vol. 1. Cd.2175. London: HMSO, 1904.

29 Boyd Orr J. *Food, Health and Income: Report on Adequacy of Diet in Relation to Income*. London: Macmillan & Co., 1936.

30 Spring Rice M. *Working Class Wives: Their Health and Conditions*. Harmondsworth: Penguin, 1939.

31 Vernon J. *Hunger: A Modern History*. Cambridge, MA: Harvard University Press, 2007.

32 Agee J, Evans W. *Let us Now Praise Famous Men*. New York: Houghton Mifflin, 2001.

33 Boyd Orr SJ. *Food and the People: Target for Tomorrow No. 3*. London: Pilot Press, 1943.

34 Keys A. Coronary heart disease in seven countries. *Circulation*, 41(suppl. 1) (1970): 1–211.

35 Keys A. *Seven Countries: A Multivariate Analysis of Death and Coronary Heart Disease*. Cambridge, MA: Harvard University Press, 1980.

36 Ewin J. *Fine Wines and Fish Oil: The Life of Hugh Macdonald Sinclair*. Oxford: Oxford University Press, 2001.

37 Sinclair H. Preface. In: McCarrison SR. *Nutrition and Health*. 3rd edn. London: Faber & Faber, 1963.

38 WHO. Environmental health. <http://www.who.int/topics/environmental_health/en> [accessed July 2014]. Geneva: World Health Organisation, 2014.

39 Carson R. *Silent Spring*. Boston/Cambridge, MA: Houghton Mifflin/Riverside Press, 1962.

40 Conway G, Pretty JN. *Unwelcome Harvest: Agriculture and Pollution*. London: Earthscan, 1991.

41 Bull D. *A Growing Problem: Pesticides and the Third World Poor*. Oxford: OXFAM, 1982.

42 Dinham B. *The Pesticide Hazard: A Global Health and Environmental Audit*. London and Atlantic Highlands, NJ: Zed Books, 1993.

43 Lang T, Clutterbuck C. *P is for Pesticides*. London: Ebury, 1991.

44 WHO. *Public Health Impact of Pesticides Used in Agriculture*. Geneva: World Health Organisation, 1990.

45 Jacobs M, Dinham B. *Silent Invaders: Pesticides, Livelihoods, and Women's Health*. London: Zed Books, 2003.

46 Taskforce on Systemic Pesticides. Report of the taskforce: <http://www.tfsp.info/worldwide-integrated-assessment> [accessed July 2014]. Gland: International Union for Conservation of Nature, 2014.

47 Lappé FM. *Diet for a Small Planet*. New York: Ballantine Books, 1971.

48 Leach G. *Energy and Food Production*. Guildford: IPC Science and Technology Press for the International Institute for Environment and Development, 1976.

49 Pimentel D, Pimentel M. *Food, Energy, and Society*. London: Edward Arnold, 1979.

50 Pimentel D, Pimentel MH. *Food, Energy and Society*. 3rd edn. Boca Raton, FLA: CRC Press, 2008.

51 Meadows DH, Meadows DL, Randers J, *et al. The Limits to Growth: A Report for the Club of Rome's Project on the Predicament of Mankind*. New York: Universe Books, 1972.

52 OECD, FAO. *Agricultural Outlook 2011–20*. Paris and Rome: Organisation for Economic Co-operation and Development and Food and Agriculture Organisation., 2011.

53 Lipton M, Longhurst R. *New Seeds and Poor People*. Baltimore, MD: Johns Hopkins University Press, 1989.

54 Mazoyer M, Roudart L. *A History of World Agriculture*. New York: Monthly Review Press (London: Earthscan), 2006.

55 Hazell PBR. *Green Revolution: Curse or Blessing?* Washington, DC: International Food Policy Research Institute, 2002.

56 Conway G. *The Doubly Green Revolution. Food for All in the 21st Century*, London: Penguin, 1997.

57 Commoner B. *The Closing Circle: Confronting the Environmental Crisis*. London: Cape, 1972.

58 Ehrlich PR. *The Population Bomb*. New York: Ballantine, 1968.

59 Smil V. *Feeding the World: A Challenge for the Twenty-First Century*. Cambridge, MA, and London: MIT, 2000.

60 WHO. *Preamble to the Constitution of the World Health Organization as adopted by the International Health Conference, New York, 19 June–22 July 1946; signed on 22 July 1946 by the Representatives of 61 States*. Official Records of the World Health Organization, 2. Geneva: World Health Organisation, 1946.

61 WHO. Ottawa Charter. <http://www.euro.who.int/AboutWHO/Policy/20010827_2> Geneva: World Health Organisation, 1986.

62 McKeown T. *Medicine in Modern Society*. London: Allen & Unwin, 1965.

63 McKeown T. *The Role of Medicine: Dream, Mirage, or Nemesis?* Princeton, NJ: Princeton University Press, 1979.

64 Fogel RW. Technophysio evolution and the measurement of economic growth. *Journal of Evolutionary Economics*, 14(2) (2004): 217–21.

65 Davies SC. *Annual Report of the Chief Medical Officer: Surveillance Volume, 2012: On the State of the Public's Health*. London: Department of Health, 2014: 121.

66 Shaw DJ. *World Food Security: A History since 1945*. London: Palgrave Macmillan, 2007.

67   FAO. Voluntary Guidelines to support the progressive realization of the right to adequate food in the context of national food security. Adopted by the 127th Session of the FAO Council, Nov. 2004. Rome: Food and Agriculture Organisation, 2004.

68   Eide A, Eide WB, Goonatilake S, *et al.*, eds. *Food as a Human Right.* Tokyo: United Nations University, 1984.

69   Eide WB, Kracht U, eds. *Food and Human Rights in Development,* vol. 1, *Legal and Institutional Dimensions and Selected Topics.* 1st edn. Antwerp: Intersentia, 2005.

70   Special Rapporteur. *Reports of the UN Special Rapporteur on the Right to Food.* <http://www.srfood.org> Geneva: UN Economic and Social Council, 2014.

71   De Schutter O. *Final Report: The Transformative Potential of the Right to Food.* Report of the Special Rapporteur on the right to food, Olivier De Schutter. report to Human Rights Council Twenty-fifth session, Agenda item 3. Geneva: Human Rights Council, 2014.

72   Pearce F. *Green Warriors: The People and the Politics behind the Environmental Revolution.* London: Bodley Head, 1991.

73   Clapp J. *Food.* Cambridge: Polity, 2012.

74   Morgan K, Marsden T, Murdoch J. *Worlds of Food: Place, Power and Provenance in the Food Chain.* Oxford: Oxford University Press, 2006.

75   McMichael P. *The Global Restructuring of Agro-Food Systems.* Ithaca, NY, and London: Cornell University Press, 1994.

76   Lang T. Going public: food campaigns during the 1980s and 1990s. In: Smith D, ed. *Nutrition Scientists and Nutrition Policy in the 20th Century.* London: Routledge, 1997: 238–60.

77   Lang T. Food, the law and public health: Three models of the relationship. *Public Health,* 2006: 30–41.

78   Lang T, Rayner G. Corporate responsibility in public health. *BMJ* 341 (2010): c3758.

79   Kinley D, Nolan J, Zerial N. The Politics of corporate social responsibility: reflections on the United Nations human rights norms for corporations. *Company and Securities Law Journal,* 25(1) (2007): 30–42.

80   WHO. Public private partnereships for health. <http://www.who.int/trade/glossary/story077/en> [accessed July 2014]. Geneva: World Health Organisation, 2014.

81   IFBA. The International Food and Beverage Alliance. <https://ifballiance.org> [accessed July 2014].

82   WHO. Global strategy on diet, physical activity and health. 57th World Health Assembly. WHA 57.17, agenda item 12.6. Geneva: World Health Assembly, 2004.

83   Lang T, Rayner G, Kaelin E. *The Food Industry, Diet, Physical Activity and Health: A Review of Reported Commitments and Practice of 25 of the World's Largest Food Companies.* Report to the World Health Organisation. London: City University Centre for Food Policy, 2006.

84   Kraak VI, Harrigan PB, Lawrence M, *et al.* Balancing the benefits and risks of public-private partnerships to address the global double burden of malnutrition. *Public Health Nutrition* (2011): 1–15.

85   European Commission. *Making Europe a Pole of Excellence on Corporate Social Responsibility.* COM(2006) 136. Brussels: Commission of the European Communities, 2006.

86   European Commission. *European Platform for Action on Diet, Physical Acivity and Health.*   <http://ec.europa.eu/health/ph_determinants/life_style/nutrition/platform/platform_en.htm> Brussels: European Commission DG Health and Consumers/Public Health, 2005.

87   London Food Commission. *Food Adulteration and How to Beat it*. London: Unwin Hyman, 1987.
88   Van Zwanenberg P, Millstone E. *BSE: Risk, Science, and Governance*. Oxford and New York: Oxford University Press, 2005.
89   Kjaernes U, Harvey M, Warde A. *Trust in Food: An Institutional and Comparative Analysis*. Basingstoke: Macmillan/Palgrave, 2007.
90   Commission of the European Communities. *White Paper on Food Safety*. Brussels: Commission of the European Communities, 2000.
91   Trichopoulou A, Millstone E, Lang T, *et al. European Policy on Food Safety*. Luxembourg: European Parliament Directorate General for Research Office for Science and Technology Options Assessment (STOA), 2000.
92   Avery N, Drake M, Lang T. *Cracking the Codex: A Report on the Codex Alimentarius Commission*. London: National Food Alliance, 1993.
93   International Organization of Consumers' Unions. Time to put consumers first: Statement on the Uruguay Round. The Hague: International Organization of Consumers Unions, 1993: p. v.
94   Davies SC, Grant J, Catchpole M. *The Drugs Don't Work: A Global Threat*. London: Penguin, 2013.
95   Lymbery P, Oakeshott I. *Farmageddon: The True Cost of Cheap Meat*. London: Bloomsbury, 2014.
96   Zhu Y-G, Johnson TA, Su J-Q, *et al*. Diverse and abundant antibiotic resistance genes in Chinese swine farms. *Proceedings of the National Academy of Sciences*, 110(9) (2013): 3435–40.
97   Hvistendahl M. China takes aim at rampant antibiotic resistance. *Science*, 336(6083) (2012): 795.
98   WHO. *The Evolving Threat of Antimicrobial Resistance: Options for Action*. Geneva: World Health Organisation, 2012.
99   Wallinga D, Burch DGS. Head to head: does adding routine antibiotics to animal feed pose a serious risk to human health? *British Medical Journal*, 347(f4214) (2013).
100 FDA. *2011 Summary Report on Antimicrobials Sold or Distributed for Use in Food-Producing Animals*. Washington, DC: Department of Health and Human Services, Food and Drug Administration, Center for Veterinary Medicine, 2013.
101 FDA. *Drug Use Review*. Washington, DC: Department of Health and Human Services Public Health Service Food and Drug Administration Center for Drug Evaluation and Research Office of Surveillance and Epidemiology, 2012.
102 National Academies of Science (USA), National Research Council (USA). *First International Conference on the Use of Antibiotics in Agriculture*. Washington, DC: National Academies of Science/National Research Council, 1956.
103 Swann MM, Baxter KL, Field HI, *et al. Report of the Joint Committee on the Use of Antibiotics in Animal Husbandry and Veterinary Medicine*. London: HMSO, 1969.
104 Department of Health EaW, Food and Drug Administration. Notices. Docket 77N/0230. Diamond Shamrock, Chemical Co. penicillin-containing premixes: opportunity for hearing. *Federal Register*, 42(168) (1977): 43772–93.
105 Department of Health EaW, Food and Drug Administration. Notices: Pfizer, Inc. *et al*. Docket no. 77N: 0316. Tectracycline (chlortetracycline and oxytetracycline in animal feed)-containing premixes: opportunity for hearing. *Federal Register*, 42(204) (1977): 56264–89.

106 Gustavsson J, Cederberg C, Sonnesson U, *et al. Global Food Losses and Food Waste: Extent, Causes and Prevention*. Rome: Food and Agriculture Organisation, 2011.

107 House of Lords EU Committee. *Counting the Cost of Food Waste: EU Food Waste Prevention*. 10th Report of Session 2013–14. London: The Stationery Office, 2014.

108 WRAP. Love food hate waste: UK Household food waste campaign facts. <http://www.wrap.org.uk> Banbury (Oxon): Waste Resources Action Programme (WRAP), 2008.

109 WRAP. The Courtauld Agreements. <http://www.wrap.org.uk/content/what-is-courtauld> Swindon: Waste Resources Action Programme, 2014.

110 WRAP. *Household Food and Drink Waste in the UK 2012*. Swindon: WRAP, 2013.

111 WRAP. Reports – food waste from all sectors. <http://www.wrap.org.uk/content/all-sectors> [accessed June 2014]. Swindon: WRAP, 2014.

112 WRAP. WRAP and the circular economy. <http://www.wrap.org.uk/content/wrap-and-circular-economy> Swindon: Waste Resources Action Programme, 2014.

113 European Commission. *Towards a Circular Economy: A Zero Waste Programme for Europe*. Brussels: Commission of the European Communities, 2014.

114 Maxwell S. The evolution of thinking about food security. In: Devereux S, Maxwell S, eds. *Food Security in Sub-Saharan Africa*. London: ITDG Publishing, 2001: 13–31.

115 Pearce F. *The Land Grabbers: The New Fight over Who Owns the Earth*. London: Transworld Publishers, 2012.

116 Duncan J, Barling D. Renewal through participation in global food security governance: implementing the International Food Security and Nutrition Civil Society Mechanism to the Committee on World Food Security. *International Journal of the Sociology of Agriculture and Food*, 19(2) (2012): 143–61.

117 Cassidy ES, West PC, Gerber JS, *et al.* Redefining agricultural yields: from tonnes to people nourished per hectare. *Environmental Research Letters*, 8 (2013): 034015 (8pp).

118 Stark O, Lloyd JK. Some aspects of obesity in childhood. *Postgraduate Medical Journal*, 62 (1986): 87–92.

119 Royal College of Physicians. Obesity: report of the Royal College of Physicians. *Journal of the Royal College of Physicians of London*, 17(5) (1983): 5–65.

120 CDC. US obesity trends among US Adults 1985–2009. <http://www.cdc.gov/nccdphp/dnpa/obesity/trend/maps/index.htm> Atlanta, GA: Centers for Disease Control and Prevention, 2010.

121 US Surgeon General. *A Call to Action to Prevent and Decrease Overweight and Obesity*. Rockville, MD: US Department of Health and Human Services, Public Health Service, Office of the Surgeon General, 2001.

122 National Council on Nutrition and Physical Activity. *Strategy*. Oslo: National Council on Nutrition, 2003.

123 WHO-Europe. *Nutrition, Physical Activity and the Prevention of Obesity: Policy Developments in the WHO European Region*. Copenhagen: Regional Office for Europe, 2007.

124 Uauy R, Albala C, Kain J. Obesity trends in Latin America: transiting from under- to overweight. *Journal of Nutrition*, 131(3) (2001): 893S–899S.

125 Tsomondo E, Jones J. Obesity: a disease of indolence and affluence. *Central African Journal of Medicine*, 20(1) (1974): 1–4.

126 Health Committee of the House of Commons. *Obesity*. Third Report of Session 2003–04, HC 23-1,vol. 1. London: The Stationery Office, 2004.

127 Foresight. *Tackling Obesities: Future Choices*. London: Government Office of Science, 2007.

128 Kopelman PG, Caterson ID, Dietz WH. *Clinical Obesity in Adults and Children*. 3rd edn. Oxford: Wiley-Blackwell, 2009.

129 Lang T, Rayner G. Overcoming policy cacophony on obesity: an ecological public health framework for policymakers. *Obesity Reviews*, 8(suppl.) (2007): 165–81.

130 Milio N, Helsing E, eds. *European Food and Nutrition Policies in Action*. Copenhagen: World Health Organisation Regional Office for Europe, 1998.

131 Burkitt D. Some diseases characteristic of modern Western civilisation. *British Medical Journal*, 1 (1973): 274–8.

132 Burkitt DP, Trowell HC. *Refined Carbohydrate Foods and Disease: Some Implications of Dietary Fibre*. London: Academic Press, 1975.

133 Trusswell AS. The evolution of dietary recommendations, goals and guidelines. *American Journal of Clinical Nutrition*, 45 (1987): 1060–72.

134 Eurodiet. Nutrition and diet for healthy lifestyles in Europe: the EURODIET evidence. *Public Health Nutrition*, 4(2A–2B) (2001).

135 Garnett T. *What is a Sustainable Diet? A Discussion Paper*. Oxford: Food & Climate Research Network, 2014: 31.

136 Smith A, Mackinnon JB. *The 100-Mile Diet: A Year of Local Eating*. New York: Random House, 2007.

137 Kinross E, Small K, Small M, *et al*. The Fife diet: about us. <http://www.fifediet.co.uk/about-us> Burntisland (Fife): The Fife Diet, 2012.

138 FAO, Bioversity International. *Final Document: International Scientific Symposium: Biodiversity and Sustainable Diets – United against Hunger. 3–5 November 2010, Rome.* <http://www.eurofir.net/sites/default/files/9th%20IFDC/FAO_Symposium_final_121110.pdf> Rome: Food and Agriculture Organisation, 2010.

139 Burlingame D, Dernini S, eds. *Sustainable Diets and Biodiversity: Directions and Solutions for Policy, Research and Action. Proceedings of the International Scientific Symposium 'Biodiversity and Sustainable Diets United against Hunger', 3–5 November 2010, FAO Headquarters, Rome*. Rome: FAO and Bioversity International, 2012.

140 Lang T. *Sustainable Diets: A Report to Friends of the Earth*. London: Centre for Food Policy City University London, 2014.

141 Monteiro CA, Levy RB, Claro RM, *et al*. Increasing consumption of ultra-processed foods and likely impact on human health: evidence from Brazil. *Public Health Nutrition*, 14(1) (2011): 5–13.

142 National Food Administration, Sweden's Environmental Protection Agency. *Environmentally Effective Food Choices: Proposal Notified to the EU, 15 May 2009.* Stockholm: National Food Administration and Swedish Environmental Protection Agency, 2009.

143 German Coucil for Sustainable Development (RNE). *The Sustainable Shopping Basket – A Guide to Better Shopping*. <http://www.nachhaltigkeitsrat.de/en/projects/projects-of-the-council/nachhaltiger-warenkorb> Berlin: Rat für Nachhaltige Entwicklung/German Council for Sustainable Development, 2014: 93.

144 Health Council of the Netherlands. *Guidelines for a Healthy Diet: The Ecological Perspective*. The Hague: Health Council of the Netherlands, 2011.

145 DEFRA. Green Food Project. <http://engage.defra.gov.uk/green-food> London: Department for Environment, Food and Rural Affairs, 2012.

146 Ministry of Health (Brazil). *Guia Alimentar para a População Brasileira*. Brasilia: Ministério da Saúde, 2014.

6 The Food Wars business

Take care to drive your cow gently, if you want to milk her comfortably.

Catherine the Great, Empress of All Russia (1729–96)

Core arguments

The previous chapters discussed whether and how public bodies can address the challenges of health and environment for the food system. In this chapter we review the role of business in food supply. It is food commerce which ultimately feeds most people on the now-urbanised planet; the exceptions are those consumers mostly in low-income countries who live self-sufficient and often poor lives or are otherwise excluded and marginalised. Since the turn of the millennium, food industries across the world have gradually been made to engage more with health and the environment. This has happened partly due to external pressures and partly out of self-interest. We see three main approaches emerging from food business in this dynamic. The first approach is to ignore or downplay or subvert the challenges. The second is to engage but in a light way. The third is to begin the process of re-imagining and re-engineering how food businesses work and to put in place new commercial goals and metrics. These approaches each carry risks and are complicated by external dynamics such as global pressures, competition, geo-politics, over which not even giant food corporations have sufficient power to ensure smooth transitions. Food business is thus locked into a double-bind. On the one hand, its material resources are becoming less controllable, commodity prices increasingly volatile, and subject to externalities beyond corporate control. And on the other hand, consumers and governments are slowly waking up to the urgent need to reduce and alter food consumption in rich societies while improving diets without environmental damage in the majority of the world. Cutting across this double bind is the continued pressure to make profits from food and to operate in highly competitive markets where margins can be tight. There is an ongoing battle over which sector in food supply chains makes the most money and has the most control. Whereas this dynamic used to occur almost entirely within national boundaries, today it occurs across continents and the world. In this new map of food business, there are giant corporations whose turnover is vast, alongside hundreds of millions of tiny enterprises from farms to cafés. Business risk and volatility is compounded by commerce seeking solutions within the different paradigms of the Food Wars.

The food industry

The food industry or food business is at the heart of the Food Wars as we describe them in this book. Whatever paradigmatic shift prevails will create a particular type of food economy linking agriculture and farming through to consumption in specific ways made up of many 'food systems'. But despite the importance of food business, it cannot sort out the Food Wars singlehandedly. The food industry is too complex and big; power and concentration is uneven within and between food supply chains that continue to lengthen and become more obtuse. This was publicly exposed by the horsemeat scandal that hit Europe and the UK in 2013.[1] The scandal revealed the tangled web of international suppliers and subcontractors needed for a relatively simple activity – the provision of meat to make basic products such as burgers and lasagna. Even big companies did not know what was going into their products.

In this chapter our task is to give an overview of the food industry and an insight into the scale of the challenge in reshaping food systems that might address sustainability, nutrition and health concerns simultaneously. Even though the food industry exhibits certain dominant characteristics – such as increasing corporate concentration – it is wrong to think of it as homogeneous. It is a complex and fragmented industry and it is often problematic to analyse the consequences of change, or proposed policies, in one part of a food supply chain without seeing altered dynamics elsewhere. For example, the advice from public health is for consumers to eat more fish yet the future sustainability of many fish stocks is uncertain. The fishing industry is confronted with a mix of problems: controlling over-fishing, the need to become more sustainable, as well as making changes to become a viable and profitable business.[2] Can this be done?

How food businesses address environmental and human health concerns is framed by commercial realities, and managing business and supply chain risks. They are also influenced by think-tanks or powerful trade organisations on specific issues such as on nutrition, animal welfare, support for particular technologies, or the role of markets and trade in global food supply. In fact one major study on food and beverage corporations found it was only towards the end of the 2000s that sustainability started to be taken on board by much of food industry as a serious topic that needed a strategic response.[3] Parts of the food industry also have a long history of actively resisting change or have fought against what they see as interference in their operations, particularly over food, nutrition and health; at the same time, other businesses show a record of innovation and being far-sighted – sometimes these contrasting positions can be found in different business units of the same corporation.

Compared to the late 20th century, by the early 21st century there is more agreement about the scale, nature and urgency of the health and environmental pressures on food business. Initially such pressures were seen as dangerous, radical undermining of business itself. But gradually, there has been growing recognition among food business leaders and analysts, as evidence and science advanced, that health and the environment must be addressed or else they would threaten the very existence of some businesses and collectively threaten the entire food system. Based on hard science, a clear sustainability challenge for food business has gained strength during the 2000s and been recognised to have profound consequences. This was set out starkly in the UK's 2011 Foresight report on the future of food and farming which stated: '[n]othing less is required than a redesign of the whole food system to bring sustainability to the fore'.[4: p 12]

Even though some elements want to, the food industry cannot stand aside from the weight of evidence about the food system's impact: the coincidence of overweight and underweight; the scale and depth of food problems; and deliberate distortion of markets. How could any responsible food industry ignore the environmental problems discussed in Chapter 4, or that, in 2014, 2–3 billion people, out of a global population of just over 7 billion, were deemed to be mal-nourished, in one form or other, as outlined in Chapter 3. According to the high-profile *Global Nutrition Report* launched in November 2014, the scale of this global malnutrition had become a 'new normal' in the sense it had become so widespread as to be almost taken for granted rather than the consequence of a dysfunctional food system. The report authors call for a 're-imaging' of nutrition, especially in relation to the accountability of nutrition interventions and actions. They wrote that this is needed across sectors and across stakeholders, including business as well as government, civil society, the research community, and international development partners that need to work together across sectors and in broad alliances.[5]

The need to set a new framework for food business in efforts to tackle this widespread malnutrition throughout food supply is not a new policy approach. The policy makers of the 1930s and 40s knew it and were tough on what they wanted from business change. More recently, the language has been more deferential. In 2004, for example, as part of WHO's global strategy on diet, physical activity and health, the role of the private sector in addressing problems such as NCDs was being put in softer 'stakeholder' terms, something to be deferred to rather than reframed.[6] For the food and beverage today, what is perhaps needed is not to produce ever more food, but changes such as reducing fat, sugar and salt content of many processed foods, reducing portion sizes and being more innovative in developing healthy and nutritious food choices. It includes developing, undertaking and practising responsible marketing, advertising, sponsorship and promotional activities and helping to provide clear and consistent information to consumers. As the 2004 WHO DPAS report said: 'the private sector can be a significant player in promoting healthy diets and physical activity'.[6] The issue is what will make this happen? At the core of the Food Wars is the issue of power within food systems – that is, with whom and where power lies and therefore who has the ability to act – when it comes to addressing both environmental and human health.

The uniqueness of food as an industrial sector

What exactly is the food industry? One common approach is to visualise a simplified linear food supply chain starting off at one end with agriculture (and not forgetting the input industries to agriculture) – the food and drink industry (processing and manufacturing) – wholesalers of agricultural and food products – food and drink retailing and foodservice (supermarkets, catering, restaurants, etc.). At the end of food chain, of course, are 'consumers' (see Chapter 7). In Chapter 2 we presented a typical food supply chain within a wider systemic representation, and suggested that while each of the links of the food supply chain are somehow separate and have their own unique set of economics and dynamics, they are, at the same time, operating as structurally bound or integrated with other parts of the supply chain. Linear models of the food supply chain tend to underplay or simplify these structural, lateral, web-like and feedback connections. For these reasons, most social scientists no longer hold the view that there is not a single food system as captured by a simple 'food supply chain', but many food systems that cut across geographies and space. Each 'food system' or sub-system is potentially structured differently and has a distinct historical chronology.[7] It is often difficult analytically to capture the balance of shifts along a food system, say between agriculture and industry, industry and retailing or foodservice or with the influence or otherwise of consumer markets and consumerism.[8] A simple 'food supply chain' approach tends to obscure, lose or downplay critical elements in food supply, such as the role of financial markets, ethical issues, the role of labour and food workers, even ignoring major industry sectors such as food ingredient suppliers, or producing work, such as on SMEs, that appears isolated from wider food system dynamics.

It is all very complex. Nevertheless, in this book, like many academics, we talk of 'the food system' when attempting to portray how business relationships operate, who and what controls this, and how it affects needs and wants, health and the environment. In evolutionary terms, the change has been remarkable. The modern food system that has evolved over the past 100–150 years has opened up entirely new consumer markets for processed and industrially produced food products, seen the expansion of eating outside the home such as fast-food outlets, and revolutionised what, how and where we eat.[9] The 21st-century problems associated with the Food Wars all stem from this revolution in food systems.

The food industry as an economic sector can be seen as distinct from other industrial sectors such as the textile, automotive, electronics or other significant areas of economic activity. In particular the relationship between the food system and consumption and the fact that food is consumed (eaten) means its 'organic' properties are of crucial importance. 'Organic' is used here to mean the physical and biological properties of food and how they are created by, and relate to, the socio-economic conditions within which they function.[8 9] For example, we focus a great deal on the nutritional properties of food, on the consequences of under- or over-consumption, but this would also include issues of perishability (crucial in securing food supply chains), aspects of food waste and the effects of weather.

The 'organic' nature of food systems lends itself to other analytical approaches. For example, Hansen picks up this theme of the biological and physical conditions of agriculture and food from a food economics perspective. He links this to two further themes that make food markets unique, namely their structural conditions and market and economic conditions.[10] In Hansen's scheme biological and physical conditions include, among a number of things, the seasonality of agricultural production, suitable land, geographical and country-specific differences. Structural conditions include the competition and price setting within the food sector and market-related conditions like trade and subsidies, through to consumers' incomes and the relative importance of the food sector in relation to the total economy. Importantly, he argues such 'conditions' are often interconnected or linked from an economics perspective.

Many of the now-dominant characteristics of food industry are not new, but the scale and rate of change are. There have been massive upheavals in:

- global consumer markets which can be summarised as little growth in developed world countries or markets, but fast-expanding markets of high-income consumers to be found in BRICS especially China;
- the growth in retail concentration;
- creative use of distribution and logistics;
- exploiting relationships with suppliers;
- global strategies but adapted for local markets; and
- product differentiation and innovation especially in processed foods and foodservice.

In 1981, the OECD identified in a seminal study on food industry a range of pressures on resources employed in the food economy and outlined the then future choices facing the food industry; these were:[11]

- improving efficiency (lowering unit costs);
- diversification of product ranges and costs;
- remaining within the food economy, but accepting lower returns on resources;
- transferring resources to activities providing higher returns;
- persuading governments to subsidise the various parts of the food economy.

These OECD 'predictions' have proved remarkably prescient. In our view, the key components of business strategy led by food manufacturers today now include in addition to these:

- restructuring and concentration (including Foreign Direct Investment in different markets);
- cost savings especially through reduction in numbers of employees or employing new technologies;
- increasing scale of production (more from fewer factories);
- mergers and acquisitions and other intra-firm alliances;
- focusing on core brands (with substantial marketing and advertising support);
- focusing on key product categories (including product innovations and new product development).

What the 1981 OECD study could not address in detail, as these were then emerging tendencies, was the role globalisation would play in shaping food industry strategies and activities; the impact of nutritional concerns especially in the developed world since the obesity 'crisis' was barely underway, nor the manner that environmental and sustainability factors would emerge onto global agendas under the banner of sustainable development – the Brundtland report 'Our Common Future' was only published in 1987 and the world had to wait until 1992 for the first Rio Summit to set out a global agenda for sustainability.[12 13] Basically, the case for a new framing for the world of food has emerged from the mid-1980s into the 2000s, increasing uncertainty and presenting unprecedented future risks for business, but at the same time offering unprecedented opportunities.

Since 2007–8 there has been a renewed focus on the financial aspects of agribusiness and the food industry as commodity prices rose significantly This has helped to turn policy and business interests towards a renewed focus on 'food security' when defined as fully functioning, resilient and efficient markets – the next section explains this price crisis in more detail.

The food commodity price crisis

The most common message cited from science to business today is the need to produce 'more from less'. This is being couched in neo-Malthusian terms as a squeeze of rising population on land use and food growing capacity – that is, how to feed a global population of more than 9 billion people by 2050. Taken to the logical conclusion, this argument and data behind it suggest that food businesses will be key actors in managing the heightened risks confronting humanity, health and eco-systems. At the same time, such risks suggest that previous models of business, their strategies and practices, might need to change. The debate is over how extensive this change might need to be, what shape the food system might ultimately take, and what drives this change.

While this battle of how food business responds to the Food Wars continues, there is a parallel and interwoven battle over profits and who makes the money from food. Over the 20th century a radical transformation of food business occurred, in different phases in countries at different times. In some countries, transnational corporations arrived threatening national businesses. In others, already powerful sectors competed for internal power. But across the world as a whole a shift in power was detected from farmers and growers on the land to sectors off the land such as processing, logistics, retailing and catering, and even resource companies feeding into farming and growing. This dynamic put a squeeze on primary producers everywhere.

However, the vulnerability of food supply with a renewed emphasis on primary production, was brought into sharp focus across political and business interests in the 2007–8 commodity and food price crisis. After decades of relatively declining prices of both raw commodities and food purchased by consumers, suddenly prices went up sharply, down and then gradually rose again to remain at historic highs. Even those analysts who sought long-term price drops had to accept that the new norm was volatility and uncertainty.[14] Figure 6.1 gives a half century look at global food commodity prices, through the FAO's Food Price Index. While nominal (to the purchaser) prices have risen slowly, 1961–2014, in real terms they slowly dropped, until the crisis of 2007.[15]

On past precedents, economists were confident that sharp rises in food prices would incentivise farmers to grow more, and that thus the long-term slow decline would continue. This, they have suggested, might squeeze primary producers but was good news for consumers in that it brought food prices down for the increasing numbers of urbanised consumers. In fact, something else happened. Prices did fall but then went up and went volatile. New explanations were thus needed, and the 'new fundamentals' argument began to win ground. This argued that new conditions were at last being recognised in market prices, the combination of much that has been summarised in the previous chapters: resource squeeze, power shifts, feedbacks, and costly consequences. Other factors also emerged – such as the rise in using land and crops for biofuel production or a new type and breed of financial speculation in food commodities lie behind the food price crisis.[16-20]

In response, some policy-makers and industrialists sought comfort from the prospect that new technologies could once more produce more, for example,

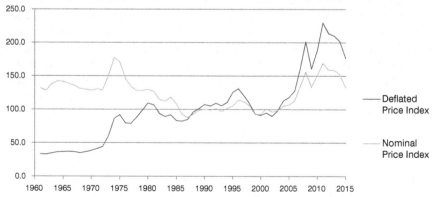

Figure 6.1 FAO global Food Price Index in nominal and real terms, 1961–2014

Source: FAO 2014 [15]

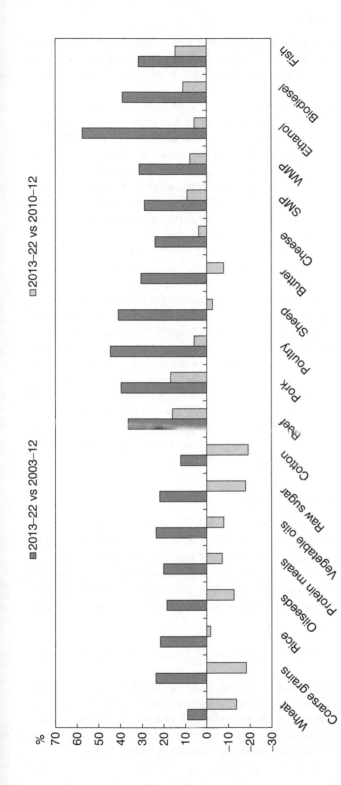

Figure 6.2 Likelihood of all commodity prices rising in 2013–22 relative to prices in 2010–12 and 2003–12

Source: OECD-FAO 2013[21]

from using resources more efficiently, and that new management efficiencies and systems could kick in, thereby keeping the 'ever cheaper food' paradigm in place. Others in business and beyond argued that this new situation was a wake-up call.

After years of subscribing to (indeed even championing) the view that prices could and should gently fall, the OECD began to recognise the new realities. Joining forces with the FAO, its prognoses began to chart the new futures. Whereas Figure 6.1 gives their forecast for all world food commodities for 2013–22, they also asked if the prognoses might vary according to which benchmark time period was taken. Figure 6.2 shows that, whether the comparison time point was taken as 2003–12 (i.e. covering pre- to just post- the 2007–8 price spike) or more narrowly as 2010–12 (immediately post-spike), the general picture is of future rises on average, but there will be sharp variations by commodity.[21]

These are forecasts, of course, and there are many commodity speculators who thrive on such forecasts and make billions betting for or against price variations. But our point here is that this new commodity price volatility casts a long shadow over the whole food system. In particular the environmental picture set out in Chapter 4 suggests factors such as climate change or water stress are also compounding the problem. The OECD-FAO report from which Figure 6.2 comes summarised soberly the tricky price situation in relation to environmental uncertainties. For example, it reported that a 'wide spread drought', like the one in 2012 in the United States and CIS countries, on top of low stocks, could raise crop prices by 15–40 per cent (p. 12). Potential price vulnerabilities such as these will have real consequences for food consumers as well as affecting the competitiveness and profitability of different food industries. The next sections attempt to situate these pressures within the broader context of the role of food industry and how sectors have evolved.

The origins of industrial food supply

Without modern food production and processing industries it would not be possible to enjoy the sort of society and lifestyles many of the global middle and upper classes can now lead. At the heart of this food economy is the way food and beverages are manufactured and processed. These transform simple raw foods and liquids into food and drink products. Today, when strong criticisms of processed food are commonly made, it is easy to lose sight of how useful processing can be; it can:

- make food edible and palatable: for example, wheat is milled into flour and then bread;
- preserve foods: using canning, drying, pickling, smoking, fermentation, refrigeration and freezing and other preserving techniques to even out vagaries in crop and food availability, to store food and reduce food waste;
- overcome climatic and seasonal problems: people in dry areas can still be fed;
- make food safe: preventing pathogens and protecting public health;

- enable transportation of food: from food-surplus to food insecure areas;
- make food more interesting: mixing combinations of crops and bringing unusual tastes and flavours to the table, creating new products, meeting the needs of different consumer groups.

Finding, producing and processing food has been central to the whole saga of human survival and cultural evolution. Homo sapiens has been foraging for food for more than 100,000 years; has farmed it for over 10,000 years; and has probably manipulated the environment (not farming per se) for considerably longer. But people have been farming and processing food industrially on a mass scale for mass markets across a high proportion of total diet for only 200–300 years. Now human society is gearing up to change the ancient and historic genetic make-up of both farm animals and crops in the belief this will secure future food supplies. Table 6.1 gives a schematic outline of the last ten millennia. It summarises some major land-based technical and social revolutions and presents these in a periodic series.

Within the historical trajectory set out in Table 6.1 the 20th century witnessed arguably one of the most significant food revolutions since settled agriculture began around 10,000 years ago – the creation of the Productionist paradigm. Although the roots of this revolution lay in the 18th and 19th centuries, it was in the 20th century that the application of chemical, transport, breeding and energy technologies transformed food supply. Beginning in the European and US agricultural heartlands, radical changes to how food was grown, processed, distributed, and consumed were experimented with, applied and marketed.[23 24] These countries entered the 20th century already well endowed with processing technology, such as the giant roller mills used in grain milling, and with industrialised baking machinery enabling, for example, the production of biscuits by the million. But these changes were nothing compared to what was to follow. Giant machinery soon began to replace human labour and 'Fordist' thinking was applied to both plant and animal production.[25] Large-scale experimentation was expended on trying to reduce nature's unpredictability.[26]

On the 20th century farm, agro-chemicals replaced the hoe; feedlots or thousands of animals confined inside buildings in so-called factory farms replaced grazing; vast crop monocultures replaced smallholdings. This growth owed much to the spread of fossil fuel culture, in particular the use of oil to drive machines and transportation, but also in the form of fertilisers.[27]

If the first half of the 20th century was marked by the industrialisation of both agriculture and processing, the second half will surely be enshrined as the decades of retailing industrialisation.[28 29] New ways of packaging, distributing, selling, trading and cooking food were developed, all to entice the consumer to purchase. These power shifts in the food economy have contributed to the contemporary conflict within the food system between the 'productivism' and 'consumerist' sectors: agribusiness versus consumer business; primary producers versus traders; food processors versus food retailers; and even production interests versus public health goals. Other changes saw year-round varieties of foods available from all

Table 6.1 10,000 years of agricultural and food revolutions, and their links with farming, culture, and food-related health

Era/revolution	Date	Impact on Farming	Culture	Implications for Food-related health
Settled agriculture	From 8500 bce on	Decline of hunter-gathering; greater control over food supply but new skills needed	Fixed human habitats; division between 'wild' and 'cultivated'	Risks of crop failures dependent on local conditions and cultivation and storage skills; diet entirely local and subject to self-reliance; food safety subject to herbal skills
Iron age	5000–6000 bce	Tougher implements (plows, saws)	Emergence of technology; spread of artistic expression	New techniques for preparing food for domestic consumption (pots and pans); food still overwhelmingly local, but trade in some preservable foods (e.g. oil, spices)
Feudal and peasant agriculture (not in some regions, e.g. North America)	Variable, by region/continent	Spread of enclosed land (parceling up of formerly common land by private landowners); use of animals as motive power; marginalisation of nomadic practices	Division of labour; settlement around land-based production and village systems	Food insecurity subject to climate, wars, location; peasant uprisings against oppression and hunger
Industrial and agricultural revolution in Europe and U.S.	Mid-18th century	Land enclosure; rotation systems; rural labour leaves for towns; emergence of mechanisation	Growth of towns; emergence of industrial working class with no access to land; rise of democratic demands	Transport and energy revolutions dramatically raise output and spread foods; improved range of foods available to more people; emergence of commodity trading on significant scale; emergence of industrial working-class diets
Chemical revolution	Begins in 19th century in developed world, spreads thereafter	Fertilisers; later pesticides; emergence of fortified foods (e.g. Liebig's beef extract)	New applications such as packaging; emergence of large-scale food processing; population gradually increases with wealth	Significant increases in food production; beginning of modern nutrition; identification of importance of protein; beginnings of modern food legislation affecting trade; opportunities for systematic adulteration grow; scandals over food safety result

Mendelian genetics	1860s; applied in early 20th century	Plant breeding gives new varieties with 'hybrid vigor'	Beginnings of biological science in everyday life, e.g. enzymes	Plant availability extends beyond original 'Vavilov' area; increased potential for variety in the diet, in turn increases chances of diet providing all essential nutrients for a healthy life
The oil era	20th century	Animal traction replaced by the tractor; spread of modern, intensive agricultural techniques	Car use and supermarkets rise; emergence of large-scale food processors; modern mass consumerist food culture and brands take off	Less land used to grow feed for animals as motive power; rise of impact of excess calorie intake leading to diet-related chronic diseases; discovery of vitamins stresses importance of micronutrients; increase in food trade gives ever wider food choice
Green Revolution in developing countries	1960s and after	Systematic plant breeding programs on key regional crops (rice, potatoes) to raise yields	Concentration of farming in larger holdings and more commercialised, intensive agriculture	Transition from underproduction to global surplus with continued mal-distribution; over-consumption continues to rise
Modern livestock revolution	1980s and after	Growth of meat consumption creates 'pull' in agriculture; increased use of cereals to produce meat	Rising incomes as more low-income countries achieve affluence; meat consumption rises (in meat-eating cultures); food suitable for humans (e.g. soya) is redirected to animals	Rise in meat consumption associated with Nutrition Transition; global evidence of simultaneous under-, over-, and mal-consumption; beginning of the end of the 1940s production-focused policy consensus that increased output will, if guided by science and if distributed fairly, end most food-related health problems.
Biotechnology	End of 20th century	New generation of industrial crops; emergence of 'biological era': crop protection, genetic modification, genomics	Debate about drivers of progress, patent ownership; consumer information becomes central to management in 'risk society'	Uncertain as yet; debates about safety and human health impacts and whether biotechnology will deliver food security gains to whole populations; investment in technical solutions to degenerative diseases (e.g. nutria-genomics)

Source: Lang 2009[22]

corners of the world and the virtual elimination in the seasonality of fruits and vegetables by global sourcing.[30 31] However, when reduced to its bare essentials, the global food economy still has a relatively simple base: most consumers eat foods from a core group of about 100 basic food items, which account for 75 per cent of our total food intake.[32]

Thus, the 20th century saw a food supply chain revolution characterised by integration, control systems and astonishing leaps in productivity, and scale in terms not only of size but deployment of capital and infrastructure. Key elements can be summarised as restructuring:

- how food is grown, such as mass use of agrochemicals and hybrid plant breeding;
- how animals are reared: factory farms, intensive livestock rearing, prophylactic use of pharmaceuticals to increase weight gain;
- the emergence of biotechnology applied to plants, animals and processing;
- food sourcing: a shift from local to regional and global supply;
- the means of processing, such as the use of extrusion technology, fermentation and cosmetic additives;
- the use of technology to shape quality and to deliver consistency and regularity;
- the workforce: labour-shedding on developed world farms; a retention of cheap labour and a strong push to 24-hour work; large increase in off-farm labour such as in retailing and foodservice operations
- marketing: a new emphasis on new product development, innovation, branding and marketing;
- consolidation of the retailers' role as the main gateways to consumers;
- distribution logistics, such as the use of airfreight, regional distribution systems, heavy lorry networks and satellite tracking;
- the methods of supply chain management – centralisation of ordering and application of computer technology;
- the moulding of consumer tastes and markets – mass marketing of highly processed foods and beverages, the use of product placement methods, investments in advertising and marketing and the targeting of particular consumer types;
- the level of control over markets – rapid regionalisation and moves towards globalisation, and the emergence of cross-border concentrations;
- the growing importance of tough, legally backed intellectual property rights.

It is important to remember that much of food business is very consumer-focused and has worked to respond to changes in society and among consumers as well as help shape consumer tastes and behaviours. One manifestation of this is the way the food industry continually launches new products. According to Neilsen, the data company, 65,725 new products were introduced into western European markets alone in 2011–12. But only 55 per cent lasted six months on shelves, whilst 73 per cent were dumped after a year.[33] Shifts in consumer lifestyles are reflected in dietary structures and increasing urbanisation – in developed world countries up to 70 per cent of people live in cities or urban areas and in 2011 for the first time in human history more people lived in urban centres than rural areas.

As incomes rise and lifestyles change, people tend to move from a traditional cereal-based diet to a more protein rich, diversified diet – consumption tends to be processed and prepared foods together with substantial increases in meat and dairy consumption – widening the spread between farmgate and retail prices. And it is the nature of these 'processed' foods and drinks that is often singled out in relation to nutritional problems and as another factor in the unsustainability of current food systems.

The nature of food processing itself has become problematic. Does the nature or type of ingredients in processing affect health and the risk of disease? Monteiro and colleagues have proposed that classifying foods according to the extent and purpose of the industrial processing applied can be used to assess trends in the quality of people's overall diet.[34 35] They propose foods should be classified in three ways:

Group 1: unprocessed and minimally processed foods. Examples would include fresh meat and milk, grains, legumes, nuts, fruits and vegetables, and roots and tubers.

Group 2: processed culinary or food industry ingredients. By this they mean substances extracted and purified from unprocessed or minimally processed foods in order to produce kitchen and food industry ingredients. This would include products such as flour, starches, oils, fats, salt, sugar and also industrial ingredients such as high fructose corn syrups or milk and soy proteins.

Group 3: ultra-processed food products. These are food products that are ready to eat or ready to heat with little or no preparation. In this category would also fall products that have experienced processing baking, frying, curing, smoking, pickling and canning. In other words, the food products most of us usually refer to as convenient or 'fast' foods.

One of the implications of this proposed classification is that dietary advice and analysis should consider more the type of processing or what sort of processed product and not simply be discussed in terms of types of nutrients (such as fats, sugars and so on). Such classifications offer a useful connection to the realities consumers meet on supermarket shelves and might provide insights into how consumers' buying patterns change over time. In the next sections we consider the structures of particular food systems, what factors drive change and the consequences.

Food business as part of a 'food system'

In the last two to three decades of the 20th century there was a rebirth of academic interest in food studies.[24] Here we do not summarise this extensive literature,[36 37] but signpost one rich seam which helps inform our approach to the Food Wars. This is the political economy and 'food-systems' perspective pioneered by rural sociologists from the 1980s onwards, trying to understand the deepening rural and farm crisis, the squeeze on farming, and why different parts of that supply chain have an impact on one another.[26 38 39] Social scientists interested in food systems have focused on four major areas.[8 40]

1   How agrarian structures and state agricultural policies developed over time in both the developed and developing world and in the growth of globalised food 'regimes' formed by the tendencies and structures of advanced capitalism.
2   Detailed empirical analyses of particular agricultural commodity regimes (such as for beef or tomatoes), with an emphasis on the structures and strategies of multinational firms.
3   The role of regulation: how state practices and rules governing food systems are changing and how they shape agri-food systems.
4   How key players or actors and networks of interest work together to formulate policy and define the workings of the food supply chain.

It is the last point – the role of different 'actors' or 'networks', known as Actor Network Theory – that has become a prominent method of food chain analysis in the 2000s. This approach to food systems emphasises the role of human agency or the social as part of the economic structuring and restructuring of food chains.[37]

Although these four areas have been identified as distinct areas of theoretical and empirical study, in practice there are areas of overlap. From our perspective a weakness of some of this work, as in other areas of food studies, is that it too rarely acknowledges or includes health and nutrition as either an outcome of the food supply chain or as a separate topic worthy of research and analysis. In this sense nutrition, diet and health is not seen as central to the workings of food value chains or at best is marginal, such as producing a crop with a nutritional profile that could be marketed on the basis of a supposed health benefit. A further theoretical and academic challenge is to integrate different meanings and practices of food and agricultural 'sustainability' into a more profound understanding and interpretation of food systems change and the behaviours of key actors.

But food system studies provide a rich vein of insight and detail into the food industry and how the activities of food business impact of other parts of society and a broader food policy environment. The following three case studies – on olive oil, fish and local food economies – provide examples of different types of food systems analysis that also illustrate how modern food businesses operate.

Food and land use: olive oil

The Mediterranean diet is widely recognised as a model for 'healthy eating' and the type of diet that if followed can help reduce the risk of NCDs such as heart disease.[41] The Mediterranean diet is one based mainly on olive oil, bread and pasta, garlic, red onions, aromatic herbs, vegetables and not much meat. As such it has been promoted by public health nutritionists internationally and some industry sectors have used this as an opportunity to promote their products on the back of the Mediterranean diet. An irony of the public health promotion of the Mediterranean diet is that Mediterranean country consumer food culture itself has started to succumb to industrializing dietary tendencies.

A study published in 2009 using the Mediterranean Adequacy Index (MAI) found that the majority of the 41 countries included in the study between 1961–5 and 2000–3 have tended to drift away from Mediterranean-like dietary patterns.[42] But one of the most striking results from this study was that the European group of Mediterranean countries experienced some of the greatest decreases in their MAI value as they moved away from traditional Mediterranean foodstuffs and towards products such as sugar, meat and edible oils other than olive oil. As a result the diet of Mediterranean countries has become much higher in both sugars and saturated fat and childhood obesity rates are now higher in some Mediterranean countries compared to Northern Europe.[43]

The olive oil industry in particular has benefited from the promotion of the Mediterranean diet as a way to increase sales and production. While from a health and nutrition perspective this might seem a good thing, a study by Scheidel and Krausmann has shown how this has unexpected impacts on the environment.[44] They have produced a detailed case study of the European olive oil system and its impact on land use. They write – and this is perhaps surprising – that, to their knowledge, the case of olive oil has not been looked at in such a comprehensive way before. Their case study shows how between 1972 and 2003 olive oil developed from a niche product that could hardly be found in food stores outside of the producing countries to become a regular component in the diets of industrial countries. They discuss the impacts of the promotion of the 'healthy Mediterranean diet' on land use and agro-eco-systems in the producing countries. Their study focuses on 15 EU countries – five producing countries, Spain, Italy, Greece, Portugal and France, and ten non-producing (NP) countries Austria, Belgium, Denmark, Finland, Germany, Ireland, Luxembourg, the Netherlands, Sweden and the UK.

Global olive oil production is concentrated in the Mediterranean region of which three countries Greece, Italy and Spain dominate. The olive oil sector, with 2.5 million producers, or roughly a third of all EU15 farmers (at the time of the study), is an important economic component of rural Europe. Spain is the country with the largest olive area, with production especially concentrated in the Andalusia region. Until relatively recently olive oil markets were predominantly for local consumption. But promotion campaigns for the healthy Mediterranean diet, especially since the 1980s, increased demand in non-traditional markets – with a more than ten-fold increase for the ten NP countries in their study.

So that over the period of three decades (1970s–2000s), production and consumption almost doubled. The EU15 countries held a share of 79 per cent of global production and 70 per cent of global consumption (using FAO data from 2008). However, consumption is also concentrated in the producing countries – Spain (59 per cent of total production), Italy (24 per cent), Greece (15 per cent) – who together consume 86 per cent of total olive oil. Portugal and France consume a further 8 per cent, whereas the non-producing countries together end up with only 6 per cent of total olive oil supply. Consumption is therefore dependent on distinct regional geographies – producing countries continue to have high levels of consumption. For example, Greeks consume on average 20 litres per person per year, compared to an average of 0.6 litres/person/year in the ten non-producing countries – despite the ten-fold increase over the period!

Scheidel and Krausmann show some of the environmental consequences of these production-consumption changes. First is the impact on local and traditional olive groves. Many of these were abandoned and modern, intensive, production plantations set up which rely upon irrigation systems, agro-chemicals and mechanisation. This has enabled much higher productivity (and often better quality products and widening consumer choice), but has meant major structural changes in land use. This intensification process has been especially pronounced in Andalusia, Spain. As Scheidel and Krausmann write: 'While traditionally rain fed olive trees were grown mainly on marginal soils, industrial olive groves expanded primarily into agricultural land with high quality soils' (p. 51). The promotion of olive oil as a 'healthy eating' product has led to an unexpected and structural transformation of Mediterranean landscapes.

This has come about as a result of the active marketing of the 'healthy Mediterranean diet' – promotion driven and undertaken by the EU, national food trade promotion organisations and producer bodies such as the International Olive Oil Council. In other words consumption was influenced by productivist interests, not least the EU's Common Agricultural Policy market-based policies to support products and producers, for example, EU olive subsidies rose from €160m in 1975 to €2.3bn between 1998 and 2003.

Scheidel and Krausmann conclude that, in the case of olive oil, supply has driven demand, with both helped by the institutional measures of the EU CAP with the accompanying transformation of the production system over a 20-year period from traditionally adapted crop to an industrial cash crop. The case of olive oil 'system' also serves as a lesson in how it is difficult for consumers in a globalised food system to connect to the environmental consequences of their consumption patterns.

Global Value Chain analysis of food systems: fish

Global Value Chain (GVC) analysis starts with the premise that the global economy is increasingly structured around global value chains that account for a rising share of international trade, global GDP and employment.[45] This type of analysis can be used to examine how global industries are organised and explore the dynamics of different actors in a particular industry. One important focus is on 'lead firms', that is, those companies that have critical marketing, technological or financial edge that permits them to set standards or specifications for the other companies they deal with – in food systems this sums up how the major food retailers of food processors have developed their food system influence in relations to their suppliers, especially those supplying bulk products.

The GVC methodology is described as a useful tool to trace the shifting patterns of global production, link geographically dispersed activities and actors of a single industry, and determine the roles they play in developed and developing countries alike to providing a holistic view of global industries. The example below considers the global fish value chain to show how this framework enables a historical analysis of complex food industries.

Fish is the world's biggest food commodity with a value of US$217.5bn in 2010.[2] However, global fisheries (ocean capture) are increasingly recorded as facing problems of unsustainability, with fish being taken out of the sea in such quantities that not enough fish are of the right age or numbers to reproduce future stocks. By some estimates something like two-thirds to three-quarters of ocean fishing stocks are in trouble with differing levels of exploitation. The exact nature of the fishing 'crisis' is subject to scientific debates – over the methodologies used by different studies and challenges of over-simplifications, errors and gaps in data resulting in inappropriate use and interpretations of data.[7] But it appears clear that the majority of global fisheries are overfished, despite these scientific disagreements on where and to what extent fish resources are being unsustainably used and about how well fisheries might be recovering under good management.

From a nutritional perspective, in developing countries fish is an important source of protein – on average providing 20 per cent of protein for 3 billion people – and in developed world countries people are urged to consume more fish as a good source of essential fatty acids (such as omega-3s). In addition to diet, fisheries are important globally for poverty alleviation, employment, income generation and economic growth.

So how can we make sense of the current state of global fisheries, not least in terms of diet and health and environmental concerns? In an outstanding paper John Wilkinson explores the 'fish system' from the 1950s to the early 2000s using a global value chain methodology.[46] His concern is to analyse the fish system through the process of production and product development, including intra-firm organisation and the relationship between firms to help inform policy-making on ways in which poor countries and producers might connect with producers and consumers in a global economy.

Using a global value chain approach he is able to start to unravel the complex relationships and intra-relationships between production and consumption and show how a range of factors influence these; from his study these factors can be summarised as:

- technology
- regulation and governance
- marketing/consumers

- north/south dynamics
- geographies of supply and demand
- externalities: health, environment, socio-economic
- industry structure and concentration
- policy.

One of the important findings from Wilkinson's analysis over the 40-year period he surveys are the implications of trade flows. For the fish system these have completely reversed, with developing world countries becoming the major exporters of fish for developed world markets where previously developed world countries had been net exporters. He shows the importance of consumption drivers shaping consumer tastes, with major supermarkets and foodservice operators creating new consumer markets for products such as shrimp and salmon and how this changed industry dynamics, particularly with respect to the rise in acquaculture. Another key finding from his work is how the geographies of food value chains change over time with certain countries, such as Chile, with hardly a presence in global fish markets at the start, becoming major players in a relatively short time.

He also weaves in some of the causes of the sustainablity challenges fisheries now face, such as the introduction of massive factory trawlers from the 1960s onwards and their impact on creating overfishing, together with how this changed industry structures including new country entrants into fish markets. The analysis shows how the growth of aquaculture was seen, as we would describe, as a 'technical fix' to this overfishing of ocean capture, but came with a package of its own separate evironmental challenges – such as disease, the over-use of chemicals such as veterinary drugs and the 'achilles heel' of aquaculture that farmed fish are fed pellargic fish that come from ocean capture, exacerbating overfishing.[46]

Studies such as this serve a number of important purposes, not least in helping us understand the dynamics of food business and industries and how these interact and connect with other parts or stakeholders with an influence on particular food systems, especially as new entrants or countries become integrated into an existing food system. They also show, as in the olive oil case study, how food industry activities – or production-led initiatives responding to consumer health and nutrition trends – can have unexpected environmental consequences even while production is promoted or developed on the back of supposed nutritional benefits. Both examples also illustrate the importance of understanding the impact of food systems change over longer time-scales.

Local food economies

A further important contribution 'food system' analysis has made is to an understanding of growth in demand for 'local food'. The local food economy (LFE) can be described as a system in which foods are grown, produced, or processed and then distributed or sold within a similar area – be it a particular distance from producer to food retailer or consumer, or defined by a geographic area such as a municipality, state. From this perspective, viable local food economies are proposed as a solution to revive flagging rural economies and to improve the incomes of small-scale farmers and producers. Local food economies are seen as a process that relinks food producers to food consumers through 'short food supply chains' (SFSCs) or 'alternative food networks' (AFNs). Local food economies and businesses are also discussed as a counter-trend to the globalizing tendencies and the logistical remoteness of the conventional, large-scale food business.

Key characteristics of local food economies, from a consumer perspective, are that products are perceived as being fresh, tasty and healthier. In addition, consumers want to help local farmers, support their local food economy, and they want the trust and knowledge of a face-to-face encounter with producers through their food purchases. In some of the academic literature this multi-faceted consumer relationship with local food economies is described as a 'quality-turn' in food consumption behaviour. By the 'quality-turn' is meant that LFEs fulfil a new type of food 'quality' dynamic driven by consumers looking for a different relationship with their food which seeks to combine 'healthy' with 'sustainable' food. The 'quality-turn' is being pushed by the urban consumer in particular and creative and innovative food producers pursuing urban market opportunities.[47] More recently local food economies have been analysed as being more 'reflective', in the sense that people or consumers are concerned about where their food comes from, how it is grown and processed and looking for greater transparency and openness in the way food businesses operate.

For food producers and farmers operating in local food economies this implies developing a different business mindset: such as producing crops for food, not for commodities or animal feed, growing and innovating with a greater diversity of produce, working towards ecological and sustainable agricultural goals, and wanting to engage directly with consumers. Business goals are not to increase yields of monoculture crops per acre, but increase the value of crops per acre through diversity of production and from producing 'quality' food.

Currently key markets are direct farm sales, farmers' markets, agricultural tourism, selling direct to local restaurants: all areas of future business growth opportunity (a further component is the increasing interest in urban agriculture). For most local food economies, the urban–rural linkage is critical. The advantages of local food economies have been described as:[48]

- *Proximate* – originating from the closest practicable source or the minimisation of energy use
- *Healthy* – as part of a balanced diet and not containing harmful biological or chemical contaminants
- *Fairly or co-operatively traded* – between producers, processors, retailers and consumers
- *Non-exploiting* – of employees in the food sector in terms of rights, pay and conditions
- *Environmentally beneficial* – or benign in its production (e.g. organic)
- *Accessible* – both in terms of geographic access and affordability
- *High animal welfare standards* – in both production and transport

Table 6.2 Attributes associated with 'global' and 'local'

GLOBAL	LOCAL
Market economy	Moral economy
An economics of price	An economic sociology of quality
TNCs dominating	Independent artisan producers prevailing
Corporate profits	Community well-being
Intensification	Extensification
Large-scale production	Small-scale production
Industrial models	'Natural' models
Monoculture	Bio-diversity
Resource consumption and degradation	Resource protection and regeneration
Relations across distance	Relations of proximity
Commodities across space	Communities in place
Big structures	Voluntary actors
Technocratic rules	Democratic participation
Homogenisation of foods	Regional palates

Source: adapted from Hinrichs (2003)[51]

- *Socially inclusive* – of all people in society
- *Encouraging knowledge and understanding* – of food and food culture

AFNs and SFSCs attempts to establish 'closer' or more 'connected' relationships between food producers/production and consumers/consumption, and represent modes of food provisioning which in various ways different from, or alternatives to, the prevalent, supermarket mode of provisioning.[49] The AFN issue tries to address what are observed as attempts in developed world countries to: 'reconfigure relationships between food producers and food consumers'.[50] Key characteristics of local food economies from this type of academic investigation emerge:

- 'quality' attributes are central;
- there is a growing tendency towards the 're-localisation' of food production and marketing;
- there are new socially 'embedded' relationships developing between and within these new food supply connections; and
- there is a need to establish 'transparency' in the transactions between local food producers and food consumers.

AFNs have not been free from criticism. These include the fact that much local food is not that 'alternative'; it can be on an industrial scale and the same as or a hybrid form of conventional food systems; they are elitist, serving better off consumers; and because they are very niche local food economies are unable to scale up to provide food for wider populations. Within the Food Wars paradigm, framing 'local' food acts as a counter-tendency to the Productionist and Life Sciences paradigms, with the research on local food suggesting a set of attributes that are more at home with our model of the Ecologically Integrated paradigm. These contrasting attributes are summarised in Table 6.2.

These three case studies – olive oil, fish and local food – demonstrate that there are sophisticated and different analytical perspectives on food systems from those of conventional food supply chain analysis alone. It is from these approaches that we see the Food Wars paradigms being played out and that there is a need for a more holistic and integrative approach to food policy to address the major challenges we identify here, particularly those that engage with social outcomes beyond narrowly defined economic assumptions. Nothing less than a societal redefinition of what is meant by food business 'success' is needed.[52]

Characteristics of today's food industry

In this section, in keeping with a Global Value Chain framework, we focus on the 'lead firms' of the food system to provide an overview of the key actors who are increasingly important in either transforming the food system or in hindering or even obstructing such food system redesign when it comes to nutrition and sustainability. A common theme in all areas of food supply chains is the dominance and market concentration held by these leading corporations. For example, international grain trade has been dominated historically by four multinational corporations – Cargill, Archer Daniels Midland (ADM), Louis Dreyfus and Bunge – although more recent entrants such as Glencore and a growing number of mainly Asian-based trading giants appear to be changing the market dynamics.[53] As McMahon reports, the 2000s have been good years for the historic four, with combined revenues of these firms rising from \$150bn in 2005 to \$318bn in 2011.[53] Companies such as these play a critical role in enabling the physical flow of food in the world food system, yet often operate in ways that are difficult for analytical or public scrutiny despite their importance for fully functioning food markets.

Many of these large commodity trading companies have long histories – a peculiar feature of the food processing and manufacturing part of the food chain is its longevity and relative conservatism. Many of today's leading manufacturers have their origins in the 19th century or early 20th century. Nestlé, the world's largest food manufacturer, was founded in 1866, Coca Cola was invented in 1886, Pepsi first appeared in 1898, and even the icon of fast-food modernity McDonald's began in 1940. Compare this to today's social media mega corporations that now shape people's behaviours and lives globally – Facebook began in 2004, YouTube in 2005 and Twitter in 2006, while Google is a relative oldie launching in 1996. With food and beverage brands 50, 60 or even more than 100 years old, some companies could be said to be anachronistic, but like today's modern social media corporations, the food industry as a combined sector is formidable in its range, political clout and influence on both individuals and society.

Most food in rich countries is, if not local, certainly national or regional; despite the undoubted influence and importance of globalisation, much food continues to be produced, processed and consumed within domestic borders. Only around 10 per cent of food and agriculture is traded globally, although this can vary considerably between type of foodstuff, so, for example, 80 per cent of coffee is traded on international markets, around 37 per cent of fish, but in packaged food and beverage products, in terms of global trade, this is less than 5 per cent. But more and more, food business, particularly its inputs and ingredients (such as animal feed), are reliant on global connections, whether through trade, investments, or the diffusion of technologies or specific food cultures such as a taste for branded fast foods.

While the food system can be painted as dominated by global corporations, in terms of absolute numbers much of food business consists of tens of thousands of small-scale producers. Together these giants and small players engage in a global food and beverage processing industry worth around $4 trillion in 2012.[54] The nature of this fragmented industry structure can be illustrated through an analysis of the European Union food and drinks manufacturing industry (thus excluding other parts of the food chain) – the EU being the world's leading food and drink producer, ahead of the USA and China. The following sections are based on data produced by FoodDrinkEurope (FDE, known as the Confederation of Food and Drink Industries – CIAA – until June 2011) which speaks for and represents the European food and drink industries.[55]

Food and drink is the EU's largest manufacturing sector, with a turnover of €1,048bn in 2012 and employed directly around 4.24 million people. To serve this market the EU is home to some 286,000 food and beverage companies. Of these, 283,000, or 99.1 per cent of all food and drink companies in the EU, were small and medium-sized companies (SMEs). These accounted for 51.6 per cent of turnover, therefore the remaining 0.9 per cent of EU food and drink manufacturing companies made up the remaining 48 per cent of sales. The EU defines a small company as having 10–49 employees; a medium company 50–249; and a large one more than 250 employees. Surprisingly, the average number of people employed in an EU food and drink company in 2012 was 16.

While large companies are important employers, SMEs are also significant, representing 64.3 per cent of all EU food and drink employment. Medium-sized companies contribute 29 per cent of EU food and drink turnover, employ 26 per cent, but represent only 4 per cent of EU food and drink companies. In other words, less than 5 per cent of all EU food and drink producers (out of 286,000 companies) create approximately 77 per cent of turnover and provide 61 per cent of food and drink employment.[55]

It is also important to understand the relative importance of core sub-sectors. The major EU sub-sectors are: bakery, meat, dairy products, drinks, and 'various food products' (this includes cocoa, chocolate and confectionery; tea and coffee; prepared meals; drinks; and sugar) which represent 76 per cent of total EU food manufacturing turnover and 80 per cent of employees. Meat is the largest sub-sector by value, representing 20 per cent of total turnover, but bakery comes out first in terms of number of companies and employment. The €1 trillion plus EU market, in turn, is dominated by five countries. These are Germany, France, Italy, the UK and Spain who between them account for €650bn of all food and drink produced within the EU. The average household food expenditure across the EU in 2011 was 14.6 per cent, although this varies between EU Member States. Another interesting piece of data produced by FoodDrinkEurope is on the top 15 food innovation trends in Europe. It is perhaps surprising to see that, as far as food and drink innovation goes, varieties of hedonism prevail, with 'pleasure', which includes sensory and sophistication aspects, being the leading driver of innovation trends in Europe in 2013. 'Ethical' factors are by a long way the least important and in a poor fifth place behind the 'physical' attributes of food, 'convenience', 'health' and then 'pleasure'.[54]

Finally, the FoodDrinkEurope data situate food and drink production within the context of the overall European food supply chain of: agricultural holdings, food and drink industry, wholesale of agricultural and food products, and the food and drink retail sector. Agricultural holdings are by far the largest sector by number of companies and employees, reflecting the large number of small farms across Europe, yet have the least turnover compared to the other links in the supply chain. Collectively the supply chain employed 24 million people with a turnover of more than €3.5 trillion in 2011. This is broken down by agricultural holdings with a turnover of €392bn and 11.9 million employees; the food and drink industry €1,016bn and 4.2 million employed; wholesale agricultural and food products €1,100bn and 1.8 million employed; and food and drink retail €1,110bn and 6.1 million employees.[54]

Digging down further, we can look at a country-specific example. Figure 6.3 details the UK food chain and its various links. The food system as a whole is the UK's biggest employer, and food and drink production is the biggest manufacturing sector. Trends in employment has for decades been ever further away from the land. In 2012 there were only 480,000 people working on UK farms, many part-time,[56] but 1.2 million people worked in UK food retail and 1.5 million in UK catering.[57] Food epitomises the service economy. Money is made off the land but food comes from the land – somewhere. People other than primary growers add the value. Figure 6.3 gives the current UK government's overview of the agri-food economy, showing each sector, its employment, number of enterprises and Gross Value Added, and so on.

Of the £196bn spent in 2013 by the UK's 64 million consumers, the returns by sector were:[58]

- £9.2bn to farmers and primary producers
- £24.1bn going to food and drink manufacturers
- £9.6bn going to wholesalers
- £26.7bn going to caterers
- £27.7bn to retailers.

It is perhaps sobering to note that farming these days represents less than 5 per cent of the total turnover of the UK food system. This pattern of market-dominant large companies with numerous small food producers is emerging globally. Fights occur off the land as to which sector will dominate produce from the land. Table 6.3 lists the world's top food and drink processors in 2012.[59]

As can be seen the top 100 is dominated by US and European based food companies and is a mix of branded food and beverage companies, conglomerates

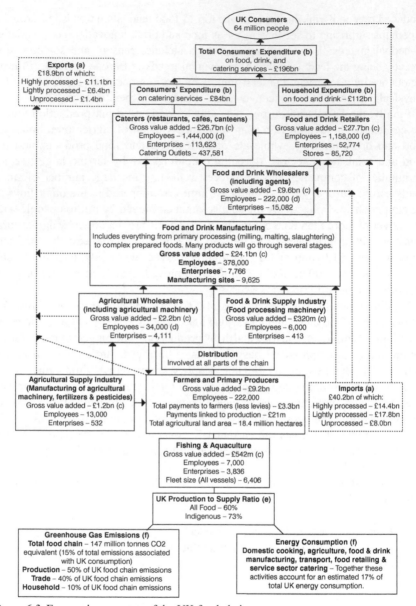

Figure 6.3 Economic summary of the UK food chain

(a) Overseas trade data provisional for 2013 from H M Treasury

(b) Consumers' expenditure provision for 2013 from the Office for National Statistics

(c) Gross value added (GVA) is the difference between the value of goods and services produced and the cost of raw materials and other inputs used up in production. Source: UK Annual Business Survey for 2012

(d) Employee data for retailers is for Great Britain in Q4 2013 from the Office for National Statistics

(e) UK Production to Consumption ratio (formerly know as the 'Self-Sufficiency' ratio)

(f) UK greenhouse gas emissions (GHG) do not relate to 2013. The figures here do not take into account embedded energy in exports or imports; therefore the 17% of energy consumption cannot be directly compared to the 15% of GHG emissions.

Source: DEFRA (2014) [58]

Table 6.3 Top 100 global food and beverage companies

Rank	Company	Food Sales ($m)	Year Ending	Rank	Company	Food Sales ($m)	Year Ending
1	Nestlé	87,977	Dec. 12	20	Diageo	17,005	June 12
2	PepsiCo, Inc.	65,492	Dec. 12	21	General Mills Inc.	16,658	May 12
3	The Coca-Cola Company	48,017	Dec. 12	22	Brf Brasil Foods	14,928	Dec. 12
4	Archer Daniels Midland Co.	46,829	June 12	23	Kellogg Company	14,197	Dec. 12
5	Anheuser-Busch InBev	39,758	Dec. 12	24	Ajinomoto	14,037	Mar. 12
6	JBS	38,675	Dec. 12	25	Fonterra	14,020	July 12
7	Mondelez International	35,015	Dec. 12	26	Carlsberg	13,388	Dec. 12
8	SABMiller	34,487	Mar. 12	27	Royal FrieslandCampina	13,300	Dec. 12
9	Tyson Foods	33,278	Sept. 12	28	ConAgra Foods Inc.	13,263	May 12
10	Cargill	32,500	May. 12	29	Vion	13,190	Dec. 11
11	Unilever	31,180	Dec. 12	30	Grupo Bimbo (Mexico)	13,164	Dec. 12
12	Mars	30,000	Dec. 12	31	Smithfield Foods Inc.	13,094	Apr. 12
13	Danone	26,920	Dec. 12	32	Marfrig Group	12,825	Dec. 12
14	Heineken	23,715	Dec. 12	33	Nippon Meat Packers	12,815	Mar. 13
15	Kirin Brewery Co.	22,130	Dec. 12	34	Arla Foods	12,575	Dec. 12
16	Lactalis	20,255	Dec. 12	35	Meiji Holdings	12,548	Mar. 13
17	Suntory	19,370	Dec. 12	36	CHS Inc.	12,500	Aug. 12
18	Kraft Foods Group	18,339	Dec. 12	37	Associated British Foods	11,825	Sept. 12
19	Asahi Breweries	18,850	Dec. 12	38	HJ Heinz Company	11,649	Apr. 12

Continued

Table 6.3 continued

Rank	Company	Food Sales ($m)	Year Ending
39	Dean Food Company	11,462	Dec. 12
40	Femsa	11,395	Dec. 12
41	Bunge	11,305	Dec. 12
42	Yamazaki Baking	10,970	Dec. 12
43	Pernod Ricard	10,598	June 12
44	Danish Crown	10,310	Sept. 12
45	Sudzucker	10,165	Feb. 13
46	Ferrero	10,055	Aug. 12
47	Maruha Nichiro Holdings	9,955	Mar. 13
48	Hormel Food Corporation	8,231	Oct. 12
49	Campbell Soup Company	7,707	Aug. 12
50	Grupo Modelo (Mexico)	7,550	Dec. 12
51	Morinaga Milk Industry	7,245	Mar. 12
52	Saputo	7,000	Mar. 12
53	Parmalat	6,743	Dec. 12
54	Oetker Group	6,702	Dec. 12
55	The Hershey Company	6,644	Dec. 12
56	Nissui	6,590	Mar. 13
57	Ingredion Inc.	6,532	Dec. 12
58	McCain Foods Ltd	6,455	Jun. 12
59	Red Bull	6,360	Dec. 12
60	DMK Deutsches Milchkontor	6,255	Dec. 11
61	Dr Pepper Snapple Group	5,995	Dec. 12
62	Yili Group	5,800	Dec. 11
63	Sodiaal	5,675	Dec. 12
64	China Mengniu Dairy Co.	5,638	Dec. 12
65	Bacardi	5,600	Mar. 12
66	Sapporo Holdings	5,548	Dec. 12
67	The JM Smucker Company	5,526	Apr. 12
68	Itoham Foods	5,498	Mar. 13
69	LVMH	5,338	Dec. 12
70	Bongrain	5,270	Dec. 12
71	Nisshin Seifun Group	5,265	Mar. 12
72	Barry Callebaut	5,218	Aug. 12
73	Barilla	5,155	Dec. 12
74	Tate & Lyle	4,940	Mar. 12
75	QP Corporation	4,800	Nov. 12
76	Nissin Food Products	4,775	Mar. 12
77	Maple Leaf Foods	4,760	Dec. 12
78	Ralcorp Holdings	4,741	Sep. 11
79	Ito En	4,630	Apr. 12
80	Maxingvest/Tchibo	4,605	Dec. 12

Rank	Company	Food Sales ($m)	Year Ending
81	Japan Tobacco International	4,505	Mar. 13
82	Coca-Cola Amatil	4,250	Dec. 12
83	Dole Food Company Inc.	4,247	Dec. 12
84	Land O' Lakes Inc.	4,156	Dec. 12
85	Muller Group	4,125	Dec. 11
86	Hillshire Brands	4,039	Jun. 12
87	McCormick Corporation	4,014	Nov. 12
88	J R Simplot	4,000	Aug. 12
89	Glanbia	3,920	Dec. 12
90	Molson Coors Brewing Co.	3,917	Dec. 12

Rank	Company	Food Sales ($m)	Year Ending
91	Agropur Cooperation	3,805	Dec. 12
92	Schreiber Foods	3,680	Sep. 12
93	Del Monte Foods Company	3,676	Apr. 12
94	Brown-Forman Corporation	3,614	Apr. 12
95	DE Master Blenders 1753	3,605	Dec. 12
96	Tine Group	3,463	Dec. 11
97	Groupe Bel	3,420	Dec. 12
98	E & J Gallo Winery	3,400	Dec. 11
99	HK Scan	3,285	Dec. 12
100	Perfetti Van Melle	3,210	Dec. 12

dominated by a particular processing expertise (such as diary companies) and global alcoholic beverage and brewing companies. These companies are very prominent in sustainability and health issues, but are often overlooked or not integrated into more general discussions on the wider food industry in academic analysis (as are some of the major food ingredients companies).

Food business concentration

The overview of the EU food and drink sector above shows that food production in terms of value is highly concentrated. The dominance or influence by 'lead' firms in global supply chains is replicated in specific food sectors. Below we give examples to illustrate how this plays out in different parts of the food system from a commodity chain, food retailing and foodservice.

Bananas

A more recent feature of agricultural trade has been the development of new food commodity chains, especially for fresh produce, built upon production in developing countries for consumption in rich countries. This trend is giving rise to a different form of competition not only between 'logistic' chains such as bananas and exotic fruits, but also whole food production systems. An example of such export orientated logistic chains are those for bananas – the most popular fruit in the world. In a market such as the UK, bananas are the single most profitable item sold in UK supermarkets and they sell 80 per cent of the 5 billion bananas British consumers get through each year.

The banana business is dominated by key corporate players, with trade or distribution of bananas controlled by a handful of major multinationals. Global banana exports were worth an estimated US$7bn in 2012, with multinationals Chiquita, Dole and Del Monte controlling half of the world's banana exports. However, in terms of market share (as opposed to trade) the top firms have been losing out as the industry has restructured and the big multinationals have looked to free themselves from their previous direct ownership of plantations, thus allowing national growers' companies to enter the market. So in 2013 these top three companies had a market share of 36.6 per cent, down from 65.3 per cent in 2002.[60]

Even big players can feel the squeeze in competitive commodity markets. In the UK, for example, the major supermarkets starting buying their bananas direct from the producers, cutting out the distributors like Chiquita. Such competitive pressures squeeze operating margins and profitability, leading to many mergers and acquisitions in food sectors, with bananas being no exception. In 2014 a merger was proposed between US-based Chiquita and Dublin-based Fyffes, reportedly because of the squeeze on retail prices.[61] It was estimated that the merged company would account for 14 per cent of the global market, becoming the largest banana company in the world. But a counter-bid by Brazilian family-based businesses Cutrale and Safra saw the deal come undone and the Brazilians snatched Chiquita.

4% workers

20% growers

23% transport (farm to European port)

12% EU tariff

12% ripener/distributor

29% retailer

Figure 6.4 Who makes the money from UK bananas?
Source: Banana Link (2014)[60]

Markets such as bananas today rely on sophisticated logistics chains to work efficiently and each section of the chain looks to get its cut. The price squeeze comes with who gets what from the particular commodity supply chain – often it tends to be food workers who receive least. In the case of banana exports, it can be seen from Figure 6.4 in the example of banana exports from Costa Rica to the UK. The figures give 2010 estimates by the NGO Banana Link for who makes the money from a Costa Rican banana sold in a UK supermarket. The banana workers receive the least from the banana export trade and the retailers the lion's share. The banana agribusiness is an industry based on large plantations and monoculture production methods which is reliant on massive inputs of agrochemicals to function (only cotton uses more agrochemicals) and has a reputation for using poorly paid labour with limited workforce protection, such as in health and safety, workers benefits and rights.[60]

Food retailing

As illustrated by the banana example, supermarkets have grown to become the lead actors in many food markets with increasing concentration and influence. Table 6.4 details the market concentration of supermarkets across a selection of countries, indicating the extent of the oligopolistic power of these corporations, and of course many supermarkets operate across many countries and formats.

In the international domain, many competition authorities are studying or have studied the retail sector and the vertical relations in the food supply chain. Notable amongst these is the research done by the British Competition Commission in April 2008, titled *The Supply of Groceries in the UK Market Investigation*.[63] Countries such as the US, Australia, Romania, France and Sweden have also conducted comprehensive or specific topical research into the grocery retailing sector, and many others have embarked on similar analyses.

In the European Union, these issues have been the object of debate and analysis in several institutions, including the European Commission and Parliament, with

Table 6.4 Examples of supermarket/grocery market national food market shares

Country	Year	% of national food market	Concentration ratio
European Union			
Austria	2009	82	3
Belgium	2011	71	5
Denmark	2009	80	5
Finland	2011	88	3
France	2009	65	5
Germany	2011	85	4
Greece	2009	50	5
Italy	2009	40	5
Netherlands	2010	65	5
Portugal	2011	90	3
Spain	2009	70	5
UK	2011	76	4
Rest of World			
Australia	2011	71	2
Canada	2011	75	5
Norway	2011	81	3
Switzerland	2011	76	3

Source: Consumers International 2012[62]

emphasis on the need for competition authorities of the Member States to intensify their pursuit of coordinated action in this area. After the European Commission's publication in 2009 of its communication *A Better Functioning Food Supply Chain in Europe*,[64] the High Level Group on the Competitiveness of the Agro-Food Industry, which reports to the Commission, created a European Forum composed of various agents from the food manufacturing and retail sectors. One of the Forum's aims was to determine the most problematic contractual practices in the commercial relationships within the food supply chain and to explore possible solutions. Could these sectors really self-regulate?

Despite high concentration levels and the market power of supermarkets, in practice very little has been done to rein them in or force through regulatory measures to open up retail markets. In fact, in many cases, this supermarket power is regarded as serving consumer interests by providing 'cheap' food and enhanced food choices.

Foodservice

More fragmented and diverse than many other areas of the food industry, foodservice such as fast-food outlets, catering, restaurants and so on is a huge part of how the global food system works and how people buy and consume food. Foodservice is often a neglected area in academic analysis compared to other parts of food business, especially in relationship to food systems or even value chain analysis, because of this fragmentation and diversity. But foodservice – food eaten outside the home – accounts for large percentages of the income people spend on food, nearly 50 per cent in the US, around 40 per cent in the UK, but is also an important part of developing world food expenditure. In 2012 the global market for consumer foodservice was around US$2.5 trillion, with demand declining in Western Europe, but growing in Asia and China.

Even more than food manufacturing, foodservice is a business sector dominated by small players but with high-profile global corporations shaping the perceptions of foodservice – such as fast-food chains. Consumer food service covers:

- cafés/bars
- full-service restaurants
- fast food
- 100 per cent home delivery/takeaway
- self-service cafeterias
- street stalls/kiosks (street vendors).

Table 6.5 details the largest global foodservice markets by value.

China represents 25 per cent of global restaurant transactions, and the Asia/ Pacific region 40 per cent.[65] Full service restaurants dominate global sales (by value) followed by fast food and then cafes and bars. Foodservice is ubiquitous and in a global age includes airports, shopping malls and events catering. As the above figures suggest, foodservice is highly localised or culturally dependent with many stand-alone or independent operations (which in part is what makes it academically hard to investigate using traditional methodological approaches). One area of foodservice that escapes from this fragmentation is the fast-food sector with its highly visible branded global offerings – the global top ten fast-food chains in 2012 are listed in Table 6.6 and as can be noted nine out of the ten are US based (Seven & I Holdings/7-Eleven is Japanese).

In addition to consumer foodservice companies, there is another layer of foodservice operation corporations – contract catering. These are the food businesses that provide catering services to institutions, organisations, events and workplaces. Prominent companies here include Compass, Sodexo and Aramark.

Another aspect of foodservice that is important is public food procurement – namely food produced for public welfare or public institutions such as hospitals and other healthcare facilities, nurseries, kindergartens, nurseries, schools and universities, prisons, the armed forces, day-care centres, to care homes for vulnerable groups such as the elderly or those with mental health problems. Public

Table 6.5 Largest foodservice markets, 2012 (US$bn)

USA	466.3
China	459.0
Japan	259.9
Brazil	146.2
Spain	99.2
Italy	96.4
India	93.6

Source: Euromonitor 2013[65]

Table 6.6 Top ten global fast-food chains, 2012

Company name	Value US$bn	No. of outlets
McDonald's Corp	88.29	38,964
Yum! Brands Inc. (KFC, Pizza Hut)	42.83	38,798
Doctor's Associates (Subway)	18.33	37,599
Seven & I Holdings (7-Eleven)	17.99	49,085
Burger King	16.35	12,651
Starbucks Corp	16.10	18,480
The Wendy's Co. (Arby's)	9.7	16,995
Dunkin Brands Inc.	8.64	7,621
Darden Restaurants Inc.	8.26	2,080
Domino's Pizza	7.54	10,318

Source: Euromonitor 2013[65]

food procurement has been regarded as a potent policy tool that can be deployed to support the sustainable development of local food economies. Thus public food procurement can be an instrument for environmental and sustainability objectives *and* nutritional quality and health outcomes.[66-68] Public food procurement is also portrayed as a driver for food business development, especially for local, rural or farmer-based enterprises; for example, in the UK £2bn per year spent is on public food – 7 per cent of the £26bn total UK market for food and catering services. In this manner public food procurement policies, as well as addressing sustainable development, are regarded as a means to 'reconnect' food consumers to local food producers, with the key driver to this process being supply-side factors.

Remarkable changes in the business of agriculture

With the 2007–8 food price crisis there has been a renewed focus back down the food chain to agriculture and on the future of food and farming.[4] Food security in this sense is now being defined as a sustainability problem of ensuring the environmental resources for resilient food markets and to adapt or mitigate against threats such as climate change (see Chapter 4). Agriculture once again rose up the policy agenda after years of relative complacency through the 1980s and 1990s. The developed world's concern was how to keep cheap, mass commodities fuelling the processed food industries, whereas the development agenda focused mainly on industrialisation and urbanisation and rural matters and agriculture were often seen as 'backward'.[69]

Agriculture and agribusiness is one area where battle has been truly engaged in the Food Wars, with competing visions set out by our paradigmatic models to attract minds as much as mouths and markets. The renewed interest in developing world agriculture, while in part sparked by the food price crisis from 2007, is also due to the influence of agribusiness in promoting their particular model of 'sustainable agriculture' and of policy-makers towards opening up global trade to support an export-market driven agricultural system.

Yet one of the tragic consequences of the industrial model of agribusiness is that there is less need for farmers. It is a model that sees fewer farmers producing more – driven by economic efficiencies, scaling-up operations and increasing farm concentration as more and more farms are merged to create large-scale operations. These agricultural monocultures (such as vast tracks of soybean and maize production) give rise to what has been termed the agricultural 'treadmill': this is when there is downward pressure on prices for farmers and upward pressure on the inputs needed for production; farmers are forced to adopt new technologies and increase their scale of production or go out of business; over time, the geographical concentration of production is into narrower and narrower locations or centres of production.[70] So by 1994, 50 per cent of US farm products came from 2 per cent of farms, while 9 per cent of produce came from 73 per cent of these farms; such trends can be seen across the world.

Intense concentration at every other link in the food chain – from farm inputs (such as fertilisers and pesticides) to food processing and retailing – has seen different links in the food chain vying to capture more of each consumer dollar spent on food. In the US, for example, the share of the consumer's food dollar that gets back to the farmer has dropped from around 40 cents in 1910 to 7 cents in 1997. The fall in the monies received by US farmers was graphically illustrated in March 2000 at a farm rally in Washington, DC, when farmers served legislators a 'farmers lunch'. Typically an $8 lunch – of barbecued beef on a bun, baked beans, potato salad, coleslaw, milk and a cookie – they charged only 39 cents, reflecting what farmers and ranchers actually received to grow the food for such a meal.

The economic pressures of the treadmill have led to a sharp decline in the numbers of so-called inefficient farms, with smaller family farms being particularly badly hit. For example, in the US there were close to 7 million farms in the 1930s, but less than 1.8 million by the mid-1990s; in France 3 million farms in the 1960s, yet fewer than 700,000 in the 1990s; 450,000 farms in the UK in the 1950s, half that number in the 1990s. Over the past 50 years the number of actual farmers has declined by 86 per cent in Germany, 85 per cent in France, 85 per cent in Japan, 64 per cent in the US, 59 per cent in Korea and 59 per cent in the UK.[71]

However, there is a major global battle going on over what model of agricultural development works best to 'feed the world' and will lead sustainable development. In this section we set out two strongly competing visions for the future of agriculture and the future of developing world agriculture to illustrate how our paradigmatic model is playing out on international stages. There are strong forces for industrializing agriculture further along Western-style models, but this is pitted against a growing body of research advocating for policy to be directed towards agro-ecological systems that support and engage more directly with smallholders, are more localised and use less agri-inputs. The agro-ecological approach is underpinned by a renewed interest in 'peasant'-based production, since it is peasant agriculture that continues to feed the majority of people and it is argued has the best prospects of providing more food in the future that meets the criteria of classic food security – i.e. enabling food that is accessible, affordable, is useable and can be sustained over time.

Work undertaken by the Canadian based NGO the ETC Group has documented in detail the importance of the peasant economy to food production. By their definition 'peasants' make up more than half of the world's population and continue to grow 70 per cent of the world's food as illustrated by Figure 6.5. The industrial model based on this analysis only contributes to 30 per cent of world food supply.[72] ETC Group and others point to the peasant food system as the one that needs supporting and nurturing to feed the world, with policies that are both relevant to this sector and originate from within it. This approach would conflict with an 'export-led' strategy, for example, since this peasant food system is largely defined as food grown and consumed at the very least within national borders or the same eco-regional area.

The peasant system is large-scale in terms of absolute numbers: it is estimated that there are 1.5 billion peasants working on 380 million farms. Seventeen million peasant farms in Latin America grow between half to two-thirds of staple food crops, Africa's 33 million peasant farms, as well as often being female-led, make up around 80 per cent of all farms and most of domestic food consumption.

The argument that smallholder and peasant production is vital for food security and feeding people has gained traction through the growing promotion of agro-ecology, a method that gained widespread coverage based on the agricultural techniques developed in Cuba after it lost support from the old USSR with the collapse of the Berlin Wall in 1989. Up to this time Cuba relied on heavy support from the Soviets and had an economy based on industrial agricultural models such as the production of sugar cane to supply Eastern Europe. After 1989 this support ended and Cuba was left to fend for itself and turned to small-scale agro-ecological methods to feed its population.

Agro-ecology is agriculture and food production based on diverse, biodiversity-rich, smallholder farming, often co-operatively run farm, operating as part of a local food economy.[73] Agro-ecology aims to minimise environmental impact, maximise the use of renewable energy, enhance human well-being and food justice and has been defined as the application of ecological science to the study, design and management of sustainable agro-eco-systems.[74] In development terms it is regarded as a method to tackle poverty, especially that of small farmers, and

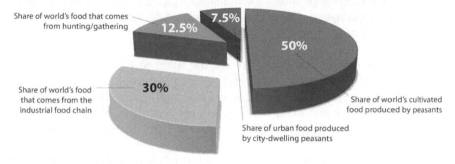

Peasants Feed at Least 70% of the World's Population

Share of world's food that comes from hunting/gathering — 12.5%

7.5%

50%

Share of world's food that comes from the industrial food chain — 30%

Share of world's cultivated food produced by peasants

Share of urban food produced by city-dwelling peasants

Figure 6.5 Who grows the food?

Source: The ETC Group 2009[72]

at the same time address environmental challenges through mimicking natural processes and creating biological synergies.

Apart from the case study of Cuba, other agro-ecological approaches have been shown to improve yields and diversity of food being produced. One study often cited is that led by Professor Jules Pretty, which studied 286 interventions of resource-conserving technologies in 57 countries. Their analysis found the average crop yield increase was 79 per cent; in 25 per cent of cases the crop yields increased 100 per cent.[75]

Agro-forestry provides another example of a sustainable alternative to industrial models. As well as increasing scientific evidence on the viability of agro-ecological systems,[76] agro-ecology has emerged as a cornerstone of new political movements confronting dominant narratives of food security via agri-industrialisation. One of the most important of these social justice movements is La Via Campesina which emerged in 1990s which has sought to reassert the role of 'peasant' farming. La Via Campesina (which translates as: 'the way of the peasant') has championed the concept of 'food sovereignty' which has grown to become a powerful 'grassroots' alternative to that of industrial agribusiness and even traditional notions of food security.[77]

Officially founded in 1993, La Via Campesina has grown from grassroots activism to being a movement which claims to embrace 200 million people across the globe under its food sovereignty umbrella. La Via Campesina movement includes small-scale producers, fisherfolk, women, youth, the landless, migrants, indigenous peoples, and farm and food workers. Food sovereignty is defined as:

> the right of peoples to define their own food and agriculture; to protect and regulate domestic agricultural production and trade in order to achieve sustainable development objectives; to determine the extent to which they want to be self-reliant; to restrict the dumping of products in their markets; and to provide local fisheries-based communities the priority in managing the

use of and the rights to aquatic resources. Food Sovereignty does not negate trade, but rather it promotes the formulation of trade policies and practices that serve the rights of peoples to food and to safe, healthy and ecologically sustainable production.[78]

The former UN Special Rapporteur on the Right to Food, Prof Olivier De Schutter, linked agro-ecology to the right to food because it has proven results as a method to help vulnerable groups and can contribute to broader economic development. From this perspective he argued that the scaling up of agro-ecological methods is the best policy route forward. To this end he recommends:

- putting in place appropriate public policies to create an enabling environment for such sustainable modes of production;
- investing in knowledge by reinvesting in agricultural research and extension services; investing in forms of social organisation that encourage partnerships, including farmer field schools and farmers' movements innovation networks;
- investing in agricultural research and extension systems;
- empowering women; and
- creating a macro-economic enabling environment, including connecting sustainable farms to fair markets.

In addition he has argued for the need to rebuild local food economies in support of developing more broadly the right to food.[79]

In summary, the particular small-scale, 'peasant' and localised approaches to the business of agriculture set out above are what might be termed as regenerative models of agricultural food business. Their core characteristics are:

- integrate biological and ecological processes (such as nutrient cycling, soil regeneration);
- minimise inputs from outside the system;
- apply principles of agro-ecology;
- make productive use of indigenous knowledge and skills of farmers;
- make use of the collective capacities of people and communities to work together;
- apply principles of food sovereignty: the right for people to define their own food, agriculture, livestock and fisheries systems.

De Schutter summed up such an agenda: '[p]ublic action is needed, not in order to "feed the world", as stated in the food security policies of the past century, but rather in order to "help the world feed itself"'.[80]

GM foods: is this the answer to sustainable agriculture?

While there is a strong emerging narrative around agro-ecology which relates strongly to the Ecologically Integrated paradigm, there is a competing agricultural business system being championed as the diffusion of biotech crops and practices. Although this sits within the Life Sciences Integrated paradigm, proponents claim it shares a lot of the same 'narrative' threads and terminology as agro-ecology. For example, in its 2014 briefing the ISAAA (International Service for the Acquisition of Agri-Biotech Applications) – which claims to be the most quoted publication on biotech crops globally – argues how biotech crops contribute to food security, sustainability and to addressing environmental and climate change mitigation. The ISAAA stresses how biotech crops support 'small resource-poor farmers in developing countries'.[81] So what is going on?

Biotechnology – in all its forms – aspires to be one of the defining sciences of the 21st century. In food systems biotechnology has been developed and applied in particular through the commercial introduction in the mid-1990s of genetically modified organisms (GMO) into agricultural production. GMO technology is proving to be one of the most controversial issues in the future of food supply and its introduction has been met with widespread resistance.

GMOs should not be confused with conventional biotechnologies, such as breeding techniques, tissue culture, cultivation practices and fermentation or the development of high-yield varieties that formed the basis of the so-called 'green revolution' of the 1960s onwards.[82] GMOs use a distinct technology which, put simply, combines the genetic material from different sources that would not be possible in nature or normal reproductive processes. GM biotechnology allows science to manipulate the natural world at its most fundamental level in that it refers to biotechnological techniques for the manipulation of genetic material and the fusion of cells beyond normal breeding barriers.[76] In other words, GMOs are organisms in which the genetic material (DNA) has been altered in a way that does not occur naturally, such as taking a gene from one species and inserting into another – an event that only happens through a process of human-led genetic engineering.

The core of the biotech revolution has been seeds and the genetic alteration of these to be resistant to the application of particular agri-chemicals such as a specific brand of herbicide, insecticide or a combination of these traits. GMOs have proven controversial because of this control over the intellectual property of seeds, the regulatory approval needed (especially in terms of their environmental or human health impacts) in global markets and finally the corporate control over the GMOs and hence market power.

GMOs have also been troubled by competing claims for their potential benefits or problems. On the positive side GMOs are promoted because they are said to

- improve crop yields;
- reduce the vulnerability of crops to environmental stresses, e.g. drought, floods, pests;
- provide environmental benefits, e.g. reduction in agri-chemical inputs;
- increase the nutritional qualities of food crops;

- improve taste, texture or appearance of food;
- contribute to livestock and aquaculture through the development of diagnostics and vaccines for infectious disease resistance, transgenes for disease resistance and feeds lower in content of nitrogen and phosphorous in waste.

Critics have argued that biotech crops themselves can cause unforeseen issues and might become weeds, i.e. plants with undesirable effects. Unintended gene flows from GMOs may assist wild relatives and other crops to become more tolerant to a range of environmental conditions threatening sustainable production. Plants engineered to express potentially toxic substances could present risks to other organisms, e.g. birds, or biotech crops might threaten centres of crop diversity (while proponents argue that GMOs can help preserve biodiversity by reducing land use for agricultural crops).[76]

A major difference between GMO proponents and those who have tried to support agro-ecological schemes is that GM is associated with powerful agribusiness corporations, the industrial model of agriculture and the policy, scientific and political interests behind large-scale farming. In terms of food business this has heralded a flurry of mergers and acquisitions to secure substantial stakes in the inputs needed to make a GMO system work – particularly the ownership of GM seeds and the chemical inputs needed to enable these to flourish. The ETC Group has documented the extent of this consolidation in seed and agrochemical companies over the years; their findings are summarised in Tables 6.7 and 6.8.

As can be seen from Tables 6.7 and 6.8, the world's top three seed corporations control over half (53 per cent) of the world's commercial seed market and six firms control 76 per cent of the global agro-chemical market (such as for herbicides and pesticides) – often the very same corporations who can benefit from, or are actively working for, the diffusion of GMOs. Also noteworthy from the ETC Group research is their analysis of other, and often overlooked, areas of corporate control, namely animal pharmaceuticals and livestock genetics. Here, again, there is significant global corporate concentration. In animal pharmaceuticals three companies account for 46 per cent of the market (and the top seven – all subsidiaries of multinational drug companies – 72 per cent of the market). In livestock genetics, four global firms are responsible for 97 per cent of poultry genetics R&D, and in pig (swine) genetics, four companies account for two-thirds of global industry R&D.[83]

From an agricultural point of view the commercialisation of GM crops – supported by the major seed and agrochemical companies – has so far has been spectacular, rising from 1.7m ha in 1996 (the start of widespread commercialisation) to more than 175m ha planted world wide by 2013. The ISAAA reports that 18 million farmers grew biotech crops in 2013, of which 90 per cent (16.5 million) were based in developing countries – the vast majority of these being in China, India and the Philippines.

More than 89 per cent of all biotech crops planted are accounted for by just five countries:[81]

Table 6.7 World's top ten seed companies in 2011

Rank	Company	Country	Seed sales, 2011, US$m	% market share
1.	Monsanto	USA	8,953	26.0
2.	DuPont Pioneer	USA	6,261	18.2
3.	Syngenta	Switzerland	3,185	9.2
4.	Vilmorin (Groupe Limagrain)	France	1,670	4.8
5.	WinField	USA	(est)1,346	3.9
6.	KWS	Germany	1,226	3.6
7.	Bayer Cropscience	Germany	1,140	3.3
8.	Dow AgroSciences	USA	1,074	3.1
9=	Sakata	Japan	548	1.6
9=	Takii & Co.	Japan	548	1.6
	TOTAL		**25,951**	**75.3**

Source: The ETC Group 2013[83]

Table 6.8 The world's top 11 agrochemical companies in 2011

Rank	Company	Country HQ	Sales, US$m	% market share
1.	Syngenta	Switzerland	10,162	23.1
2.	Bayer CropScience	Germany	7,522	17.1
3.	BASF	Germany	5,393	12.2
4.	Dow AgroScience	USA	4,241	9.6
5.	Monsanto	USA	3,240	7.4
6.	DuPont	USA	2,900	6.6
7.	Makhteshim-Agan Industries	Israel (acquired by China National Agrochemical Co., Oct. 2011)	2,691	6.1
8.	Nufarm	Australia	2,185	5.0
9.	Sumitomo Chemical	Japan	1,738	3.9
10.	Arysta LifeScience	Japan	1,504	3.4
11.	FMC Corporation	USA	1,465	3.3
	TOP 10		**41,576**	**94.5**
	TOP 11		**43,041**	**97.8**

Source: The ETC Group 2013[83]

- United States – 70.1m ha (40 per cent)
- Brazil – 40.3m ha (23 per cent)
- Argentina – 24.4m ha (14 per cent)
- India – 11m ha (6.3 per cent)
- Canada – 10.8m ha (6.2 per cent)

The major crops grown are maize, soybean and cotton (all of India's biotech crop is Bt cotton) and the most important GM trait is herbicide tolerance (i.e. a crop like soy can be sprayed with a branded weed-killer like Monsanto's Round-Up and they themselves will not be affected). In 2013, the global market value of biotech seeds was estimated at US$15.6bn, around 35 per cent of the US$45bn global commercial seed market.

Many in the biotechnology industry and policy world, while acknowledging that part of the public concern about GM crops is that they are presented as mainly delivering benefits to farmers only (agronomic traits), express great faith that GM crops will be a major part of the sustainable intensification agenda that is needed to produce more from less and contribute to global food security.

Food business responses to nutrition and sustainability challenges

The problems and issues we highlight in *Food Wars* are so widespread, so serious and on such a large-scale, no sensible business leader could ignore them or the potential risks they pose to their business interests. Often, however, and especially in the case of nutrition, the response has been highly defensive, based on attacking or trying to undermine evidence and individuals they disagree with.[84]

The food industry, like other industry sectors, employs lobbyists to promote its agendas as individual corporations or as a collective industry effort through trade associations and think-tanks. Tracking such activities is not always easy, but high-profile examples in 2013–14 that have been documented include the soft drinks industry in the United States lobbying against soda taxes and different corporations collectively funding campaigns to stop the labelling of GMOs in different US states.[85 86]

Within the EU, the Corporate Europe Observatory (CEO) claimed in 2010 that the Confederation of the Food and Drink Industries of the EU (CIAA – renamed FoodDrinkEurope) spent €1bn opposing European proposals for front-of-pack 'traffic light' labels. The 'traffic light' system has a green symbol that indicates that a product contains, for example, low levels of fat, sugar or calories (thus a potential 'healthy' choice) and a red symbol when products have high levels of sugar, fat or salt, for example (thus, possibly negative health attributes). The food manufacturing industry, and some retailers, mostly preferred a system based on guideline daily amounts (GDAs), which shows, for example, how many calories a 'portion' contains as a percentage of an adult's daily needs. However, the CIAA criticised CEO's claims that the food industry spent as much as €1bn and emphasised that the CIAA was not a lobbying organisation.[87]

Australian academic Gary Sacks has highlighted a number of tactics the food industry uses to lobby on issues it tries to influence or to set agendas.[88] These include:

- funding research to confuse the evidence;
- setting up front groups to lobby on their behalf;
- promising self-regulation to prevent legislation.

One of the most damning indictments of the way business tries to subvert nutrition and health promotion came from the Director-General of the WHO, Dr Margaret Chan, in her opening address on 10 June 2013 at the 8th Global Conference on Health Promotion held in Helsinki, Finland. It is worth quoting her at length as she sums up the problem as seen from a prominent public health nutrition perspective:[89]

Efforts to prevent non-communicable diseases go against the business interests of powerful economic operators. In my view, this is one of the biggest challenges facing health promotion … it is not just Big Tobacco anymore. Public health must also contend with Big Food, Big Soda, and Big Alcohol. All of these industries fear regulation, and protect themselves by using the same tactics. Research has documented these tactics well. They include front groups, lobbies, promises of self-regulation, lawsuits, and industry-funded research that confuses the evidence and keeps the public in doubt. Tactics also include gifts, grants, and contributions to worthy causes that cast these industries as respectable corporate citizens in the eyes of politicians and the public. They include arguments that place the responsibility for harm to health on individuals, and portray government actions as interference in personal liberties and free choice. This is formidable opposition. Market power readily translates into political power. Few governments prioritise health over big business.

The point we are making here is that the Food Wars confront many vested interests, but that often the food industry comes across as inconsistent and almost a Jekyll and Hyde character – for example, as well as often resisting nutrition science as eloquently put by Dr Chan, the food industry has responded to health and nutrition concerns through highly innovative new product development and created multi-million-dollar market opportunities for 'healthy eating' products.[84]

With respect to environmental and sustainability concerns we would argue the food industry owes a debt of thanks to environmental activists and scientists and the green movement. These have alerted the food industry to its huge inefficiencies with respect to its resource use such as water, energy, waste, and many food and drink companies have publicly reported on the substantial business cost savings they have made in recent years through reducing resource use or employing greater environmental efficiencies. In the following sections we give examples from what are now wide-ranging initiatives from food and drink companies addressing both human and environmental health problems.

It should be recognised this is only a 'taster' of the range of food industry activities. In many instances it represents a substantive change in corporate strategies and how corporations engage with both nutrition and sustainability issues. But there is a lack of data, evaluation and evidence on how effective across specific food systems some of these activities have been in bringing about real food system transformation. But at least anecdotally there are significant levels of activity and engagement, especially with respect to sustainability issues. In this respect there are new technologies and innovations gaining traction, particularly in relation to the environmental challenges facing agriculture. These include:

- precision agriculture (water droplet control);
- robots (from farm to distribution centres);
- information technology (satellites to logistics);
- consumer technology: use of phones and apps;
- farm-based internet applications.

From corporate social responsibility to sustainability reporting

Since the 2000s there has been an explosion in business responses to sustainability challenges by all actors throughout food supply chain from agribusiness to food retailers. Much of this activity originated under the broad umbrella of Corporate Social Responsibility (CSR) and the subsequent strategic implementation of CSR initiatives. CSR can be defined as referring to: 'business decision-making linked to ethical values, compliance with legal requirements, and respect for people, communities and the environment'.[90]

The CSR debate, obviously not confined to the food industry alone, is complicated. Some business critics of CSR argue it is not a meaningful or workable business concept, while civil society critics say it is used as 'greenwash' – little more than a public relations exercise to achieve societal legitimacy for corporate activities. The food industry, like many other industry sectors, is still developing and evolving the implementation of CSR policies, which in some instances are now morphing into 'Corporate Sustainability'. For the food industry, CSR reporting has embraced six broad categories, these are:

1   Working conditions/employment
2   Community and philanthropy
3   Environmental issues
4   Supply chain relations
5   Financial and legal probity
6   Nutrition, health and wellness.

One striking example of this process at work is the 'sustainability' positioning of Walmart, the world's largest retailer. Starting in October 2005, Walmart's then

CEO Lee Scott painted a sweeping – and perhaps to suppliers frightening – future vision, saying: 'What if we used our size and resources to make this country [the United States] and this earth an even better place for us: customers, associates, our children, and generations unborn.'[91] This was the start of a number of far reaching sustainability initiatives from Walmart. This cumulated in October 2014 when, at a Global Sustainability Milestone Meeting, Walmart announced its latest commitment to create a more sustainable food system. At the launch Walmart said this would be framed around what is called 'four key pillars'. These are: improving the affordability of food for both customers and the environment, increasing access to food, making healthier eating easier, and improving the safety and transparency of the food chain.[92]

In a press release announcing the initiative, Walmart wrote:

> The four pillars outlined in Walmart's sustainable food system commitment aim to address major issues and threats facing today's global food system, including how to address the food needs of a growing population while reducing environmental impact; meeting an increasing consumer demand for greater food transparency; providing more options for healthier eating; and alleviating global hunger.[92]

Up until relatively recently such far-reaching and frankly, radical, statements from a major food industry would be unthinkable or very rare – now they are becoming commonplace.

Individual food and beverage companies have varied greatly in their approach to CSR or even gone beyond common approaches to CSR. A prominent example of the latter is from the world's largest food company Nestlé, which has developed its own CSR concept of 'creating shared value'– this is a concept which Nestlé pioneers using metrics relating to its business activities showing its impact in areas where it engages with society or the communities it works in as part of its business activities. The company has committed itself to communicate transparently with not only its shareholders, but its wider stakeholders, on progress in all areas where it engages with society. Nestlé positions this as offering products and services that help people improve their nutrition, health and wellness, but also on sustainability issues such as rural development and water. In its 2013 *Nestlé in Society* report the company writes: 'We continue to actively manage our commitments to environmental, social and economic sustainability needed for operating our factories and for the sustainable growth and development of the communities and countries where we have operations.'[93]

Another, and possibly one of the most significant, single company initiative has been that by Unilever which in 2010 published its *Sustainable Living* report.[94] This was committed to '(1) halve the environmental footprint of our products; (2) help more than 1 billion people take action to improve their health and well-being; (3) source 100 per cent of our agricultural raw materials sustainably'.[94] Unilever has launched a Sustainable Agriculture Programme that is working with suppliers and farmers to develop indicators to create a Sustainable Agricultural Code. The company has announced that by 2020 it aims to source 100 per cent of agricultural raw materials sustainably (up from 50 per cent by 2015). To do this Unilever will focus first on their top ten agricultural raw material groups (around 70 per cent of the company's raw material volumes). In 2013, Unilever's Chief Procurement Officer was reported as saying in a company press release:[95]

> Market transformation can only happen if everyone involved takes responsibility and is held accountable for driving a sustainability agenda. Our progress has been made possible by the commitment and efforts of a number of our strategic suppliers. We will continue to engage with our suppliers, NGOs, governments, RSPO [Roundtable on Sustainable Palm Oil], end users and other industry stakeholders to develop collaborative solutions to halt deforestation, protect peat land, and to drive positive economic and social impact for people and local communities.

Business organisation responses

Business organisations or groups have also developed a number of sustainability programmes. For example, in 2002, the Sustainable Agriculture Initiative (SAI) was launched by a group of transnationals, including Groupe Danone, Nestlé and Unilever.[96] In 2007, the EU Corporate Leaders Group on Climate Change convened by the Prince of Wales began to speak out about the need to tackle climate change in consumption.[97] It featured many top food companies including Barilla, Coca-Cola, Tesco and Unilever. In 2008, the World Business Council for Sustainable Development (WBCSD), launched a report on *Sustainable Consumption Facts and Trends*, followed by its *Vision 2050* document (2010) and *Vision for Sustainable Consumption* in 2011.[98] This quick sequence by the WBCSD indicates coherent policy development. It stressed the role of 'choice-editing', and the need to move from linear to circular economies.

In 2010 the World Economic Forum took on McKinsey to produce its *Roadmap* for agriculture, looking at how agriculture must become more sustainable if the commodities which companies rely upon to manufacture food products are to continue securely or not be a risk of shocks to supply chains.[99] This was followed by *More with Less: Scaling Sustainable Consumption with Resource Efficiency* in 2012, produced with Accenture.[100] This document went further than the EU is prepared to go and addressed how there needs to be less but better consumption, a policy point with radical implications for food.

Also in 2010, the Barilla Center for Food and Nutrition, a think-tank funded independently by the Barilla Foundation – Italian-based Barilla is the largest pasta company in the world – launched its double inverted pyramid nutrition and environment guidelines (see Figure 6.6).[101] This did what governments had fought shy of doing; it put the sustainable diet case simply. The pyramid on the left (in Figure 6.6) shows fairly standard nutritional advice to eat a diet high in those foods at the bottom and low in those at the top, while the pyramid on the right suggests those foods at the top are high in environmental impact and thus to be minimised compared to those at the bottom which have low impact.

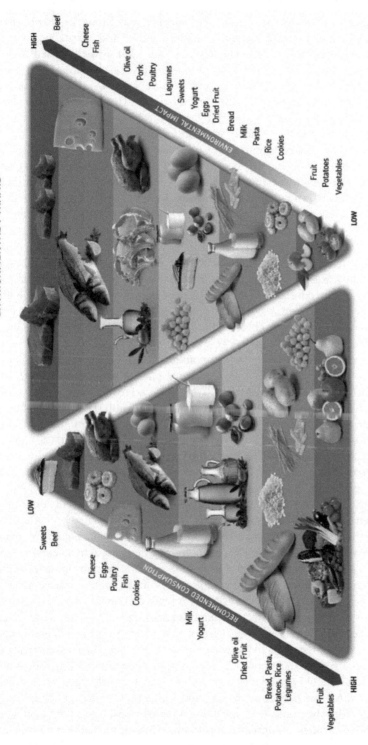

ENVIRONMENTAL PYRAMID

Beef
Cheese
Fish
Olive oil
Pork
Poultry
Legumes
Sweets
Yogurt
Eggs
Dried Fruit
Bread
Milk
Pasta
Rice
Cookies
Fruit
Potatoes
Vegetables

HIGH

ENVIRONMENTAL IMPACT

LOW

FOOD PYRAMID

LOW

RECOMMENDED CONSUMPTION

HIGH

Sweets
Beef
Cheese
Eggs
Poultry
Fish
Cookies
Milk
Yogurt
Olive oil
Dried Fruit
Bread, Pasta,
Potatoes, Rice
Legumes
Fruit
Vegetables

Figure 6.6 The Barilla double pyramid for health and environment

Source: Barilla Centre for Food and Nutrition Foundation, 2014 [102]

In November 2013 the Barilla Center launched the Milan Protocol which aims to create a new dialogue between academia, public and private institutions, companies and experts who work in the food and nutrition sector.[102] The Protocol has three main objectives: to define and reduce food waste; to explain and foster sustainable agriculture; and to support and encourage healthy lifestyles, and to halt hunger. The Protocol was launched at the 2015 Milan EXPO which had the theme of *Feeding the Planet, Energy for Life,* with the goal of getting countries and leaders to sign up to the Protocol while attending the EXPO 2015.

There have been many country-level projects. For example, in the UK, the corporate sector has been equally active. In 2007 the IGD, the UK food industry research body, launched the Food Industry Sustainability Strategy Champions Group with a focus on low carbon and ethics. In 2008 Tesco, the world's third largest food retailer, gave the largest donation yet given to social science, granting £25m to Manchester University to set up the Sustainable Consumption Institute to do long-term research. Being customer-focused, food retailers are aware of the coming crunch point where consumer behaviour change is required. The earliest to engage, arguably, was the Co-operative Group whose 1996 *Responsible Retailing* commitment signalled a renewal of some of the original 'old' co-operative belief in a values-led food system.[103] In 2009 Marks & Spencer took this further with *Plan A,*[104] an overt strategy to choice-edit to meet 100 commitments. In five years this saved M&S money and carbon.[105]

At a regional level the same sort of momentum can be found. For example, a policy paper, *Actions towards a More Sustainable European Food Chain,* was published in April 2014 by representatives from across Europe's food chain together with the NGOs which set out to encourage EU policy-makers to support a more cohesive approach to safeguarding the sustainability of food systems. [106] The document included 32 policy recommendations which it was argued could help achieve a more sustainable food chain by 2020, including improving the coherence among different food-related policy objectives and among EU stakeholder platforms, taking into account all aspects of sustainability, ranging from EU agriculture and fisheries, environmental policies, health and consumers, to waste management and energy policies.

What all these types of food industry activities suggest – and the many more not mentioned here – is that there is now a recognition that there needs to be commitment and focus to address global food problems of sustainability, nutrition and health. What at this stage is less clear is how collectively these might coalesce so as to provide the scale of food system 'redesign' or transformation urged by reports such as Foresight's with which this chapter began. Tensions remain over how extensively or fast change needs to happen – not least over workers' rights and wages. Many food workers do not receive a living wage for the work they do. There remains an issue of trust – even the most powerful companies cannot alone address the issues raised here. Why should people trust companies to resolve such urgent and difficult problems or to believe they have the solutions? Food and drink corporations are caught in a bind – they operate in hugely competitive business environments, so need to survive in often cut-throat competition, while at the same time they are increasingly aware of looming sustainability crises and a bigger picture that demands collaboration or cooperation. To accept new frameworks for long-term sustainability, however, almost certainly means actions that go against short-term financial interests of shareholders.

References

1 Elliott C. Elliott Review into the Integrity and Assurance of Food Supply Networks – interim report. <https://www.gov.uk/government/policy-advisory-groups/review-into-the-integrity-and-assurance-of-food-supply-networks> London: HM Government, 2013.
2 FAO. *The State of World Fisheries and Aquaculture*. Rome: Food and Agriculture Organisation, 2014.
3 Ionescu-Somers A, Steger U. *Business Logic for Sustainability: A Food and Beverage Industry Perspective*. New York: Palgrave Macmillan, 2008.
4 Foresight. *The Future of Food and Farming: Challenges and Choices for Global Sustainability: Final Report*. London: Government Office for Science, 2011: 211.
5 IFPRI. *Global Nutrition Report 2014: Actions and Accountability to Accelerate the World's Progress on Nutrition*. Washington, DC: IFPRI, 2014.
6 WHO. Global strategy on diet, physical activity and health. 57th World Health Assembly. WHA 57.17, agenda item 12.6. Geneva: World Health Assembly, 2004.
7 Nesheim MC, Oria M, Tsai Yih P, eds. *A Framework for Assessing Effects of the Food System. Report by the Committee on a Framework for Assessing the Health, Environmental, and Social Effects of the Food System; Food and Nutrition Board; Board on Agriculture and Natural Resources; Institute of Medicine; National Research Council*. Washington, DC: National Academies Press, 2015.
8 Fine B, Heasman M, Wright J. *Consumption in the Age of Affluence: The World of Food*. London: Routledge, 1996.
9 Goodman D, Redclift M. *Refashioning Nature: Food, Ecology and Culture*. London: Routledge, 1991.
10 Hansen HO. *Food Economics: Industry and Markets*. Abindgon: Routledge, 2013.
11 OECD. *Food Policy*. Paris: Organisation for Economic Cooperation and Development, 1981.
12 Brundtland GH. *Our Common Future: Report of the World Commission on Environment and Development (WCED) chaired by Gro Harlem Brundtland*. Oxford: Oxford University Press, 1987.
13 United Nations Conference on Environment and Development. *Rio Declaration*. Rio de Janeiro/Geneva: United Nations, 1992.
14 OECD, FAO. *OECD-FAO Agricultural Outlook 2014–2023*. Paris/Rome: Organisation for Economic Cooperation and Development, and Food and Agriculture Organisation, 2014.
15 FAO. Food Price Index. <http://www.fao.org/worldfoodsituation/foodpricesindex/en> [accessed July 2014]. Rome: Food and Agriculture Organisation, 2014.
16 OECD, FAO. *Agricultural Outlook 2008–2017*. Rome and Paris: Organisation for Economic Cooperation and Development and Food and Agriculture Organisation, 2008.
17 De Schutter O. *Final Report: The Transformative Potential of the Right to Food*. Report of the Special Rapporteur on the right to food, Olivier De Schutter. report to Human Rights Council Twenty-fifth session, Agenda item 3. Geneva: Human Rights Council, 2014.

18   Oxfam International. *Not a Game: Speculation vs Food Security: Regulating Financial Markets to Grow a Better Future*. Oxford: Oxfam International, 2011.

19   World Development Movement. Food speculation: stop bankers betting on food – our campaign to curb commodity speculation. <http://www.wdm.org.uk/food-speculation> London: World Development Movement, 2012.

20   Ajanovic A. Biofuels versus food production: Does biofuels production increase food prices? *Energy*, 36(4) (2011): 2070–6.

21   OECD, FAO. *Agricultural Outlook 2013–22*. Paris: Organisation for Economic Cooperation and Development, and Food and Agriculture Organisation, 2013.

22   Lang T. Reshaping the food system for ecological public health. *Journal of Hunger and Environmental Nutrition*, 4(3/4) (2009): 315–35.

23   Tansey G, Worsley T. *The Food System: A Guide*. London: Earthscan, 1995.

24   Atkins PJ, Bowler IR. *Food in Society: Economy, Culture, Geography*. London: Arnold, 2001.

25   Goodman D, Watts M. Reconfiguring the rural or fording the divide? *Journal of Peasant Studies*, 22(1) (1944): 1–49.

26   Goodman D, Sorj B, Wilkinson J. *From Farming to Biotechnology*. Oxford: Blackwell, 1987.

27   Jones A. Eating Oil. Secondary Eating Oil 2002. London: Sustain.

28   Seth A, Randall G. *The Grocers: The Rise and Rise of the Supermarket Chains*. 2nd edn. London: Kogan Page, 2001.

29   Gabriel Y. *Working Lives in Catering*. London: Routledge & Kegan Paul, 1988.

30   Thrupp L-A. *Bittersweet Harvests for Global Supermarkets*. Washington, DC: World Resources Institute, 1995.

31   Friedland WH, Barton AE, Thomas RJ. *Manufacturing Green Gold: Capital, Labour and Technology in the Lettuce Industry*. Cambridge: Cambridge University Press, 1981.

32   Sims LS. *The Politics of Fat: Food and Nutrition Policy in America*. New York: ME Sharpe Inc., 1998.

33   Halliwell J. The race to bring the best NPD to market. *The Grocer*. Crawley: William Reed, 2014: 28–31.

34   Monteiro C. The big issue is ultra-processing. *World Nutrition*, 1(6) (2010): 237–69.

35   Monteiro CA, Levy RB, Claro RM, *et al*. Increasing consumption of ultra-processed foods and likely impact on human health: evidence from Brazil. *Public Health Nutrition*, 14(1) (2011): 5–13.

36   Goodman D, DuPuis EM, Goodman MK. *Alternative Food Networks: Knowledge, Place and Politics*. London: Routledge, 2012.

37   Oosterveer P, Sonnenfeld DA. *Food, Globalization and Sustainability*. Abingdon: Earthscan Routledge, 2012.

38   Friedland W, Busch L, Buttel F, *et al*., eds. *Towards a New Political Economy of Agriculture*. Boulder, CO: Westview, 1991.

39   Pritchard W, Burch D. *Agri-Food Globalisation in Perspective: Restructuring in the Global Tomato Processing Industry*. Aldershot: Ashgate Publishing, 2003.

40   Buttel F. Some reflections on late 20th century agrarian economy. *Sociologia Ruralis*, 41(2) (2001): 165–81.

41   Alexandratos N. The Mediterranean diet in a world context. *Public Health Nutrition*, 9(1A) (2006): 111–17.

42 da Silva R, Bach-Faig A, Raidó Quintana B, *et al*. Worldwide variation of adherence to the Mediterranean diet, in 1961–1965 and 2000–2003. *Public Health Nutrition*, 12(9A) (2009): 1676–84.

43 Drewnoski A, Eichelsdoerfer P. The Mediterranean diet: does it have to cost more? *Public Health Nutrition*, 12(9A) (2009): 1621–8.

44 Scheidel A, Krausmann F. Diet, trade and land use: a socio-ecological analysis of the transformation of the olive oil system. *Land Use Policy*, 28 (2011): 47–56.

45 Gereffi G, Fernandez-Stark K. *Global Value Chain Analysis: A Primer*. Durham N Carolina: Center on Globalization, Governance and Competitiveness (CGGC) Duke University, 2011.

46 Wilkinson J. Fish: a global value chain driven on the rocks. *Sociologia Ruralis*, 46(2) (2006): 139–53.

47 Goodman D. The quality 'turn' and alternative food practices: reflections and agenda. *Journal of Rural Studies*, 19(1) (2003): 1–7.

48 Ilbery B, Maye D. Retailing local food in the Scottish-English borders: a supply chain perspective. *Geoforum*, 37 (2005): 352–67.

49 Holloway L, Kneafsey M, Venn L, *et al*. Possible food economies: a methodological framework for exploring food production–consumption relationships. *Sociologia Ruralis*, 47(1) (2007): 1–19.

50 Venn L. Researching European alternative food networks: some methodological considerations. *Area*, 38(3) (2006): 248–59.

51 Hinrichs CC. The practice and politics of food system localisation. *Journal of Rural Studies*, 19(1) (2003): 33–45.

52 Devlin S, Dosch T, Esteban A, *et al*. *Urgent Recall: Our Food System Under Review*. London: New Economics Foundation, 2014.

53 McMahon P. *Feeding Frenzy*. London: Profile Books, 2014.

54 USDA ERS. Global Food Industry. <http://w w w.ers.usda gov/topics/international-markets-trade/global-food-markets.aspx> [accessed July 2014]. Washington, DC: US Department of Agriculture Economic Research Services, 2014.

55 FoodDrinkEurope. *Data and Trends of the European Food and Drink Industry*. Brussels: FoodDrinkEurope, 2013.

56 DEFRA. *Agriculture in the United Kingdom 2012*. London: Department for Environment, Food and Rural Affairs, 2013.

57 DEFRA. *Food Statistics Pocketbook 2013 (April 2014 Update)*. London: Department for Environment, Food and Rural Affairs, 2014.

58 DEFRA. *Agriculture in the UK 2013 (29 May Update)*. London: Department for Enviornment, Food and Rural Affairs, 2014.

59 Rowan C. The world's top 100 food and beverage companies. *Food Engineering*. <wwwfoodengineeringmagcom> [accessed July 2014] 68 (2013).

60 Banana Link. World's top banana companies. <http://www.bananalink.org.uk/who-earns-what-from-field-to-supermarket> [accessed July 2014]. Norwich: Banana Link, 2014.

61 Financial Times. Supermarket pricing made merger of Chiquita and Fyffes necessary. *Financial Times*. 10 Mar. 2014.

62 Consumers International. *The Relationship between Supermarkets and Suppliers*. London: Consumers International, 2012.

63 Competition Commission. *The Supply of Groceries in the UK Market Investigation: Final Report*. London: Competition Commission, 2008.

64 Commission of the European Communities. *A Better Functioning Food Supply Chain in Europe*. Communication COM(2009) 591. Brussels: European Commission, 2009.

65 Euromonitor. Consumer Foodservice: Market Reports. <http://www.euromonitor.com> London: Euromonitor, 2013.

66 Kneafsey M, Holloway L, Cox R, *et al. Reconnecting Consumers, Producers and Food: Exploring Alternatives*. Oxford: Berg, 2008.

67 Morgan K, Morley A. *Sustainable Public Procurement: From Good Intentions to Good Practice*. Report to the Welsh Local Government Association, Cardiff. Cardiff: University of Cardiff, BRASS, 2006.

68 Morgan K, Morley A. *Relocalising the Food Chain: The Role of Creative Public Procurement*. Cardiff: Cardiff University Regeneration Institute, 2002.

69 Bello WF. *The Food Wars*. London: Verso, 2009.

70 Buttel FH, Magdoff F, Foster JB, eds. *Hungry for Profit: The Agribusiness Threat to Farmers, Food, and the Environment*. New York: Monthly Review Press, 2000.

71 Halweil B. Where have all the farmers gone? *World-Watch*, 13(5) (2000): 12–28.

72 ETC. *Who Will Feed us? Questions for the Food and Climate Crisis*. Ottawa: ETC Group, 2009: 31.

73 Silici L. *Agroecology: What it is and What it Has to Offer*. IIED Issue Paper. London: IIED, 2014.

74 Hotl-Giminez E, Altieri M. Agroecology, food sovereignty and the green revolution. *Agroecology and Sustainable Food Systems*, 37(1) (2013): 90–102.

75 Pretty JN, Noble AD, Bossio D, *et al*. Resource-conserving agriculture increases yields in developing countries. *Environmental Science and Technology*, 40(4) (2006): 1114–19.

76 IAASTD. *Global Report and Synthesis Report*. London: International Assessment of Agricultural Science and Technology Development Knowledge, 2008.

77 Rosset P. Food Sovereignty and alternative paradigms to confront land-grabbing and the food and climate crisis. *Development*, 54(1) (2011): 21–30.

78 La Via Campesina. About us. <www.viacampesina.org> 2014.

79 De Schutter O. *Agroecology and the Right to Food: Report to the Human Rights Council*. Geneva: Office of the UN Rapporteur on the Right to Food, 2010.

80 De Schutter O, Vanloqueren G. The new green revolution: how twenty-first-century science can feed the world. *Solutions Journal*, 2(4) (2011): 33–44.

81 James C. *Global Status of Commercialised Biotech/GM Crops: 2013*. ISAAA Brief 46. Ithaca, NY: ISAAA, 2013.

82 Conway G. *The Doubly Green Revolution: Food for All in the 21st Century*. London: Penguin, 1997.

83 ETC. *Putting the Cart(el) Before the Horse: Who Will Control Agricultural Inputs?* Ottawa: ETC Group, 2013.

84 Heasman M, Mellentin J. *The Functional Foods Revolution: Healthy People, Healthy Profits?* London: Earthscan, 2001.

85 Ruskin G. *Seedy Business: What Big Food is Hiding with its Slick PR Campaigns on GMOS*. Oakland, CA: USRTK, 2015.

86 Simon M. *Appetite for Profit*. New York: Nation Books, 2006.

87 CEO. A red light for consumer information. <http://corporateeurope.org/news/red-light-consumer-information> [accessed Jan. 2015]. Brussels: Corporate Europe Observatory, 2010.

88 Sacks G. Big Food lobbying: tip of the iceberg exposed. The Conversation 2014. <http://theconversation.com/big-food-lobbying-tip-of-the-iceberg-exposed-23232> [accessed Jan. 2015].

89   Chan M. Opening Address by the WHO Director-General to 8th Global Conference on Health Promotion, Helsinki. <http://www.who.int/dg/speeches/2013/health_promotion_20130610/en> [accessed Jan. 2015]. Geneva: World Health Organisation, 2013.

90   Zadek S. *The Civil Corporation: The New Economy of Corporate Citizenship*. London: Earthscan, 2001.

91   Heasman M. Toward an ethical system. *Harvard International Review*, 2006. <http://hir.harvard.edu/archives/1456>

92   Walmart. Walmart announces new commitment to a sustainable food system at global milestone meeting. 2014. <http://news.walmart.com/news-archive/2014/10/06/walmart-announces-new-commitment-to-a-sustainable-food-system-at-global-milestone-meeting> [accessed Jan. 2015].

93   Nestlé. Nestlé in society. 2014. <http://www.nestle.com/asset-library/documents/library/documents/corporate_social_responsibility/nestle-csv-summary-report-2013-en.pdf> [accessed Jan. 2015].

94   Unilever. Sustainable living plan 2010. <http://www.sustainable-living.unilever.com/the-plan> [accessed Dec. 2010]. London: Unilever plc, 2010.

95   Unilever. 100 percent of palm oil bought will be traceable to known source by end 2014. Press release [accessed Jan. 2015]. London: Unilever plc, 12 Nov. 2013.

96   SAI. Sustainable Agriculture Initiative platform. <http://www.saiplatform.org> Brussels: Sustainable Agriculture Initiative, 2014.

97   University of Cambridge. The Prince of Wales's EU Corporate Leaders Group. <http://www.cisl.cam.ac.uk/Business-Platforms/The-Prince-of-Wales-Corporate-Leaders-Group/EU-CLG.aspx> Cambridge: Institute of Sustainability Leadership, University of Cambridge, 2014.

98   World Business Council for Susainable Development. A vision for sustainable consumption. <http://www.wbcsdpublications.org/cd_files/datas/capacity building/consumption/pdf/AVisionForSustainableConsumption.pdf> [accessed July 2014]. World Business Council for Sustainable Development. Conches-Geneva, 2011.

99   World Economic Forum, McKinsey & Co. *Realizing a New Vision for Agriculture: A Roadmap for Stakeholders*. Davos: World Economic Forum, 2010.

100 World Economic Forum, Accenture. *More with Less: Scaling Sustainable Consumption with Resource Efficiency*. Geneva: World Economic Forum, 2012.

101 Barilla Center for Food & Nutrition Foundation. *Double Pyramid: Health Food for People, Sustainable Food for the Planet*. Parma: Barilla Centre for Food & Nutrition Foundation, 2014.

102 Barilla Center for Food and Nutrition. The Milan Protocol. <http://www.milanprotocol.com> Parma: Barilla Center for Food and Nutrition, 2014.

103 Co-operative Wholesale Society. *Responsible Retailing*. Manchester: Co-operative Retail Group, 1996.

104 Marks & Spencer plc. About Plan A: Plan A is our five year, 100 point plan. <http://plana.marksandspencer.com/about> [accessed May 2012]. London: Marks & Spencer plc, 2009.

105 Marks & Spencer. *Plan A Report 2014*. London: Marks & Spencer plc, 2014.

106 FDE. Press Release: Europe's food chain partners working towards more sustainable food systems. <http://pr.euractiv.com/node/106496> [accessed Jan. 2015]. Brussels: FDE, 2014.

7 The consumer culture war

We are not consumers. For most of humanity's existence, we were makers, not consumers: we made our clothes, shelter, and education, we hunted and gathered our food.

Matthew Fox [1]

Core arguments

In an ideological world where the consumer is ostensibly sovereign, the notion of a 'food culture' is a robust way of understanding people's food beliefs and behaviour. The concept of food culture is central to the food wars. Business spends huge sums of money trying to mould and respond to consumer aspirations in a highly competitive way. At the same, food culture is inherited from the past; it is not solely dictated by present commerce. There are also counter-trends and contradictory meanings, behaviours and beliefs. Governments are also involved in shaping food cultures, particularly around health or what they allow supply chains to do. They might appeal to consumers to behave sensibly while actually spending little money on helping them to do so. The 'consciousness' industries compete with public health education over the shape of food culture and food policies. Food culture is not homogeneous, despite globalisation. Part of the battle for minds is about who and what shapes choice.

The battle for mouths and minds

Food and beverage companies today all want to describe themselves and their activities as 'consumer-led', and the Food Wars are played out in the arena of public consumption and the imagery and media that help shape it. The battles here are not just for the mouths of consumers, but increasingly for their minds. This is happening at a global scale, as food brands seek to engage with developing countries and relatively new mass consumer markets such as in Asia, Africa and Latin America. Branded food and beverage companies, in particular, seek to influence consumers so they form an emotional bond with products or even companies.

A struggle is underway between food processors selling their branded dreams, caterers luring consumers to their food offerings often through strategies that owe more to the entertainment industries than to food (a strategy of 'eatertainment' which ranges from product placement in TV soaps and films to web-based games), and supermarkets trying to outdo both. Many ironies have emerged, such as when brand managers present their products with smiling faces of 'real' farmers and growers, or when supermarkets promote their sustainability credentials, when too often the reality is that they are food warehouses with giant car parks. There is also a class food war with companies and marketers appealing to or developing products aimed at different population segments.

This chapter maps out the evolving consumer landscape and how that has become a key battleground. This is highly contested terrain. Food culture is never static, but the 20th century witnessed unprecedented creation of methods by which powerful food supply chain forces could intervene in consumer understanding of food. At the same time, policy-makers became aware of the implications of consumers as powerful forces in the food system. Consumer choice has become a totem pole around which many dance. The reality, however, is that consumer choice means different things to different sectors. An old philosophical – and marketing – distinction remains important: the difference between needs and wants.

By 'food culture' we mean the totality of food beliefs and behaviours. These are socially framed. The term refers to a constellation of socially produced values, attitudes, relationships, tastes, cuisines and practices as exhibited through food. They may and do vary internationally.[2] As one definition put it, food culture is the 'shared practices and meanings relating to food'.[3] To explore food culture requires us to acknowledge the meanings that individual people and population groups attach to food.[4] When people sit down to a meal together, every action they take – sitting on chairs, using plates and cutlery, the order of food, the type of food they choose, where and what they eat and how much – is an indicator of food culture.[5] Food culture comprises both social 'cement' – binding groups together with shared assumptions – and opportunities for difference and distinction.[6 7] People express their identities and classes through food and derive cultural meanings from it.[8]

Consumers can hold opposing food beliefs: they may welcome some innovations but resist others; they say that they are concerned about health, but do not always want to change diet-related behaviour in a healthy direction; they judge food by its price but also by its quality. Consumer food culture is never simple, being multi-dimensional and constantly changing. Food patterns, tastes and beliefs should be assumed to be in a state of flux.[9] Existing and older food cultures are semi-permanently under assault, partly driven by willing change and partly by changes stemming from the restructuring of the food economy and its major players, and processes such as globalisation and the internet. This has opened up mass-market shopping online and the globalisation of brands.[10] Yet despite this, market research suggests consumers still have local and regional characteristics to their tastes. Values are important.[11] So are identities. Modern consumerism is riven by inequalities and fragmentation.[12] The language of the consumer in markets must be taken with a pinch of salt. The studies by academics of consumers and consumerism suggest the importance of inequalities. Not all consumers are created equal. They do not behave the same. In this sense, when politicians talk of 'consumer power', this can be more rhetorical than real. Indeed, consumer organisations – who champion consumer interests – are wary of the myth of consumer sovereignty.[13]

More than any other factor including science, we see the current food culture as driving the demise of the Productionist paradigm and shaping the future success or otherwise of the Life Sciences Integrated and the Ecologically Integrated paradigms. Food culture is in tension with business philosophies and state policy, as well as with the ways people have traditionally purchased, prepared and consumed. Our thesis here is that culture is a key to understanding the dynamics between the supply chain and issues such as health or sustainability; but culture is not a consistent entity in its influx.[14 15]

Food and health: a done deal for the consumer?

According to some market research, a majority of consumers see the connections between food and health, and in many cases make choices based on their understanding of diet and health. But it should be noted this is not a recent phenomenon. For example, in the 1990s, a large study for the European Union of 14,331 people in 15 countries found that, while overall, respondents saw quality, price and taste as the main factors affecting their food choice, health was close behind.[16] That survey found that the majority of EU citizens were happy with their diets and saw no need to change them. So why is there such an ongoing struggle to persuade populations to eat more healthily and why the endless efforts of health campaigns and interventions? Our view is that, in part, this current tension over behaviour change is to do with pro-health forces not understanding food cultures and how these are shaped and moulded.

That is not to say consumer attitudes don't change, they do. Studies of EU-wide attitudes to food undertaken in the 2000s portrayed Europeans as developing more complex values. Food security had risen up their agenda: 71 per cent were concerned. But the highest supported value was for quality (96 per cent supporting) and price (91 per cent), while a large majority (71 per cent) thought that the origin of their food was also important, although there were big differences within the EU on this. For example 94 per cent of respondents in Greece were concerned about national food production whereas only 11 per cent of people surveyed in Denmark or the Netherlands were. This variation reflected the economic realities of the countries. Greece was in a dire economic state, the northern EU member states much less so.[17]

Consumer wants and needs

The relationship between the consumer and food can be fraught. Affluent societies are now understandably concerned about developing obsessions with body images.[18 19] Media messages extol the beauty of waif-like fashion models on the one hand, and the dangers of anorexia on the other. A new food and health *zeitgeist* emerged as an eternal triangle linking 'health', 'food' and 'beauty', but this is now being reshaped by constant battles with overweight and obesity.[20] As a result food, beverage and cosmetic companies have become skilled at dreaming up new products for consumers that will deliver the perceived benefits and imagery represented by this new consciousness. Debates have, of course, raged for decades about the morality of consumerism,[21] its impacts on global divisions and the environment,[22] its myths and so on. Yet consumerism drives the world's economies, including for food. The growth model is dominant, but the question is now countered in relation to health and the environment as represented by the concept of sustainable diets as set out in earlier chapters. From analyses in those respects, food culture is tipping over from addressing primary needs to servicing excessive wants. John Maynard Keynes, anticipating such tensions, famously asked in a 1930 essay: 'how much is enough?',[23] a question returned to by others recently.[23-27]

Global food marketing is putting before people an awesome offer: endless all-year round food choice for very little effort, and now at last, the new global food machines – 'intellectual property' as well as actual techniques – are meeting resistance. after decades of fast food, there is a 'slow food' movement;[28] after decades of the legal adulteration of food, there is now a burgeoning market for more natural foods; after decades of enticing consumers to eat world cuisines, there is now a counter-move to return to localism, regional foods and real cooking. It is not fanciful to see the revolt over GM foods, for instance, as a popular uprising against imposed new technologies, the patenting of 'life', environmental risks, and corporate power, rather than human health concerns alone.[29] There is a discernible consumer resistance to the perceived corporate and scientific arrogance of companies patenting a rice, for example.[30] With the food companies' global brands so ubiquitous, it is not surprising that they are becoming sensitive to consumer concerns and now want to 'help' them behave more sustainably.[31] What that means, however, is still open to interpretation.

Today's consumption is tomorrow's waste, both in the form of excrement and packaging, and in actual disposal of food waste. As we showed in Chapter 4, globally, consumers in low-income societies waste very little food – there the waste happens near or post farm production (through poor storage facilities or technologically weak distribution infrastructures). In affluent societies, on the other hand, consumers throw away large amounts of food they have bought to consume, up to a third of all food purchased in rich countries is jettisoned.[32] FAO estimated that globally this is 1.3bn tonnes per year. European and North American consumers waste an estimated 95–115kg of food per person per year. In sub-Saharan Africa and South East Asia, by contrast, only 6–11kg per person per year are wasted.

In the UK, in the first report by the national Waste Resource Action Programme (WRAP), published in 2008, they estimated that the country threw away 6.7m tonnes of food every year, roughly a third of everything consumers bought.[33] Most of this was avoidable and it could have been eaten if only there had been better household planning, storage and management. Less than a fifth of that waste was truly unavoidable – things like bones, cores and peelings. This shocked the UK policy-makers and politicians, helped by the fact that these figures were published in the middle of a volatile time in global and national finance – the 2007–8 commodity crisis when banks went bust. Public awareness campaigns, a big media pick-up, rising food prices and a squeeze on consumer expenditure led to significant behaviour change in subsequent years. In a review five years later in 2012, UK food waste had fallen by 21 per cent but UK consumers were still throwing away a staggering 4.2m tonnes of household food and drink annually, the equivalent of six meals every week for the average UK household.[34]

Energy used in food production is also astonishingly wasteful in itself. Indeed, it can be plausibly argued that modern food systems are intrinsically wasteful. 'Just-in-time' logistics for food supply chains may be efficient in economic terms, but they inevitably create waste; so argues Tristram Stuart in a powerful critique of food waste in the modern world.[35] The point here is that food culture in affluent societies has somehow become detached from the consequences of consumption. Food has become so cheap it is almost devalued. This new cultural aspect of waste is sensitive. Politicians do not want to criticise affluent consumers whose spending is critical in 'advanced' economies. Consumers feel embarrassed about waste but continue to produce it. This is a major new dynamic in food culture. No civilisation ever before has had or expressed this capacity to waste the means of its own survival.

The rise of modern 'burgerised' food politics

Food is intensely personal. It is our identity and what we consume sends signals to everyone – family, friends, neighbours and strangers. This is why these meanings are so carefully monitored and tracked by food companies, building on the thinking first developed by Sigmund Freud (an explanation of why so many psychologists are employed in the consciousness and marketing industries). Everyone likes to think that they control their own diet, that it is theirs by choice and tradition. We like to think that the tomato is Italian and that rhubarb is English, for example. Yet even a cursory review of food history reminds us that there is nothing new about food globalisation. Food culture follows trade, which, in turn, follows people and culture. So much of what we think of as traditional food is not.

Take the potato, which came from the Andes.[36] It met a combination of resistance and apathy when introduced to Europe by Sir Walter Raleigh in the 17th century, but now the chip or 'fries' which drive global fast-food production are closely associated with the USA. In fact, the US love affair with fries dates from World War II when returning troops brought the taste for 'French fries' back to the USA, hence the name.[37] It is due to trade that the potato spread into the mass food culture worldwide. The British introduced the potato to India where it is now a staple alongside rice. They also introduced it to Ireland, where its crop failures due to mono-cropping and bad weather engendered the dreadful Irish Famine of 1845–6.[38 39] The humble potato was the culinary 'technical fix' for Irish smallholders who lived off tiny smallholdings and took to the potato as a high-yielding crop which would feed large families off relatively little land. Meanwhile absentee British landlords, who milked rents from the land and put too little back, relied on them doing so. When grown in mono-cultures, the potato is highly vulnerable to decimation by the spread of a virus, the 'blight'.[40] This happened in the mid-1840s. Mono-cropping thus caused disaster for both growers and food culture, leading to famine and emigration.

Few major food commodities are heavily cultivated and eaten solely where they originate, their so-called 'Vavilov' centres (named after the Russian scientist Vavilov who first proposed the theory of botanic origins of species and mapped them to regions). Figure 7.1 gives the world map of Vavilov's theory of the origins of plants. He proposed eight centres with some sub-divided. Group 1, the Chinese centre, he saw as home to 138 distinct species, of which probably the earlier and most important were cereals, buckwheats and legumes. Group 2, the Indian subcontinent, was mainly rice, millets and legumes, with a total of 117 species. Wheats, rye and many herbaceous legumes came from Group 3 and 4. And so on. Although some subsequent scientists question the neatness of Vavilov's

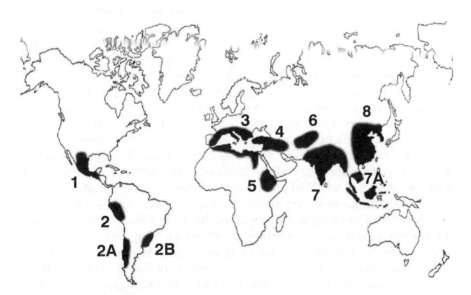

Figure 7.1 The eight Vavilov centres of crop origin

Note: The Vavilov centres of origin are (1) Mexico-Guatemala, (2) Peru-Ecuador-Bolivia, (2A) Southern Chile, (2B) Southern Brazil, (3) Mediterranean, (4) Middle East, (5) Ethiopia, (6) Central Asia, (7) Indo-Burma, (7A) Siam-Malaya-Java, (8) China

Source: Peterson, 2015 [42]/Redwoodseed, Wikipedia

original thinking,[41] the schema is a reminder of how human activity – trade, plant breeding, agriculture – have shaped biological diversity since its 'natural' origins.

The importance here is that food culture is the 'glue' linking consumption and production, as created by evolution and then altered by human activity, first in trade, then in plant genetics. This can be illustrated by cane sugar: a native of Polynesia, it spread to India, China and elsewhere but only became a global commodity and staple ingredient and food in the industrial diet made possible through the plantation system built upon the slave trade in the notorious 'golden triangle' between the Caribbean, Africa and Europe.[40 43] Wheat originated in Africa, but it was US and European surpluses, aided via mechanisms set out in national and international trade agreements like the Global Agreement on Tariffs and Trade (GATT) and later the World Trade Organisation, which enabled global markets for wheat to grow. Wheat 'consumption' has now gradually won over the rice-eaters of Asia to its charms. Food in this sense can be both cultural ambassador and 'fifth columnist', a 'below the radar' and overt change agent.

What is new today about the role of food and this cultural dissemination is the pace and scale of the change and the corporate influence over that process of change. Colas and burgers are now synonymous with the USA, and 'McDonaldisation' – what we call 'burgerisation' – used as shorthand for the modern bureaucracies of society.[44] The burger has become a symbol of modernity and the triumph of a North American mode of eating, a metaphor for how rational, bureaucratic society can command the production of any good and the fulfilment of any need. The burger symbolises a society on the move, constantly seeking new markets, travelling large continents, forced to eat 'on the hoof' and at speed.[45] In cultural terms, sweet, fatty and bland foods are usurping gustatory and culinary diversity. Palates are kept at the pre-adolescent stage of development. 'Comfort' foods sell. Sugary drinks triumph. Cokes and soft drinks usurp water and teas.

Food's capacity to be changed – its 'plasticity' – is why it has been such a perennial threat to those in control, whether governments, companies or landowners. It can be a method of control, but also of cultural resistance. The Slow Food movement took off in Italy in reaction to a perceived takeover of US fast food in the heartlands of Italian culinary culture.[46] Ethical consumers have won mass support demanding improved information about their food. The relationship between consumers and food culture can be uneasy: partly identity but partly oppression. Over the long-term, however, some trends dominate. The rise of meat culture is a notable example. As people become more affluent, they eat more of foods which were previously rare due to cost and for 'feast days' only.[47] For the already affluent, meat has become routine, but behind this ubiquity there lies an industrialised machine of production, now much criticised by animal welfarists and those anxious about the environmental footprint of meat production.[48] Meat and dairy consumption are now in a state of cultural uncertainty and anomie for some, for others they mean symbols of dietary modernisation or essential ingredients of what constitutes a 'real meal'.

Companies know only too well that consumers can be fickle but customer loyalty is the dream of any firm; governments know that it is worth paying more than mere lip-service to consumer opinion. At the same time, if consumers led opinion, there would be no room for politics. The legacy of the French Revolution in which all citizens vote, not just the free as in Athenian democracy under which slaves had no rights, is replaced by a marketplace 'ballot box' which occurs every time the consumer buys and pays for a good or service. The notion of citizenship is thus being redefined as a purchaser whose '(purchasing) ballots ... help create and maintain the trading areas, shopping centres, products, stores, and the like'.[49 50] Buying becomes tantamount to voting, market surveys the nearest thing that exists to expressions of the collective will. It follows that the more wealth or purchasing power the consumer has, the more 'votes' she or he gets. Indeed the parallel status has been many times noted between consumers who are individualised in the marketplace and workers who without the organizing muscle of trade unions are easily outmanoeuvred in the wage market. The French sociologist Jean Baudrillard, for instance, stated acidly in 1970:[51]

> Overall ... consumers as such are lacking in consciousness and unorganized, as was often the case with workers in the early nineteenth century. It is as such that they are everywhere celebrated, praised, hymned by 'right-thinking' writers as 'Public Opinion', that mystical, providential, sovereign reality. Just as 'the People' is glorified by Democracy provided that it remains the people (and does not intervene on the political and social stage), so consumers are recognized as enjoying sovereignty ... so long as they do not attempt to exercise it on the social stage. The People are the workers, provided they are unorganized. The Public and Public Opinion are the consumers, provided they content themselves with consuming.

Is that the role of consumers? To consume and not to intervene? We think their role has developed significantly and that, certainly in the food world, consumers have begun to exercise their muscles, to threaten vested interests and to raise their voices.

The new consumer web and competing models

Most countries have had a long struggle over who controls food: between 'democratizing' food culture and the imposition of a top-down control system.[52] Much depends upon whether one is rich or poor, living in a developed or developing country; and on how equal that society is.[53] Relative equality/inequality within societies has been shown to be a key determinant of health and well-being.[54] The reality of modern food consumption is that the dominant food culture is pulling one way, while reason and caution urge other policy goals. The mantra of 'choice' pulls food consumption towards eating more and more wastefully, while the case for behaving responsibly and consuming within environmental limits and for health pulls another way. The dominant policy language is of consumer sovereignty; the reality is of increasing market concentration.

Food culture has contrary patterns and dimensions. Table 7.1 contrasts the 'dominant' patterns with 'alternatives'. Even though we use the term 'alternative' here, these are just as real as the 'dominant' models. Both have policy ramifications. The purpose of the table is to set out how there are definable tensions along linked meanings. History will show which pattern emerges as dominant overall, or indeed which particular tensions triumph. The political challenge is to reconcile these contradictory demands and to ensure policy coherence. We see this list as key indicators over which the Food Wars paradigms compete. At its most basic level, consumerism offers a dream of total choice. The neo-liberal economic model has ushered in an era in which consumers, rich ones in the main, can browse the world in their local supermarket without even having to travel; the world's food is literally coming to their shopping basket and table. The reach of retail giants, as we showed in the last chapter, is operating almost like a new food colonialism: instead of owning the land, as the great European imperialists did, the 21st-century food empires own the trade routes, logistics and marketing channels. This process is accelerated by tourism which opens up people to new food tastes and by the marketing executives who constantly search for new products, recipes and niches, and by celebrity chefs who parade the globe on TV cooking shows.

Modern food culture offers huge choice for some, but also little for many;[56] it promises information on labels but provides little education with which to interpret them.[57] Wants are being subtly shaped while needs remain unmet. Local foods become culturally suspect while foreign and particularly Western food becomes chic.[58] Food is becoming a commodity which allows whims and fashion to determine its consumption; meanwhile, consumer culture is deepening food's environmental footprint.[59 60] Elite food consumer culture is pressing deeper and deeper imprint in the sands of history. Bottled water is an example of how a simple product can be loaded with cultural connotations of chicness, cleanliness and become a 'must have' consumer accessory, when in fact it is a hugely energy-wasteful way of imbibing liquid, creates a problem of unnecessary packaging and distorts economic efforts to promote clean water for all.[61-63] The paradoxes and idiocies of the emerging global food culture are legion.

Contradictory consumers?

In theory, modern consumers can now choose to remain victims or to be active citizens, finding meaning and enchantment from food in an otherwise disenchanted world. In reality, however, only affluent consumers have significant food choice; middle-income consumers have rather less, and the poor next to none. Table 7.2 gives a conceptual classification adapted and expanded from Alan Durning's original analysis from the early 1990s that continues to be compelling. The reality of there being global consuming classes remains stubbornly constant even though hundreds of millions can move category across the table in just a few years. It should be noted that low-income consumers can have a considerable impact on the environment by burning wood to cook, but it is affluent consumers

Table 7.1 Contrasting patterns of food culture

The dominant pattern	The alternative model
globalisation	localisation
urban/rural divisions	urban–rural partnership
long trade routes (food miles)	short trade routes
import/export model of food security	food from own resources
intensification	extensification
fast pace and scale of change	slow pace and scale of change
non-renewable energy	re-usable energy
few market players (concentration)	multiple players per food sector
costs externalised	costs internalised
declining rural workforce population	vibrant rural population
monoculture	biodiversity
one-track agriculture	multi-functional agriculture
science replacing labour	science supporting nature
industrial-scale 'scientific' farms	ecological sciences/sustainable farming
technocratic knowledge	indigenous knowledge
processed (stored) food	fresh (perishable) food
food from factories	food from the land
hypermarkets	markets
de-skilling	up-skilling
standardisation	'difference' and diversity
niche markets on shelves	real variety on field and plate
people to food	food to people
homogenised food culture	diverse/fragmented food cultures
seasonality eliminated	eating seasonally
created wants (advertising)	real wants (learning through culture)
brands	local distinctiveness
'burgerisation'/McDonaldisation	local food specialities/artisanal foods
microwave reheated food/meals	cooked meals
fast food	slow food
individualised food	commensality/shared food
private intellectual property	common goods
neo-liberal food economies	local food economies
footloose production	fixed location for production
food from anywhere	bio-regionalism
rapidly changing artificial nature	slowly changing nature
global decisions	local decisions
top-down controls	bottom-up controls
dependency culture	self-reliance
health inequalities widening	health inequalities narrowing
social polarisation and exclusion	social inclusion
consumers	citizens
de- or self-regulation	state/public regulation
food control	food democracy
policy segmentation	policy integration

Source: adapted from Lang [55]

whose impact is greatest on resource use. Companies entering low-income markets apply the 'bottom of the pyramid' strategy, making small or individual items available to reduce purchasing costs so some consumption can begin.[64] In this way, the poor are levered onto the treadmill of consumerism. Or that is the argument. (Others argue that the role of the poor is to remain cheap labour in production!) The purpose of Table 7.2 is to highlight how the more affluent, by dint of purchasing and consuming more, have the greater environmental impact. It should be noted that even small increases in the purchasing power (as opposed to consumerism) of the poor can have a profound, and positive, impact on poor people's life opportunities.[65]

Immigration is another well-evidenced mechanism of food cultural change. Immigrants bring their food tastes and cultures with them. It is why setting up restaurants selling their food culture and cuisine is such a common first- or second-generation commercial enterprise. It is why one could eat good Ethiopian food in Washington, DC, not long after the 1983–5 famine.[66] France has long had a range of excellent African food. The UK is a second home to marvellous Indian sub-continental foods. In fact, the ethnic food market in the UK grew to a market value of £1.4bn by 2011,[67] and today English food is defined as much by 'chicken tikka masala' as by 'fish and chips', once a British staple and still viewed as a national dish.[68]

Table 7.2 World consuming classes

Category of consumption	High	Middle	Poor
Population	2 bn	3 bn	2 bn
Diet	More meat & dairy, packaged foods, soft drinks	Grain and plant-based, clean water	Insufficient grain, unsafe water
Transport	Private cars, air	Bicycles, bus	Walking
Access	Long-distance foods, hypermarkets + delicatessen/specialist shops	Some long-distance, local shops and markets	Local food, local shops and markets
Materials	Throw-away	Durables	Local biomass
Choice	Wide choice, global horizons	Sufficient, regional horizons	Limited or absent local horizon
Health impact	Endemic diet-related non-communicable diseases	Rising diet-related non-communicable disease	Under-nourishment, hunger, food insecurity
Environmental impact	Deep footprints, globalised externalities	Increasing carbon, water and land use	Low but variable, localised immediate impact

Source: extended from Durning 1992[25]

In the public sphere, the food industry may appear to bow before the consumer, yet this facade is thin: the marketers, opinion pollsters and focus group consultants who daily probe consumers' tastes, fetishes and aspirations will privately admit that there is no one such being as 'the consumer '; market research highlights how, over and above demographic givens such as gender, age and ethnicity, consumers are divided by income, social class, ideology and desires; they are fragmented in their tastes in ways that make mass markets hard to corral. The public wants good safety and environmental practices in theory but is often unprepared in practice to pay the extra costs that tougher standards entail, until there is a crisis – as happened in the UK with mad cow disease (BSE) in the late 1980s or the horsemeat scandal of 2013.[69]

All medium to large food companies ceaselessly research the consumer and incorporate findings into their product planning and strategy. Most large companies conduct and/or buy such research on a continuing basis. Market researchers and advertisers are very good at (re)packaging such work, producing ever-more-clever categories of consumers in market-led groups to tailor products to suit their differences. They assume the standard socio-economic differences and overlay aspirational group differences to yield categories. Gabriel and Lang have argued that part of the complexity of modern consumer culture is that there are many models of being a consumer now available, and that consumers can and do adapt and enter these role models, often in sequence and sometimes at the same time.[70] They might approach food as an explorer one day but another day seek repeat comfort food. They make citizenship demands but also behave as hedonists.

Food advertising and marketing: moulding food culture

One of the most remarkable changes in consumer food culture is the sources people use for their information about food. Home experience may still be the initial 'fount', but there are now many competing channels of information: advertising, sponsorship, formal education, the media and the internet, and peer groups. Table 7.3 lists the myriad of different media used to influence and mould people's food consumption behaviours.

When the food industry is portrayed in a not so favourable light, the media are often singled out for blame – such as for biased or inaccurate reporting, especially in the case of over-simplifying science and technology issues. However, as Table 7.3 shows, the industry itself is deeply engaged with the media and the global food and beverage industry spends hundreds of millions of dollars, euros, pounds, yen, etc., annually on advertising, public relations and other marketing activities. In 2012, the US fast-food industry alone spent $4.6 billion to advertise mostly unhealthy products, much of it with children and teenagers as the audiences.[72] In the UK, food was the biggest sector for advertising in 2013, with £0.6bn being spent. Much of this is for sugary, fatty, 'fun' foods, and for supermarket bargains.[73]

But the tension continues not least as from the late 2000s the media landscape has been changing fundamentally. In the post-World War II boom years, especially

Table 7.3 Range of media used to advertise and market food and beverages

Broadcast media
- Television/radio advertising and sponsorship
- Television – product placement

Stores
- On-shelf displays
- Displays at check-out tills
- Special offers and pricing incentives
- Purchase-linked gifts, toys, collectables
- Free samples and tastings

Product packaging or content
- Product formulation, e.g. colour/shape
- Product portions, e.g. king size
- In-pack and on-pack promotions, e.g. gifts, games, puzzles and vouchers
- Packaging design, e.g. colours, playshapes

In school
- Sponsorship, educ. materials/equipment
- Vending machines
- School participation – promotion/sampling schemes

Other media
- Print media, e.g. magazines, comic books
- Cinema advertising
- Film product placement
- Posters and advertising boards
- Branded books, e.g. counting books
- Internet: email clubs, chat rooms, free ring tones
- Websites – puzzles/interactive games
- Promotional sales by telephone
- Text messaging to mobile telephones
- Direct marketing, e.g. home catalogues
- Sponsorship – events, venues, teams, sports heroes
- Cross branding on household goods
- Branded toys
- Branded computer games and product placement in computer games

Source: Adapted from WHO Europe, 2007 [71]

from the 1960s onwards, manufacturers produced standardised products and then bought a lot of TV ads to shift the product. TV ads led to retail distribution and to sales, and sales of course created profits. Marketing analyst Seth Godin calls this process the TV-industrial complex and shows how this model is dying.[74] The principal reason, he argues, is that people have ceased to pay attention to the type of advertising that interrupts our lives – he includes magazines and newspapers as well as TV in this process. The TV-industrial complex – and how it worked for industry – has been forever disrupted by the internet and new social media. In 2009 the UK became the first major economy where advertisers spent more on the internet than they did on TV – 23.5 per cent in the first quarter of 2009 compared to 21.9 per cent on television.[75]

The new media landscape is leading to changing business practices, marketing and advertising techniques and technologies. In the UK total advertising expenditure across media was forecast as £18,517m in 2014 (up from £17,510m in 2013). While TV spot advertising continues to be important, accounting for 26.8 per cent of the total, internet ad spending was 38.5 per cent, with a fast-growing part of this being mobile advertising, around £1.6bn in 2014. By contrast direct mail was £1.82bn and radio £570m.[76] For food and beverage companies, TV remains the most important advertising route within traditional media in terms of total expenditure. Table 7.4 shows examples of TV advertising spend in the UK by major food and beverage companies compared to expenditure on press, radio, cinema and outdoors – but what these data do not show is how much these companies are spending on social media and the internet or the effectiveness of different advertising and marketing channels. As

Table 7.4 Examples of UK food and beverage company advertising expenditures within traditional media, 2014 (as % of total £m spent)

Advertiser	TV	Press	Radio	Cinema	Outdoor	TOTAL
Kellogg's	94.7%	3.3	0.3	1.7	0.0	£35.6m
Coca-Cola	54.9%	7.3	0.9	0.4	36.5	£29.0m
Muller Dairy	93.0%	5.0	1.3	0.2	0.5	£25.2m
Cadbury	54.1%	7.2	0.3	13.2	25.3	£19.4m
Nestle	84.6%	5.2	–	0.1	10.1	£19.4m
Danone	76.4%	3.3	2.7	–	17.6	£18.1
Ferrero	90.3%	2.9	1.6	–	5.2	£16.7m
Arla Foods	70.1%	8.5	3.5	3.2	14.8	£16.6m
Birds Eye	90.1%	0.7	0.6	3.4	5.2	£15.5m
Walkers	82.7%	10.2	2.6	–	4.5	£15.3m
Innocent Drinks	81.9%	1.4	–	–	16.7	£6.5m
Weetabix	38.3%	0.7	–	1.3	59.7	£6.3m

Source: *The Grocer* [77]

can be noted within traditional media, for UK food companies the cost of TV dominates ad budgets.

In the United States the Yale Rudd Center for Food Policy and Obesity calculates advertising and marketing expenditure for 25 of the largest fast-food restaurants, ranging from TV, websites, social media and Spanish-language TV advertising to TV ads being viewed by black children and teens. Their data show that TV again takes the biggest chunk of advertising expenditure as a percentage of the total budget (around 80–98 per cent) – Table 7.5 shows the top 13 fast food companies (those spending over $100m) ranked by total advertising spend in 2012 (it should be noted companies can often spend a lot more or less each year).

The Yale Rudd Center also gives some insight into the digital marketing activities of major fast-food restaurants in the US. Table 7.6 details this activity for five leading companies based on their Facebook 'likes', Twitter followers and YouTube upload views, comparing 2010 to 2013. The incredible increases over this time suggests the newness and novelty of this type of marketing activity.

The marketing activities highlighted by Table 7.6 are what Seth Godin means by consumers moving away from being 'interrupted' by marketing messages – such as a TV ad in the middle of our favourite programme – to social media where we give an element of permission or actively engage with an advertising company, such as watching a YouTube video or playing an online game devised by the company aimed to support one of its brands.

But despite the newness of media it is another manifestation of the long-term moulding of consumer food cultures, particularly in relation to brands. The food industry is serviced by an army of professionals, from psychologists to home economists, dedicated to studying, analysing and squeezing consumers into 'pigeon holes' – or 'niches' in business parlance – better to predict and manage

Table 7.5 Top US fast food advertising spenders, 2012 ($m/% of total spent on TV)

1. McDonald's	$971.8 (79%)
2. Subway	$595.3 (85%)
3. Taco Bell	$274.7 (91%)
4. Wendy's	$274.5 (87%)
5. KFC	$258.1 (98%)
6. Pizza Hut	$245.8 (99%)
7. Burger King	$236.4 (93%)
8. Domino's	$191.1 (93%)
9. Sonic	$173.7 (96%)
10. Papa John's	$153.3 (96%)
11. Arby's	$137.8m (97%)
12. Dunkin' Donuts	$135.1m (82%)
13. Jack in the Box	$103.7m (90%)

Source: adapted from Yale Rudd Center for Food Policy and Obesity, 2013[78]

Table 7.6 Digital marketing activity for five major US fast food restaurant chains 2010–2013

Restaurant	Facebooks 'likes' (000)			Twitter followers (000)			YouTube upload views (000)		
	2010	2013	Increase	2010	2013	Increase	2010	2013	Increase
Starbucks	11,353.4	34,969.7	208%	989.2	4,215.4	326%	5,293.6	8,166.8	54%
McDonald's	2,636.8	29,202.5	1007%	39.5	1,573.1	3883%	115.6	7,749.4	6602%
Subway	3,088.1	23,651.2	666%	22.8	1,483.4	6406%	0.0	1,726.6	
Taco Bell	1,770.8	10,200.8	476%	35.2	717.6	1939%	2,073.8	13,756.3	563%
Pizza Hut	1,414.8	10,623.6	651%	31.3	439.9	1305%	16.8	3,438.0	20400%

Source: adapted from Yale Rudd Center for Food Policy & Obesity, 2013[78]

them. As Sir Michael Perry, former head of Unilever, stated as long ago as 1994 in a keynote presentation to the UK Advertising Association:

> Our whole skill as branded goods' producers is in anticipation of consumer trends. In earlier appreciation of emerging needs or wants. And in developing a quality of advertising which can interpret aspirations, focus them on products and lead consumers forward.[79]

New and emerging technologies are creating a potentially serious problem area in marketing – this is the step-change in the ability of data companies to collect data and information on individuals and how this is used (or passed on). More generally, vast amounts of information about individuals are being collected, stored and used with or without people's permission – an issue of privacy and ownership of people's personal details compounded by the potential abuse of this information using new 'Big Data' mining technologies. Society has entered an advertising, persuasion and selling revolution as modern insights from psychology, neurology and big data are applied to the advertising and marketing (dark) arts.[75] Food is a key arena for this application.

The marketing and advertising of food and beverages to children

One area where the marketing and advertising arts are seen as having a particularly dark shadow is the way these are applied to children and the selling or promotion of particular foods and beverages. Over recent decades there has been persistent concern about the targeting of young people by advertisers and arguably this is one of the few areas the food industry has consistently been put on the back foot in relation to public health nutrition concerns. This problem for the food industry has been compounded by the growth in childhood obesity and children and teenagers becoming overweight. International concerns about the marketing strategies of the food industry towards children predates the rise in social media so in this sense it is not a new problem but a long-term food war.

Trends in childhood obesity first started to worry public health experts from the 1980s. For example, the prevalence of the condition escalated in the US from 5 per cent in 1964 to 14 per cent in 1999; by 2014 one-third of US children were overweight or obese. In the WHO Europe region in 2007 (covering 880 million people) around 20 per cent of children and adolescents were overweight, a third of these were obese.[71] By 2010, worldwide, 43 million children (35 million in developing countries) were estimated to be overweight and obese, with a further 92 million at risk of overweight. By 2020 worldwide prevalence of childhood overweight and obesity has been estimated as rising to more than 9 per cent.[80]

In the UK, in 2001, one in six 15-year-olds were classified as obese and obesity-linked 'adult onset' diabetes mellitus was for the first time being reported in children and adolescents in the UK. While the prevalence of childhood excess weight appeared to be levelling off or even falling from the late 2000s, by 2013 this had still reached levels that impacted a significant minority of children living in England. For 4–5-year-old boys 23.4 per cent were either overweight or obese (9.9 per cent obese) and 21.6 per cent for girls (9.0 per cent obese). For 10–11 year olds 35.2 per cent of boys were either overweight or obese (20.8 per cent obese) and 31.7 per cent of girls (17.3 per cent obese). Obesity prevalence is strongly correlated with socio-economic status and is highest in England's most deprived areas.[81]

Such trends in childhood obesity spell danger to the marketing and advertising of food and drinks. Children are seen as particularly vulnerable to food promotion, with brand advertising influencing price, availability, information, personal taste and cultural values. Marketing can grow a food category – such as soft drinks – and not just encourage brand switching. Public health experts point out that much food marketing and advertising is in support of products and brands that are energy-rich, nutrient-poor products – in particular products with a high content of fat, sugar or salt. These can encourage rapid weight gain and exacerbate risk factors for chronic disease (see Chapter 3).[82]

During the 2000s public health voices started speaking up in unprecedented ways to set out strategies in response to the marketing and advertising of food and drinks to children. The International Obesity Task Force in its September 2002 report 'Obesity in Europe' was unequivocal in its concern about childhood obesity, calling for a European Union-wide restriction on advertising targeting the young, including pre-school children, to consume inappropriate foods and drinks. At the start of 2003 the World Health Organisation went public about its concerns that products like sugary drinks are contributing to obesity, especially in children. The WHO urged governments to consider clamping down on TV ads for 'sugar-rich items' to children. In May 2010 the World Health Assembly (the governing body of the WHO) endorsed a set of recommendations on the marketing of foods and non-alcoholic beverages to children.[83] This was followed up in 2012 by publication of a framework for implementing these.[84] The objective of these WHO initiatives is to provide guidance for countries to design or strengthen policies on food marketing. The WHO recognised that food marketing to children now has global reach and uses multiple messages in multiple media channels.

But in the main government or public policy responses have been to seek out or encourage voluntary efforts by the food and beverage companies rather than to pursue regulatory routes. The food industry has responded by introducing a series of industry-led 'pledges' aimed at setting self-imposed rules for themselves in relation to their marketing activities towards children. A prominent example is the European Union pledge programme announced at the end of 2007 as part of the EU Platform on Diet, Physical Activity and Health set up in 2005. In the EU Pledge, 11 major food beverage companies made a commitment to change the way they advertise to children. This included not to advertise products to children under 12 years old on TV, in print or the internet such as company-owned websites, except for products that fulfil specific nutrition criteria relating to accepted national and international dietary guidelines (known as 'nutrient profiling'). Companies also agreed to not engage in commercial communications in primary schools.[85]

By 2014 there were 21 EU Pledge members and together these companies accounted for more than 80 per cent of food and beverage advertising expenditure in the EU. In 2014 the scope to the EU Pledge was extended to cover additional media which reflected the changing marketing landscape, as this included interactive games, mobile and SMS marketing as well as radio, cinema, DVD/CD-ROM, direct marketing and product placement.

As might be surmised such voluntary arrangements are not without controversy and any causal link between, for example, online food and soft drink advertising and consumption and childhood overweight and obesity has been challenged. A review of the existing evidence (as written up in academic literature) commissioned by the UK's Committee of Advertising Practice (CAPr) led it to the conclusion that there was no need for further rules for 'advergames', websites and social media and that the existing rules on food and soft drinks marketing and advertising can provide the right level of protection for children. But the CAPr did set out a series of actions – such as training, advice to marketers and new guidance – aimed to monitor online marketing to children.[86] This conclusion by the CAPr was greeted with disappointment by campaign groups in the UK and saw the response as allowing advertisers to continue to target junk food marketing at children online.[87]

The issues around the marketing and advertising of food and drinks to children are part of the bigger trend of the commercialisation of childhood which see children as an increasingly lucrative market and marketers coming up with ever more ingenious methods to encourage children to consume.[88] For example, one study found the average 10 year old has internalised 300 to 400 brands and 70 per cent of 3 year olds recognise the McDonald's symbol but only half of them know their own surname.[89] But when it comes to food and beverages the assumption is that overconsumption results in diet and health problems. This overconsumption is also highly lucrative – one estimate suggests that excess food and beverage consumption by school-age children in the United States is worth \$20bn to the food industry.[82]

Consumer information and rights

One of the serious points of tension between commerce and government has been food labelling.[57] Labelling is a key demand for international consumer organisations. It features in the main Consumers Rights as outlined and agreed by the UN-affiliated Consumers International in 1975, following President John Kennedy's famous 1962 speech about the issue.[90] The UN enshrined six principles of consumer guidelines, and added another in relation to sustainable consumption in 1999. Table 7.7 gives both the UN's seven and CI's eight key principles. The role of labelling is central here. It is to provide consumers with information upon which to be able to make informed judgments and choices, thereby encouraging market efficiencies. This is fairly conventional liberal economic theory. Information flows are essential in the marketplace, yet there is little internal consistency in its application to food in the real world. In practice, what goes onto a food label can become the subject of intense politicking. One British senior civil servant went on record as saying that in effect the UK government in the 1980s and early 1990s regarded food policy as no more than what information could or could not appear on a label.[91] Within Europe, for instance, the presence of food additives has to be declared, but not pesticides. [92] There are ongoing fights in Europe and North America over the labelling of genetically modified foods and at what levels any inclusion should be declared on a label. Labelling, in short, is not a resolution to the Food Wars but becomes another battlefield within them. Yet politicians regularly cite labelling both as a principle of and a means for consumer sovereignty.

Food labelling and its effectiveness in consumer transactions are likely to continue to be contentious for as long as some critics argue that complex information on a label in tiny writing across dozens of purchased goods does not enable the consumer to deliver for him or herself a health-enhancing diet, and others argue that, without such information, the consumer remains in ignorance. Whilst vast resources are now put into this area of policy, with national and international meetings proliferating on all aspects of labelling, we remain to be convinced that labelling is an effective health or environmental protection strategy, yet it remains cited as such. As periodic food scandals show, consumer cynicism remains. In Europe in 2013, for example, it emerged that there was extensive and routine adulteration of meat products with horsemeat being passed off as beef

Table 7.7 Consumer rights and principles: UN and Consumers International

General Principles of the 1975 UN Guidelines for Consumer Protection (amended in 1999)	*Consumers International's Consumer Rights*
1 the protection of consumers from hazards to their health and safety	the right to the satisfaction of basic needs
2 the promotion and protection of the economic interests of consumers	the right to safety
3 consumer access to adequate information to enable making informed choices according to individual wishes and needs	the right to be informed
4 consumer education, including education on the environmental, social and economic impacts of consumer choice	the right to choose
5 the availability of effective consumer redress	the right to be heard
6 freedom to form consumer and other relevant groups or organisations and the opportunity for such organisations to present their views in decision-making processes affecting them	the right to redress
7 the promotion of sustainable consumption patterns (added in 1999)	the right to consumer education
8	the right to a healthy environment.

Source: UN 2003 [93]; Consumers International 2013[94]

in some processed products such as burgers, sausages and in lasagna. Too often such products were sold in public catering which no labelling at all. This mix of false and absent labelling suggests ongoing tension in the economic model which sees labelling as key. Others have argued that a high-quality food culture would not require labelling. History suggests otherwise.[95 96] Unfortunately, infringements of food labelling regulations will continue to need interventions by food law enforcement officers and labelling issues are destined to remain a bloody battlefield in the Food Wars for the foreseeable future.

Cooking: why cook?

Television chefs are not a new phenomenon; they are as old as television itself. Over the decades, TV cooking and food programmes have mushroomed almost in inverse proportion to the amount of cooking that actually occurred in the Western home, and went mass, mainstream and practically daily. Indeed, cooking TV programmes have almost become a hallmark of the supposedly 'leisure' industries. Books on cooking and cuisine make the best-seller lists worldwide, and some research has suggested that the greatest strength of the phenomenon is its entertainment value.[97] In rich societies, actual cooking is often now only an occasional or 'hobby' pastime rather than a daily necessity. The shift in the role and function of cooking is a long-term transition, a function of changed working conditions and hours, the different role of women with respect to the family and workplace, and the meaning of the home. Even back in the 1990s, a European survey was suggesting that, although the British public spent less time in the kitchen than their European neighbours, 42 per cent were already viewing cooking as an enjoyable occupation, with 14 per cent seeing it as a creative activity and 11 per cent using it as a 'de-stressing activity'.[98] A UK government-funded study in the early 1990s, when the transition away from direct cooking from raw ingredients had already taken hold and was worrying health campaigners, found that, while some young people were interested in learning to cook, they lacked actual skills.[99] A 20-year campaign then followed to reintroduce cooking skills into the national educational curriculum in England, which eventually happened following a high-profile campaign by TV celebrity chef Jamie Oliver in the mid-2000s about school dinners and a follow-up on skills, which he then repeated in the USA.[100]

The key issue here is whether there is a de-skilling transition underway in culture, globally; and whether this is associated with the spread of pre-cooked processed ready-to-eat meals. Initially this debate occurred within culinary circles but latterly it has become important for its health implications. Indeed, a powerful combination of data and argument has been presented by Brazilian public health researchers who have linked the rise of what they term 'ultra-processed' foods with the rise of diet-related non-communicable diseases in Brazil.[101 102] When the Brazilian government revised its national population dietary guidelines, published in 2014, these were formulated with cooking and everyday food culture in mind. The guide takes as its starting point what the Brazilian people from all social classes actually eat every day. The guide also considers the social, cultural, economic and environmental implications of food choices. Among the ten overall national dietary recommendations – this from the Ministry of Health – is advice to:[103]

- 'Prepare meals from staple and fresh foods'
- 'Develop, practice, share and enjoy your skills in food preparation and cooking'
- 'Limit consumption of ready-to-consume food and drink products'
- 'Buy food at places that offer varieties of fresh foods. Avoid those that mainly sell products ready for consumption'

It remains to be seen whether this powerful cultural advice becomes an important foundation for the implementation of national and governmental policy.

Cooking is a universal point at which the natural world meets food culture.[104] In this respect, the matrilineal thread to food culture in the cooking transition may be under pressure from external sources such as the ubiquity of processed, pre-cooked foods and the cheapness of eating out. Food's role in family structures alters, yet mothers and grandmothers remain important purveyors of food culture – teaching what to cook and how. In mobile and media-dominated cultures, there are other means for transferring food knowledge besides the family.

Traditionally, and almost without exception across cultures, cooking has been a predominantly female task, but the skills base, in industrialised societies at least, is changing. The rise of male labour in the food industry and in catering trades provides a source of male employment. TV chefs tend to be male in absolute numbers, although there are high-profile female celebrity chefs providing some gender balance, while in professional kitchens or catering establishments (such as restaurants or foodservice) kitchen personnel tend to be male.

For the non-professional cook, a study of UK culinary skills found that people retained the basics of cooking skills but did not necessarily apply them or have confidence to apply in complicated culinary concoctions.[105] Another study comparing French and UK approaches to cooking found this to be true in the UK but in France the family lunch, for example, remained important. People in both countries often lacked time to cook and increasingly relied on a mix of both raw and convenience-type foods but to varying degrees. The French were more willing to 'cook from scratch'. Men in both countries were more engaged with cooking in the home but saw this more as a leisure activity for the weekend or special occasions.[106]

As awareness of the importance of cooking skills in culture has grown, policy-makers have become more interested in cooking education as a means of health promotion. Young people may not cook but they have different skills from their predecessors. The question is: does it matter if cooking is marginalised? In one corner are those who argue that the arrival of pre-processed foods is a liberation for women generally.[107] In the other corner are those who argue that, by not being taught to cook, people lose a basic means for taking control of their diet.[105]

People cannot be expected to control their diet as health educators exhort them to do if they do not know what goes into food. This reframes the debate with cooking being a life skill rather than a high moral pursuit or a labour force skill. If people do not know how to cook, they are entirely dependent upon whatever others or the food industry provide for them. They cannot be informed consumers in the marketplace. Food culture without cooking becomes a dependency culture – a far cry from the rhetoric of consumer sovereignty.

Such debates might trouble policy-makers but the reality is an inexorable rise of eating out and eating what others cook. By the mid-1990s, one in five Americans ate in a fast-food restaurant each day.[104] In 1955, US restaurants took only 25 per cent of the US food dollar, but by 2014 this had grown to 47 cents of every US dollar spent on eating.[108] Fast food is only one expression of the American love of being on the move; the 150 million cars on the roads by 2000 reinforces fast eating,[45] which consolidates the US 'long hours/few holidays' work ethic and can incidentally break links with real food cultures that immigrants bring with them. Fast food is the American way. Convenience is now also regularly cited in the EU as an influence on food patterns. Yet, ironically, the concept of 'convenience' also implies leaving the responsibility for food consumption choices to someone else – the food and beverage processing industries. UK consumers, perhaps the most Americanised of Europeans, are now adopting 'grazing' – constant snacking – rather than eating at set times of the day and have adopted the 'dashboard' trend of eating in the car.

Research on the British in the 1980s suggested their cooking patterns were determined by structural factors such as longer working hours, the increase in waged female work and longer shopping travel times. Pre-prepared meals and better cleaning products mean the average UK person spent 2 hours and 41 minutes less time doing domestic chores per week by the 2000s.[109] This was, however, balanced by an increase in the time spent shopping and on associated travelling to food suppliers, which had risen by two hours and 48 minutes per week, largely because of the growth of out-of-town supermarkets.[110] In evolutionary terms, an entirely new food culture is being induced globally. Literally, people do not need to move much (or expend physical activity) to get their food; the car does the moving, but this is fossil-fuel driven. New culinary culture is generating environmental externalities. Packaging of food illustrates this. Traditionally almost all cultures have developed forms of wrapping food either to preserve it or make it transportable, but mostly using re-usable materials – cloth or pottery or metal. The rise of plastic to wrap food is a 20th-century phenomenon.[111]

Mobility, ease and fast food are not new cultural demands. The ubiquitous sandwich has long been a domestic form of food 'on the go'. It was so named after an English aristocrat, John Montague, the 4th Earl of Sandwich, who in 1762 was so addicted to his gaming table that he refused to leave it; instead he stuffed meat between two slices of bread in order to carry on gambling. [112] The aristocrat's name relabelled an existing practice of the common people. This is now a multi-billion dollar industry across many countries, with mass-scale routinised factory systems producing millions of sandwiches overnight for them to be trucked around countries and arrive 'fresh' for consumers in packages on supermarket or corner store shelves.[113] Something consumers could do themselves is commodified.

Shopping, spending and food

Although food is a basic need, it is only one of many spending areas, and food shopping culture is, like other shopping, increasingly driven by the retail giants.[114] Hypermarkets tend to have a higher turnover per square metre of sales space than do small shops, so that encouraging consumers to travel to a hypermarket in pursuit of small cost savings is a very successful marketing strategy. As a result, food shopping becomes a near-identical experience everywhere in the world: aisles in bricks-and-mortar malls with car parks and open 24 hours; instead of the consumer having at least eye contact with the retailer and perhaps at best a chat with him or her, there is an alienated exchange with a harassed, low-paid worker.

The net impact of this emerging shopping culture is externalised social and environmental costs: the bread or beans might be cheaper at the hypermarket but the food travels further and the consumer travels further to meet it. Internationally, the food system has grown around a particular model of the food retail environment. The supermarket – a self-service system of food shopping – began in the USA in the early 20th century, but it spread globally throughout the century and continues to do so in the 21st. This has altered how people get access to food and the culture surrounding that process – literally the amount of time spent, the mode of transport (car, walk, bike, bus) and who gets the consumers' money. Yet this triumphant model is now under attack. It is heavy in energy use. It is being threatened by online shopping – a modern version of home delivery that used to be a service offered by an earlier era of small town grocery stores. Consumer loyalty is thin.

While technology reshapes this battle between food coming to the consumer or the consumer going to the food, there is also a key battle over economics. When the world financial crash happened in 2007–8, the previous long-term drop in food prices in the West was altered. Energy prices and raw material prices rose, leading to a squeeze within the food system and on consumer spending. In poor countries, millions of people were tipped into absolute food poverty. The 'new austerity' sent shock waves through the developed world, whose policy-makers took cheap food for granted as an essential element of consumer society. The OECD and FAO, long champions of lower food prices as part of the Productionist paradigm – cheaper food being thus more affordable – had to alter their analysis. They began to assume rising prices or at least volatility.[115] For consumers, this reintroduced an old uncertainty. It also had major impacts within the food system, with the giant retail chains experiencing a profits squeeze and having to respond to customers with tighter budgets.

As this new dynamic emerged in the 2010s, food bank usage and other crisis responses to food poverty emerged. Food banks had been developed in North America from the 1980s, a mix of charity and alternative welfare, providing what was supposed to be a temporary stop-gap. With welfare cuts, food banks spread, championed often by churches and local community self-help. They became normalised in North America and, in the Great Recession from 2007, spread to Europe.[116 117] Political debate on food banks is highly sensitive, with right-wing sentiment approving food welfare but sometimes criticizing lifestyles of users, while more left-wing critics saw this as a return to Victorian values and an indicator of the erosion of welfare state safety nets and the refusal to address structural inequalities. Retailers and manufacturers, meanwhile, see them as an opportunity to dispose of surplus or unwanted or otherwise waste stock.

Sustainable diets: choice versus choice-editing for sustainability

In Chapter 4, we introduced the notion of sustainable diets as a challenge for how to integrate health and environment. Here we reintroduce it as a challenge for consumers. The problem can be stated simply: how can consumers eat in a way which is pleasurable, good for health and the environment, while meeting cultural expectations? The 2010 FAO definition acknowledged the importance of culture by saying that sustainable diets are those that are 'culturally acceptable'. [118] This recognised that consumers come to the table with cultural rules and norms, but with regard to health and environment, how are consumers to know what they are choosing, when they lack the tools and cultural rules by which to do this? In Chapter 6 we showed how business has altered simple foods into complex assemblies of ingredients. What is 'healthy' becomes a battleground of experts and technologies. In this chapter, we are suggesting the 20th-century model of the food system rendered food culture more 'plastic' and changed the form, content, messages and appearance of food in the transition to diets based on processed foods. Culture has been altered by supermarketisation, burgerisation, globalisation and the rise of what Monteiro and colleagues term 'ultra-processed' foods.[119] Sustainable diets are like an iceberg for consumers. Some of the issue is visible on the surface – the looks, perception, norms, even the label – yet most of the issues which need to be addressed by consumers to shift to a sustainable diet remain beneath the surface – the footprint, the long-term and mass effects.

A model of food, farming, culture, living, shopping and above all eating has been developed in rich societies which continues to be rolled out worldwide but which is unsustainable. It is grossly wasteful, misuses land, depletes resources, distorts culture. But it makes big money for some. While the developing world still experiences problems of chronic under-consumption and underperforming food markets (i.e. hunger remains), in the rich developed world, patterns of eating have been normalised which damage health and environment and exacerbate social inequalities even amidst wealth. This pattern of eating enshrines a mismatch of humans, land use and food culture. It is spreading globally. By any use of the word '(un)sustainable', this is unsustainable.

Many global and national reports have made this case, often from sectoral starting points: health or environment, biodiversity or inequalities, particular crops, dietary changes in this or that country. Taken collectively, a 'meta-picture' has emerged. Presented with such overviews, policy-makers alas have so far responded in the weakest way, as we saw in Chapter 5, at best calling on consumers to make responsible choices, without giving them the tools to judge. Information flows are scarce and remain at the scientific rather than popular cultural level. Meanwhile dominant market signals reinforce different values: 'cheap food' is deemed good as an end in itself. But it is hard for consumers to know what is 'cheap' among a cacophony of continuous supermarket promotions or what the size of the bill is for the hidden costs from the health and environmental externalities that the corporations pass onto society at large to produce 'cheap' products on supermarket shelves.

Ever more processed 'non-food'/'junk' foods flood onto supermarket shelves to titillate tastebuds. Sugary fatty salty food products continue to receive vast advertising support. Choices are framed which ignore the health or environmental consequences. Or 'health' is sold as the impact that will follow from this or that particular product. No wonder public health and environmental food scientists became increasingly frustrated by the snail-like policy change and by the failure to communicate the challenge to consumers. Sustainable diet is now a campaigning point for consumer organisations. In the UK, for instance, the Eating Better coalition has been set up, bringing together a wide range of citizens groups to coalesce their actions on sustainable diets.

The food industry is aware of these issues, as we showed in Chapter 6. In the 2000s, some giant companies began to create alliances to share approaches to resolving environmental impacts, such as the Sustainable Agriculture Initiative. [120] They began to audit their food products and supply chain carbon emissions, driven by a desire to become more efficient in resource use and to manage business risks. In Europe, this is being encouraged by the new policy framework to move towards a circular economy, and away from the linear economy. In food, this is currently translating mostly into a focus on waste. In general, companies and governments do not want to shake consumer trust. They prefer 'choice-editing' before consumers see the products. They favour a commercial version of 'nudging', soft attempts to change behaviour but without rocking the boat too much! Consumers are thus kept mostly in the dark about sustainability and have little awareness of reformulation of products, packaging and ingredients behind the scenes. A new tension thus emerged. Leading consumer organisations have argued that consumers must be involved and informed about how food has a combined health and environmental impact; they need government and companies to engage with them.[121] On the other hand, companies are more comfortable acting on behalf of consumers, but beneath the radar.[31]

Meanwhile the scientific messages about sustainable diet for affluent societies point broadly in a similar direction: consume less overall, less meat, less dairy-based products, more plant-based diets, more biodiversity awareness, and real cooking from raw ingredients, etc. Three examples of academic advice are given in Table 7.8. More detailed work is beginning to emerge about how to translate the general principles for particular consumer groups, and what is acceptable and affordable. Rich and poorer societies enter the sustainable diet debate from

Table 7.8 Some 'principles' from academics for sustainable eating compared

Source/ country	New Nordic Diet's ten principles [126]	Barsac Declaration: four principles [127]	Centre for Food Policy tips for eco-nutrition
Date	2010	2009	2007
Lead body	University of Copenhagen project	9 Research Groups of scientists	Centre for Food Policy discussions[129, 130]
Prime concerns	To combine health and environmental advice	To reduce nitrogen emissions [131]	To formulate some new cultural 'rules' (norms) for C 21st
Actual Advice	More fruit and vegetables every day (lots more: berries, cabbages, root vegetables, legumes, potatoes and herbs)	Encourage the availability of reduced portion sizes of meat and animal products, compared to the local 'normal'	Eat a plant-based diet, eat flesh sparingly, if at all
	More whole grain, especially oats, rye and barley	Eat meals containing half the amount of meat or fish compared with the normal local amount,	Eat simply as a norm and eat feasts as celebrations, i.e. exceptionally
	More food from the seas and lakes	Eat a correspondingly larger amount of other food products	Drink water not soft drinks; if you drink alcohol, use it moderately
	Higher-quality meat, but less of it	In social situations (eg conferences or public eating) always offer clear distinctions between: 'normal', 'demitarian', vegetarian, vegan	Celebrate and eat biodiversity (from inside the field to your plate)
	More food from wild landscapes		Eat locally where possible to support local suppliers and resilience
	Organic produce whenever possible		Eat seasonally if possible to keep embedded energy in the food low
	Avoid food additives		Choose your diet carefully and beware of hidden ingredients in food, especially salt and sugars
	More meals based on seasonal produce		Eat equitably: (a) Eat no more than you expend in energy; (b) build exercise into your daily life
	More home-cooked food		Eat less but better: Go for quality, not quantity; be prepared to pay the full (sometimes hidden) costs of producing and transporting the food
	Less waste		Enjoy Food ... in the short-term but think about its impact long-term

The 'Tips' are given for comparison. They were proposed as a discussion point from early 2007, and first appeared in public in 2007 in a paper by Lang.[128]

different starting points. The needs of lactating mothers may not be the same as an elderly grandparent. Some studies have suggested that environmental impacts fall if existing health guidelines are followed.[122-124] Others suggest this may not always be the case.[125] This debate will continue.

Conclusion

In this chapter we have set out the notion of a 'food culture' in order to capture the diversity of influences within the food economy that shape food consumption. Food culture is malleable and in transition. Consumers feed new ideas and demands back into food supply, but at the same time are moulded by demands and marketing coming from it. Diets continue to become less healthy and unsustainable, at the same time more 'healthy' and 'natural' products become available. Food culture is a delicate arena for 21st-century food wars. Food policy-makers will have to reconcile long-term needs of policy strategy with the short-termism of much business, media and consumer activity.

The food culture challenge is heightened further by the urgent need to link the health needs of individuals and populations to environmental well-being. The challenge of how to dovetail human and environmental health demands will reshape 21st-century food cultures.

References

1   Fox M. *Creativity: Where the Divine and Human Meet*. New York: Tarcher/Putnam, 2002.
2   I Aniiniluu C, Wmi Fintomilc R. *Food and Culture: A Reader*. 3rd edn. New York and London: Routledge, 2013.
3   Fitzpatrick I, et al. *Understanding Food Culture in Scotland and its Comparison in an International Context: Implications for Policy Development*. Edinburgh: NHS Scotland, 2010.
4   Goody J. *Cooking, Cuisine and Class: A Study in Comparative Sociology*. Cambridge: Cambridge University Press, 1982.
5   Murcott A. On the social significance of the 'cooked dinner' in South Wales. *Anthropology of Food*, 21(4/5) (1982): 677–96.
6   Mintz S. *Tasting Food, Tasting Freedom: Excursions into Eating, Culture and the Past*. Boston: Beacon Press, 1996.
7   Warde A. *Consumption, Food and Taste: Culinary Antinomies and Commodity Culture*. London: Sage, 1997.
8   Bourdieu P. *Distinction: A Social Critique of the Judgement of Taste*. London: Routledge, 1984.
9   Crowther G. *Eating Culture: An Anthropological Guide to Food*. Toronto: University of Toronto Press, 2013.
10   Shirky C. *Here Comes Everybody: How Change Happens When People Come Together*. London: Penguin, 2009.
11   Kjaernes U, Harvey M, Warde A. *Trust in Food: An Institutional and Comparative Analysis*. Basingstoke: Macmillan/Palgrave, 2007.
12   Trentmann F. *The Making of the Consumer: Knowledge, Power and Identity in the Modern World*. New York: Berg, 2005.

13   Gabriel Y, Lang T. *Unmanageable Consumer*. London: Sage, 2015.

14   Levett R, *et al. A Better Choice of Choice: Quality of Life, Consumption and Economic Growth*. London: Fabian Society, 2003.

15   Jackson T. *Motivating Sustainable Consumption: Report to the SDRN*. London: Sustainable Development Research Network, 2005.

16   IEFS. *A Pan EU Survey of Consumer Attitudes to Food, Nutrition and Health*. Dublin: Institute of European Food Studies, 1996.

17   Eurobarometer. *Europeans' Attitudes towards Food Security, Food Quality and the Countryside*. Special Eurobarometer, 389. Brussels: European Commission DG Agriculture and Rural Development, 2012.

18   Orbach S. *Hunger Strike: The Anorectic's Struggle as a Metaphor for our Age*. London: Faber & Faber, 1986.

19   Lupton D. *Food, the Body and the Self*. London: Sage, 1996.

20   Dixon J, Broom D, eds. *Seven Deadly Sins of Obesity: How the Modern World is Making us Fat*. Sydney: University of New South Wales Press, 2007.

21   James O. *Affluenza: How to be Successful and Stay Sane*. London: Vermilion, 2007.

22   Latouche S. *In the Wake of the Affluent Society: An Exploration of Post-Development*. London: Zed, 1993.

23   Keynes JM. Economic possibilities for our grandchildren. In: Keynes JM. *Essays in Persuasion*. New York: W. W. Norton & Co., 1963 (1930): 358–73.

24   Lansley S. *After the Goldrush: The Trouble with Affluence*. London: Century Business Books, 1994.

25   Durning AT. *How Much is Enough? The Consumer Society and the Future of the Earth*. 1st edn. New York: Norton, 1992.

26   Jackson T. *Prosperity without Growth: Economics for a Finite Planet*. London: Earthscan, 2009.

27   Skidelsky R, Skidelsky E. *How Much is Enough? The Love of Money and the Case for the Good Life*. London: Allen Lane, 2012.

28   Petrini C, Padovani G. *Slow Food Revolution: A New Culture for Eating and Living*. New York: Rizzoli, 2008.

29   Tansey G, Rajotte T, eds. *The Future Control of Food: A Guide to International Negotiations and Rules on Intellectual Property, Biodiversity and Food Security*. London and Ottawa: Earthscan and IDRC, 2008.

30   Shiva V. *Biopiracy: The Plunder of Nature and Knowledge*. Dartington: Green Books in association with The Gaia Foundation, 1998.

31   Arnold H, Pickard T. *Sustainable Diets: Helping Shoppers*. Letchmore Heath, Herts: IGD, 2013.

32   FAO, *Global Food Losses and Food Waste: Extent, Causes and Prevention*. Rome: Food and Agriculture Organisation, 2011.

33   WRAP, *The Food we Waste*. Banbury: Waste & Resources Action Programme (WRAP), 2008.

34   WRAP, *Household Food and Drink Waste in the UK 2012*. Swindon: WRAP, 2013.

35   Stuart T. *Waste: Uncovering the Global Food Scandal*. London: Penguin, 2009.

36   Salaman RN. *The History and Social Influence of the Potato*. Cambridge: Cambridge University Press, 1949.

37   Schlosser E. *Fast Food Nation: What the All-American Meal is Doing to the World*. London: Allen Lane, 2001.

38   Kinealy C. *The Great Irish Famine: Impact, Ideology and Rebellion*. Basingstoke: Palgrave, 2002.

39 Woodham Smith CBFG. *The Great Hunger: Ireland, 1845–9*. London: Hamish Hamilton, 1962.

40 Hobhouse H. *Seeds of Change: Five Plants that Transformed Mankind*. London: Macmillan, 1992.

41 Andersen R. *Conceptualizing the Convention on Biological Diversity: Why is it Difficult to Determine the 'Country of Origin' of Agricultural Plant Varieties?* FNI Report 7/2001. Lysaker (Norway): Fridtjof Nansen Institute, 2002.

42 Peterson G. *Ecosystems Services: Vavilov and Agrodiversity.* <http://rs.resalliance. org/2010/01/26/vavilov-and-agrodiversity> [accessed Feb. 2015]. Stockholm: Stockholm Resilience Centre, Stockholm University, 2010.

43 Mintz SW. *Sweetness and Power: The Place of Sugar in Modern History*. Harmondsworth: Penguin Books, 1985.

44 Ritzer G. *The McDonaldization of Society*. Thousand Oaks, CA: Sage, 2013.

45 Jakle JA, Sculle KA. *Fast Food: Roadside Restaurants in the Automobile Age*. Baltimore, MD: Johns Hopkins University Press, 1999.

46 Petrini C. *Slow Food Nation*. New York: Rizzoli, 2006.

47 Foer JS. *Eating Animals*. London: Penguin, 2010.

48 D'Silva J, Webster J, eds. *The Meat Crisis: Developing More Sustainable Production and Consumption*. London: Earthscan, 2010.

49 Dickinson R, Hollander SC. Consumer votes. *Journal of Business Research*, 22 (1991): 335–42.

50 Dickinson RA, Carsky ML. The consumer as economic voter. In: Harrison R, Newholm T, Shaw D, eds. *The Ethical Consumer*. London: Sage Publications, 2005: 25–38.

51 Baudrillard J. *The Consumer Society: Myths and Structures*. London: Sage, 1998 (1970).

52 Lang T, Barling D, Caraher M. *Food Policy: Integrating Health, Environment and Society*. Oxford: Oxford University Press, 2009.

53 Rosling H. The best stats you've ever seen. <http://www.ted.com/talks/hans_rosling_shows_the_best_stats_you_ve_ever_seen> Monterey, CA, 2006.

54 Wilkinson RG, Pickett K. *The Spirit Level: Why More Equal Societies Almost Always Do Better*. London: Allen Lane, 2009.

55 Lang T. Diet, health and globalization: five key questions. *Proceedings of the Nutrition Society*, 58 (1999): 335–43.

56 Galbraith JK. *The Culture of Contentment*. New York: Houghton Mifflin & Co., 1992.

57 Lawrence F. *Not on the Label*. London: Penguin, 2004.

58 Norberg-Hodge H. *Ancient Futures: Learning from Ladakh*. San Francisco, CA: Sierra Club Books, 1991.

59 Global Footprint Network. *Living Planet Report 2010: Biodiversity, Biocapacity and Development*. Gland: WWF, Institute of Zoology, Global Footprint Network, 2010.

60 UNEP, et al. *The Environmental Food Crisis: The Environment's Role in Averting Future Food Crises. A UNEP Rapid Response Assessment*. Arendal, Norway: United Nations Environment Programme/GRID-Arendal, 2009.

61 Barlow M, Clarke T. *Blue Gold: The Battle Against Corporate Theft of the World's Water*. London: Earthscan, 2002.

62 Shiva V. *Water Wars: Privatization, Pollution and Profit*. London: Pluto Press, 2002.

63 Chellaney B. *Water, Peace, and War: Confronting the Global Water Crisis*. Plymouth: Rowman & Littlefield Publishers, 2013.

64 Prahalad CK. *Fortune at the Bottom of the Pyramid: Eradicating Poverty through Profits*. Philadelphia, PA: Wharton School Publishing, 2004.

65 Banerjee AV, Duflo E. *Poor Economics: A Radical Rethinking of the Way to Fight Global Poverty*. New York: PublicAffairs, 2011

66 Appleton J, SCF Ethiopia Team. *Drought Relief in Ethiopia: Planning and Management of Feeding Programmes*. London: Save the Children, 1987: 186.

67 Mintel. *UK Ethnic Food Market – September 2012*. London: Mintel, 2012.

68 Walton JK. *Fish and Chips and the British Working Class, 1870–1940*. Leicester: Leicester University Press, 1992.

69 Van Zwanenberg P, Millstone, E. *BSE: Risk, Science, and Governance*. Oxford and New York: Oxford University Press, 2005.

70 Gabriel Y, Lang T. *The Unmanageable Consumer*. 3rd edn. London: Sage, 2015.

71 WHO Europe. *The Challenge of Obesity in the WHO European Region and Strategies for Response*. Copenhagen: WHO Regional Office for Europe, 2007.

72 Harris JL, *et al*. *Fast Food Facts Report 2013: Measuring Progress in Nutrition and Marketing to Children and Teens*. New Haven, CT: Rudd Center for Food Policy and Obesity, 2013.

73 Statista. Product categories with the highest TV ad spending in the United Kingdom (UK) in 2013 (in million GBP). <http://www.statista.com/statistics/270891/tv-ad-spending-in-the-uk-by-advertiser-industry-category> [accessed June 2014].

74 Godin S. *Purple Cow*. New York: Portfolio, 2002.

75 Gannon Z, Lawson N. *The Advertising Effect*. London: Compass, 2010.

76 AA/Warc. Advertising Association downgrades forecast and economic caution. Press release. London: Advertising Association & Warc, 13 Jan. 2015.

77 The Grocer. UK advertising. 2014. *The Grocer*. <www.thegrocer.co.uk> [accessed Feb. 2015].

78 Yale Rudd Center for Food Policy and Obesity, *Fast Food Marketing Ranking Tables 2012–13*. New Haven, CT: Rudd Center for Food Policy and Obesity, Yale University, 2013.

79 Perry M. The brand: vehicle for value in a changing marketplace. Advertising Association, President's Lecture, London, 7 July 1994.

80 De Onis M, Blössner M, Borghi E. Global prevalence and trends of overweight and obesity among preschool children. *American Journal of Clinical Nutrition*, 92 (2010): 1257–64.

81 Public Health England. Child weight data fact sheet. <www.noo.org.uk> London: Public Health England, 2014.

82 Lobstein T, *et al*. Child and adolescent obesity: part of a bigger picture. *The Lancet*, 2015. <http://dx.doi.org/10.1016/S0140-6736(14)61746-3>.

83 WHO. *Set of Recommendations on the Marketing of Foods and Non-Alcoholic Beverages to Children*. Geneva: World Health Organisation, 2010.

84 WHO. *A Framework for Implementing the Set of Recommendations on the Marketing of Foods and Non-Alcoholic Beverages to Children*. Geneva: World Health Organisation, 2012.

85 EU Pledge. We will change our food advertising to children. 2012. <http://www.eu-pledge.eu>

86 Committee of Advertising Practice. *Online Food and Drink Marketing to Children*. London: Committee of Advertising Practice, 2015.

87 Harrison-Dunn A-R. Campaigners disappointed by 'null conclusion' of UK report on online ads to children. <http://www.foodnavigator.com/Policy/

Campaigners-disappointed-by-null-conclusion-of-kids-advergame-report/?utm_
source=newsletter_daily&utm_medium=email&utm_campaign=11-Feb-2015&c=
3AEzn8XrIuocAm47uqPTlA%3D%3D> [accessed Feb. 2015]. Montpellier:
FoodNavigator.com, 2015.

88 Compass. *The Commercialisation of Childhood*. London: Compass, 2006.
89 Mayo E. *Shopping Generation*. London: National Consumer Council, 2005.
90 Kennedy JF. Special message on protecting the consumer interest: statement read to
Congress by President John F Kennedy Thursday, Washington, DC, 15 Mar. 1962.
91 Wiseman MJ. Government: where does nutrition policy come from? *Proceedings of
the Aristotelian Society*, 49 (1990): 397–401.
92 Lang T. The contradictions of food labelling policy. *Information Design Journal*, 8(1)
(1995): 3–16.
93 United Nations. *United Nations Guidelines for Consumer Protection (as Expanded in
1999)*. New York: Department of Economic and Social Affairs of the United Nations,
2003.
94 Consumers International. *Consumer Rights*. CI 50 Policy Briefing. <http://www.
consumersinternational.org/who-we-are/about-us/we-are-50/ci50-policy-briefings>
London: Consumers International, 2014.
95 Paulus I. *The Search for Pure Food*. Oxford: Martin Robertson, 1974.
96 Wilson B. *Swindled: From Poison Sweets to Counterfeit Coffee – The Dark History of
the Food Cheats*. London: John Murray, 2008.
97 Caraher M, Lang T, Dixon P. The influence of TV and celebrity chefs on public
attitudes and behaviour among the English public. *Journal of the Association for the
Study of Food and Society*, 4(1) (2000): 27–46.
98 NOP. *Taste 2000: Research Carried out for Hammond Communications on Cooking
(Geest Foods)*. London: National Opinion Polls, 1997.
99 National Food Alliance. *Get Cooking!* London: National Food Alliance, Department
of Health, and BBC Good Food, 1993.
100 Oliver J. *The Ministry of Food*. TV series and book. London: Channel 4 TV, 2008.
101 Monteiro C. The big issue is ultra-processing. *World Nutrition*, 1(6) (2010): 237–69.
102 Monteiro CA. Nutrition and health: the issue is not food, not nutrients, so much as
processing. *Public Health Nutrition*, 12(5) (2009): 729–31.
103 Ministry of Health (Brazil). *Guia Alimentar para a População Brasileira*. Brasilia:
Ministério da Saúde, 2014.
104 Fieldhouse P. *Food and Nutrition: Customs and Culture*. 2nd edn. London: Chapman
& Hall, 1995 (1985).
105 Short F. *Kitchen Secrets: The Meaning of Cooking in Everyday Life*. Oxford: Berg,
2006.
106 Gatley A, Caraher M, Lang T. A qualitative, cross cultural examination of attitudes
and behaviour in relation to cooking habits in France & Britain. *Appetite*, 75 (2014):
71–81.
107 Stitt S, *et al*. Schooling for capitalism: cooking and the national curriculum. In: Kohler
BM, *et al*., eds. *Poverty and Food in Welfare Society*. Berlin: WZB, 1997.
108 National Restaurant Association. *2014 Restaurant Industry Pocket Factbook*. <http://
www.restaurant.org/News-Research/Research/Facts-at-a-Glance>* Washington, DC:
National Restaurant Association, 2014.
109 Gershuny J, Fisher K. Leisure. In: Halsey AH, Webb J, eds. *Twentieth-Century British
Social Trends*. Basingstoke: Macmillan, 2000: 620–49.

110 Lang T, Barling D. The environmental impact of supermarkets: mapping the terrain and the policy problems in the UK. In: Burch D, Lawrence G, eds. *Supermarkets and Agrifood Supply Chains.* Cheltenham: Edward Elgar, 2007: 192–219.

111 Rothstein H. *Science and Policy in the Regulation of Food-Contact Plastics.* London: University of Sussex, 1993.

112 Davidson A. *The Oxford Companion to Food.* Oxford: Oxford University Press, 1999.

113 Sharpe R. *An Inconvenient Sandwich: The Throwaway Economics of Takeaway Food.* London: New Economics Foundation, 2010.

114 Burch D, Lawrence G, eds. *Supermarkets and Agri-food Supply Chains.* Cheltenham: Edward Elgar, 2007.

115 OECD, FAO. *Agricultural Outlook 2011–20.* Paris and Rome: Organisation for Economic Cooperation and Development and Food and Agriculture Organisation, 2011.

116 Hansard. Food banks: debate in the (UK) House of Commons, 18 Dec. 2013, columns 806–55. *<http://www.publications.parliament.uk/pa/cm201314/cmhansrd/cm131218/debtext/131218-0003.htm>* London: Hansard, 2013.

117 Lambie-Mumford H, *et al.*, *Household Food Security in the UK: A Review of Food Aid.* London: Department for Environment, Food and Rural Affairs, 2014.

118 Burlingame B, Dernini S, eds. *Sustainable Diets and Biodiversity: Directions and Solutions for Policy, Research and Action. Proceedings of the International Scientific Symposium 'Biodiversity and Sustainable Diets United against Hunger', 3–5 November 2010, Rome.* Rome: FAO and Bioversity International, 2012.

119 Monteiro CA, *et al.* Increasing consumption of ultra-processed foods and likely impact on human health: evidence from Brazil. *Public Health Nutrition,* 14(1) (2011): 5–13.

120 SAI. Sustainable Agriculture Initiative platform. <http://www.saiplatform.org> Brussels: Sustainable Agriculture Initiative, 2014.

121 Which? *The Future of Food: Giving Consumers a Say.* London: Which?, 2013.

122 Reddy S, Lang T, Dibb S. *Setting the Table: Advice to Government on Priority Elements on Sustainble Diets.* London: Sustainable Development Commission, 2009.

123 Macdiarmid J. Is a healthy diet an environmentally sustainable diet? *Proceedings of the Nutrition Society,* 72(1) (2012): 13–20.

124 WWF. *Food Patterns and Dietary Recommendations in Spain, France and Sweden: Livewell for Low Impact Food Europe.* Godalming: WWF, 2012.

125 Vieux F, *et al.* Greenhouse gas emissions of self-selected indvidual diets in France: change the diet structure or consuming less? *Ecological Economics,* 75 (2012): 91–101.

126 OPUS. Developing the new Nordic diet. <http://foodoflife.ku.dk/opus/english/wp/nordic_diet> Copenhagen: University of Copenhagen Research Center OPUS, 2009.

127 Barsac Declaration Group. The Barsac Declaration: environmental sustainability and the demitarian diet. <http://www.nine-esf.org/sites/nine-esf.org/files/Barsac%20Declaration%20V5.pdf>. Barsac: European Science Foundation Nitrogen in Europe (NinE) research networking programme, Biodiversity in European Grasslands, et al., 2009.

128 Lang T. *Choice, Power and Food: Nutrition in an Ecological Public Health Era.* London: Centre for Food Policy City University, 2007.

129 Lang T, Heasman M. *Food Wars: The Global Battle for Mouths, Minds and Markets.* London: Earthscan, 2004.

130 Rayner G, Lang T. *Ecological Public Health: Reshaping the Conditions for Good Health.* Abingdon: Routledge/Earthscan, 2012.

131 Sutton MA, *et al. The European Nitrogen Assessment Sources, Effects and Policy Perspectives.* Cambridge: Cambridge University Press, 2011.

8 Food democracy or food control?

A day will come when the only fields of battle will be markets opening up to trade and minds opening up to ideas. A day will come when the bullets and the bombs will be replaced by votes, by the universal suffrage of the peoples.

Victor Hugo (1802–85) in a speech to the Congrès des amis de la paix
universelle, Paris, 22 August 1849

Core arguments

Food policy requires different issues to be joined up rather than being dealt with in different policy silos. This is complex for governments, industry and civil society. Food is historically a sensitive issue. People want their food to be affordable, available and fit to eat but also to meet other values. Many institutions of food governance in both public and private sectors struggle with this complexity and how to maintain consumer trust. They have failed fully to address the changing nature of the food economy, the challenges to human and eco-systems health, and the rise of corporate influence. Over the last century, power over food systems has tended to move from local to global, weakening perceived democratic control over food. The modern evidence about food's impact on health, environment and society ought to be a central concern for the state since only the state and its institutions can ultimately be a democratically accountable facilitator and mediator between increasingly powerful interests and the consumer which complex modern supply chains require. Yet most official food policies are wedded to the 'hands-off' versions of the state promulgated by neo-liberalism. Productionism which was put into place with firm state investment and support in the 20th century coincided with a growth of corporate power. And now that the state is generally reluctant to be seen to be too hands-on about food, it remains the default policy position. Productionism seems unable to address these present, let alone future, challenges. A central tension for future food governance will be negotiating the imbalance between 'food democracy' and 'food control', one being publicly accountable, the other being 'top-down'. The chapter reviews issues where this tension is manifest, such as trade, food rights, institutional structures and subsidies. Different models and interpretations of food democracy exist. Some argue that democracy occurs in the form of consumer choice at the check-out. Others counter that democracy requires the setting of new frameworks to achieve coherent and long-term sustainability. At the very least, the awesome challenges highlighted in this book about food's role in health, the environment, society and economy suggest that more dynamic and inspiring leadership and open debate are needed. The stakes are high but there are signs that actors spread throughout the food system recognise the need for systemic change.

Why food democracy and good governance matter

In this chapter we argue that there is a crisis of democracy about food and the direction of the food system. The institutions that are supposed to bring democratic control and accountability over food systems are not in full command of the food system. Partly, this is because they have withdrawn from the terrain and cede responsibility to corporations in the marketplace, often citing consumer sovereignty as the guiding principle. And partly, this is because the problems cross boundaries and are intrinsically complex. Official objectives to ensure food is good for health, environment and social well-being are not quite working. Food's health toll and environmental impacts are accepted as normal and the price of progress. Some critics see this as inevitable. In societies which are shaped by the pursuit of profits this sets democracy against company power. Others argue that systems of governance – a curious English word that refers to the processes and practices of how decisions are made, whether by government or any relevant societal forces – can exert suitable mixtures of policies and can referee conflicts and can ensure that fair play occurs and that the general public good is achieved.

Others think that there is no need for government at all. Leave it to the marketplace, they cry. Let individual consumers relate to the companies through the buyer–seller relationship. This market view assumes that consumers and companies in supply chains are in equal positions of power and knowledge, when they are not. As scandals about food such as the 1980s mad cow disease episode,[1] or the 2013 horsemeat exposé have shown,[2] supply chains are amazingly long and complex to the extent that even companies don't know what is in their products. And also, the map of power has been fundamentally skewed by the emergence of hugely powerful corporations dominating the global and regional food systems. Although in theory consumers are powerful because they can shape markets by what they buy, in reality they are relatively weak because they are poorly organised, easily marketed to and manipulated, and ultimately are dependent on suppliers for their food. They might want food but others control the contracts, specifications and relationships within supply chains. That is why government matters, the critics argue.

And so the debate goes on. Debate and the right to do so are intrinsic to most notions of democracy. So too for food democracy. How is the 21st century to have a food system which is good for health, environment, society and culture? What would such a food economy look like? On whom and what can we rely to deliver this nirvana? What are the processes and mechanisms by which food systems can change? Addressing such questions suggests the value of the concept of food democracy, a term used since the 1990s to capture the struggles and aspiration for good food for all. The reality, however, is that there is tension between food governance shaped by democratic aims and that which achieves change by exerting control. There is a battle over food power: whoever has it determines whether difficulties covered in the book are resolved, fudged or denied. We thus propose a tension between food democracy and food control as a shorthand for this important debate.

This is a live debate. Examples of the high stakes at issue are the two vast trade negotiations by the USA in the 2010s, one westwards – the Trans-Pacific Partnership (TPP) – and one eastwards – the Transatlantic Trade and Investment Partnership (TTIP). These are both cast in the trade liberalisation mould: opening borders, harmonising standards, favouring business, spreading processed food products. The TPP negotiations between nine countries (Australia, Brunei Darussalam, Chile, Malaysia, New Zealand, Peru, Singapore, Vietnam and the USA) began in 2011.[3] Critics rightly pointed out that the food element of this was partly about opening Asian markets to US meat – i.e. promoting the nutrition transition with its inevitable ill health outcomes.[4] The TTIP negotiations between the USA and the EU began formally in 2013 and, like the TPP, had considerable implications for food and farm trade, an attempt to harmonise trade rules between these giant trading blocs.[5] Nominally a trade agreement, the early negotiations were being carried out in secret or with little public transparency and it became apparent that the negotiations went beyond trade and were addressing social and labour issues. The significance of these bilateral trade negotiations became clear due to dogged work by civil society organisations which forced some level of public scrutiny.[6 7] NGOs and others have become especially concerned over the features of the TTIP which would lead to so-called 'regulatory cooperation'. This would create a mechanism for the EU and the US jointly to review existing rules or standards (such as labour or food safety standards) that are seen as barriers to trade by industry, and prevent new ones in the future. The concern for critics was that this would be an excuse to reduce standards such as those for consumer protection, workers, the environment and health. While the TTIP remit goes far wider than food, it has profound consequences for food supply. For food safety, for example, this might mean EU member states having to accept lower standards for chemicals used in food production, antibiotic and hormone use in animals, and to accept controversial technologies such as genetic modification, on which the EU has been wary. Especially worrying was the proposed Investor State Dispute Mechanism which was to give the right to corporations to sue governments over regulations that undermine their expected profit and for these disputes to be settled in courts outside country jurisdictions. This would seriously weaken national democratic decision-making and legal traditions. Many large-scale food and agribusiness groups expressed support for the TTIP, wanting wider markets for processed food products and brands. In May 2013 over 20 US agribusiness industry groups, particularly from meat, dairy and grain industries, outlined support for what they saw as the benefits of the TTIP.[7] Here is a classic example of the tensions between food democracy and food control at the international level. It echoed similar tensions over the negotiation of the General Agreement on Tariffs and Trade (GATT) in 1987–94 which brought agricultural products into the world trade system.[8] Rules are set way beyond the level of the check-out or ordinary consumer consciousness to suit very powerful forces within the food economy. [9] Multinational food corporations were and are active at such global levels and can frame the discourse within which the states conduct their negotiations. Representatives of civil society find this harder, and are often excluded. This is why transparency and accountability within food governance are such important democratic issues.

Governance, according to Richards and Smith, is:

> generally a descriptive label that is used to highlight the changing nature of the policy process in recent decades. In particular, it sensitizes us to the ever-increasing variety of terrain and actors involved in the making of public policy. Thus, governance demands that we consider all the actors and locations beyond the 'core executive' involved in the policy-making process.[10]

Governance is a term used to cover the execution of power, not just what is done but how and by whom. Implementation matters; what is ignored and what is acted upon. In this sense, governance is not limited to governments. Indeed, with the rise of neo-liberalism influencing what governments do and do not do, the term governance has spread to cover how any institution disports its power and responsibilities.[11] Different political regimes each have their own traditions of, and approaches to, food governance. The European Union and the USA, for example, are federal in their food decision-making. States are in constant interplay with superior levels of governance. EU Member States decide much about food at Brussels rather than at the national level. Policies are negotiated.[12] The Common Agricultural Policy is as heated an issue in Europe, as is the Farm Bill in the USA or Five Year Plans in China, hence the significance of EU and US leaders deciding to engage on a new round of trade liberalisation through the TTIP. Where was the public engagement or consultation? There were no stakeholder roundtables or PPPs used to dampen down radical action on health or environmental challenges!

A major challenge for food democracy is how to integrate health, the environment and social justice. These are cross-cutting policy issues which too often fall between the institutions of food governance. The labyrinthine nature of local, national, regional and global institutions makes for awkward food governance. Today, more than at any time in human history, institutions central to food and sustainability are multi-level. They can be placed in at least five levels of governance, from the global to the community (see Table 8.1). There is tension between these levels, with political debate sometimes raging about whether power is sliding up to the global institutions, or whether more local institutions have lost their power and influence, or whether the global bodies are ultimately having to bow down to regional, national and more accountable local levels. The European Union has a principle of subsidiarity, by which any decision is supposed to be taken at the level most appropriate to the problem. Despite this there are endless arguments about whether decisions are being taken in distant unaccountable ways by 'Eurocrats'. Similar sentiments are expressed universally by locals about the levels 'above' across the globe.

Table 8.1 Multi-level food governance by governments and corporate sectors

Level of governance	*Some governmental institutions*	*Some roles and functions*	*Some company institutions*
Global	UN: WHO, FAO, UNEP, Codex Alimentarius Commission etc.; WTO, World Bank, IMF; G7 (G8 with Russia), G20; CGIAR	Intergovernmental negotiations; coordination of expert consultations; setting and sharing policy agenda and standards	World Economic Forum; World Business Council for Sustainable Development; Sustainable Agriculture Initiative; Global Food Safety Initiative; GlobalGAP; International Life Science Institute
Regional	North American Free Trade Agreement; Transatlantic Trade and Investment Partnership; TransPacific Trade Partnership; European Union; ASEAN; Mercosur; Organisation of African States	Set trade rules between Member States; develop regulation; cross-border food safety issues	FoodDrinkEurope; Copa-Cogeca (EU); Grupo de Pulses Productores del Sur (Latin America); Institute of Food Technologists
National	c.200 nation states exist within the UN	Legislation and regulation; healthcare, policy covering food supply chain; dietary guidelines	Grocery Manufacturers of America; National Restaurants Association (USA);
Sub-national	Regional health bodies; elected regional assemblies	Coordination of local initiatives; regional voice and policy	Chambers of commerce; regional development agencies;
Local/ community	Town or village council; health authorities; community centres	Delivery of local services such as food law enforcement, primary health care, dietary advice	Farmers groups; community business associations; village trade associations;

Tension between globalisation and localisation are visible across the food system. Food enforcement has to occur locally yet legislation is increasingly being framed, not at national or regional but at international level through such bodies as the European Union, and trade agreements such as were discussed above – TTIP and TPP – and earlier versions from the 1990s – the World Trade Organisation, the Asia-Pacific Economic Cooperation or the North American Free Trade Agreement. The TTIP is a further extension of this trend, and signals new forms of global governance: bilateral deals rather than through the UN or other global forums. These expose tensions within globalisation,[13] but they occur in a

Table 8.2 Formal global institutions involved in health

Remit	Examples of organisation/bodies
Public health	World Health Organisation (WHO), Food and Agriculture Organisation (FAO), World Food Programme (WFP); OIE (world organisation for animal health)
Children and health	UNICEF – UN Children's Fund, UNESCO
Global economic bodies with health impact	World Bank, International Monetary Fund, UN Conference on Trade & Development (UNCTAD), World Trade Organisation (WTO), World Intellectual Property Organisation (WIPO), Organisation for Economic Cooperation and Development (OECD)
Intergovernmental agreements with a health impact	Bio-safety Convention; International Conference on Nutrition, Basel Convention on hazardous waste; Bio-Safety Protocol
Emergency aid	World Food Programme, International Committee of the Red Cross/Crescent;
Environmental health	Global Panel on Climate Change, UN Conference on Environment & Development (UNCED), International Maritime Organisation
Commercial interests	International Chamber of Commerce, Transnational Corporations, International Federation of Pharmaceutical Manufacturers Associations, World Economic Forum, McKinsey
Regional bodies with health role	European Union, Regional Offices of WHO and FAO; Pan American Health Organisation
Trade associations	International Hospitals Federation
Networks to promote public health ([UN] indicates UN support)	Healthy Cities Network [WHO], International Baby Food Action Network (IBFAN), Local Agenda 21 network, Pesticides Action Network, Tobacco Free Initiative [WHO]
Professional associations	International Union of Health Education; World Public Health Nutrition Association; International Diabetes Federation; World Heart Foundation
Non-governmental organisations	Médecins sans Frontières, Médecins du Monde, World Federation of Public Health Associations; NCD Alliance; World Obesity Federation; AICR/World Cancer Research Fund; Oxfam

world with power moving inexorably 'upwards'. This marginalises local decision-making processes and the sphere of existence and 'choice' where most consumers exist, and buy or eat food.

At the global level, there are many bodies or institutions which have either a pre-eminent role in relation to national ministries and agencies or which have a coordinating role. Table 8.2 provides examples in relation to food and health only. Similar exercises could be done for institutions covering the environment or social aspects of food. Some have particular weight and power, particularly the financial institutions like the World Bank, the WTO, but the UN remains a softer global forum for debate about the present and future state of the world. At each governance level, there are initiators and think-tanks who seek to influence the policy and food agenda. They collect data and evidence, and lobby the formal decision-makers. Thus each level of governance becomes a battleground for vested interests, and what gets onto policy agenda. Over the last 30 years, there has been a growth of powerful commercial and business-oriented bodies seeking to frame the food supply chain within and between nation states.[14] Of particular note is the emergence of consultancies such as McKinsey, Bain or PwC as key players in food policy working either in the open or as advisers to other forums such as the World Economic Forum.[15-17]

Such institutions frame and tussle over existing policy commitments, constantly debating responsibility and control over the food supply chain. In theory, this is the function intended for the UN, but as the power of states has been altered, so the influence and legitimacy of the UN has been weakened, not helped by the greater influence and financial power of the Bretton Woods institutions – the World Bank and International Monetary Fund.

Civil society and food activism

If the purpose of food democracy is to improve accountability and openness in the food system, we have to monitor how the state and other institutions engage with the people and communities. The dominant market approach conceives of them simply as consumers, but over recent decades, as governance has been dissipated, it has become common for any decision-making to adopt a 'stakeholder consultation' approach. In this, the powerful seek the views of others who are either in the terrain or whose interest might destabilise the direction of travel. The growth of civil society organisations – particularly in the form of constituted NGOs – has seen them emerge as key players in such stakeholder and consultation processes. Sometimes NGOs are metaphorically 'shouting' from outside (wearing jeans and waving placards), and sometimes they are invited into the consulting room (wearing suits and carrying briefcases). One can identify NGOs across an 'inside-outside' spectrum. The growth of NGOs has been phenomenal. By the turn of the 21st century, there were 1–2 million NGOs in the USA, about 1 million NGOs in India, 210,000 in Brazil, 96,000 in the Philippines, 27,000 in Chile, 20,000 in Egypt and 11,000 in Thailand.[18] They were even growing in China but strictly under licence. One estimate is that, by the mid-2000s, there were over 1 million, perhaps 2 million, around the world working towards ecological sustainability and social justice alone.[19] NGOs vary in their approach, style and purpose. Some are focused on service delivery, while others are think-tanks. Some are small and others are vast, well-resourced membership organisations. Some seek behind the door influence, while others are public campaigners. Most are single-issue in focus.

In the 1980s and 1990s, some extraordinarily creative and effective NGOs emerged in the food policy world: in Brazil, for example, groups like IBASE (Instituto Brasil rode Análises Sociais e Econômicas) in Rio and the Global Forum on Sustainable Food and Nutritional Security in Brasilia both lent their weight to the mass campaigns for the right to eat and against hunger; in India, groups like the Delhi-based Research Institute for Science, Technology and Ecology worked with farmers on sustainable agriculture and in defence of home-bred seeds, while consumer groups like the Jaipur-based Consumer Union and Trust Societies (CUTS) worked on food standards and globalisation; and in Australia, the Australian Consumers' Association – renamed Choice – a member of Consumers International, worked on labelling, while the Eco Consumers Network linked human and environmental health and the Landcare network encouraged positive systems for protecting our fragile soils and ecosphere. In Canada, too, the Toronto Food Policy Council pioneered a new forum for stakeholders in the local food system working through the municipal health department while at a national level the Council of Canadians tackled governmental sale of Canada's ample water assets. In Malaysia, the Consumers' Association of Penang campaigned against pesticides and for the quality of chicken feed, while the Third World Network pushed back the boundaries of consumerism by being the earliest to recognise the significance of the Uruguay Round of GATT.[8] In the UK, an alliance was created in 1985 to bring together diverse food-related NGOs to have a stronger and more coherent voice in policy. It merged with another smaller alliance later, and today Sustain represents around 100 UK NGOs with an interest in food matters, from health, environment, animal welfare, social justice and international development.[20] Over many years, it has represented civil society interests and developed innovative campaigns to push for a range of neglected or civil society food issues to be put on policy and public agendas. Its Childrens' Food Campaign, for example, has a bigger membership than itself. Such alliances have become important for food democracy, because they reflect a political realism that NGOs have to be better organised, less 'single-issue' dominated, and more active in both appealing to the public and trying to influence policy processes. They wear both jeans and suits, to continue the metaphor above.

One of the key turning points for NGOs at the global level was the organising that went into the demonstrations in Seattle in December 1999 in protest at the WTO negotiations.[21] This was an unprecedented coming together of NGOs from all around the world. They were expressing deep disquiet at the form the world trade talks were taking. Most of the activists at Seattle were motivated by environmental more than public health concerns; if anything, the health concerns focused on safety and contamination rather than non-communicable disease or the nutrition transition. In the last two decades, spearheaded by alarm at obesity, this unevenness in the collective NGO position has improved. NGOs have played an important role in opening up popular discourse on issues such as labelling, the introduction of new food technologies, food adulteration, food poverty, the morality of marketing targeted at children and nutritional standards of public food.[22]

Civil society, perhaps because it has influence rather than power, has been quicker than government or companies to recognise the need for joined-up food policy. As social movements, they are the lifeblood of food democracy, part conscience, part activists, part thinkers.[23] They are influential in creating a shared view that food systems need change.[24] Although they emerged as single-issue campaigns, very often, there is a clear recognition today of the need for more integrated perspectives in food systems. An example of this was the 2014 launch of an alliance between Consumers International, the world's federation of consumer groups, and World Obesity Federation, an alliance of health scientists working on obesity. This was supported by over 300 professionals and organisations. It proposed a new Global Framework to tackle diet-related ill health, just as the UN had negotiated a Framework Convention on Tobacco Control. 'If this can be done for smoking, why not for food?' was their argument. They cited that, despite the WHO's 2004 Global Strategy on Diet and Physical Activity, deaths from diet-related ill health had grown. Global deaths attributable to obesity and overweight alone had risen from 2.6 million in 2005 to 3.4 million in 2010.[25] In 2008, they cited, 36 million people had died from non-communicable diseases, representing 63 per cent of the 57 million global deaths that year. In 2030, such diseases are projected to claim the lives of 52 million people. That a consumer federation would take such a strong public health line calling for regulation at the global level would have been unthinkable in the 1980s, when the NGO world split over positions on trade; some were for liberalisation, while others were opposed, seeing it as the route to greater commercial concentration of power.[26] Due to their capacity to generate publicity, NGOs have become powerful voices within government and company dialogue. They can and increasingly do challenge business and governments to reframe food governance and to set the rules of food engagement in a way that biases towards ecological public health.

Despite such grand ambitions, formal reviews of consumerism conclude that consumer NGOs have limited industrial 'muscle'.[27] In fact, history suggests moments where their influence can be significant. NGOs halted the Seattle world trade talks, but arguably this created the conditions for a new generation of bilateral trade deals such as TPP and TTIP. What methods do they have? The oldest is the boycotts. These were pioneered in the USA against the British as early as 1764–76. [28] They can be successful if coupled with mass political movements, as happened with Gandhi's salt tax boycott when leading his revolt against the British. The international boycott of Nestlé in the late 20th century for breaking of the UN code against the marketing of breast-milk substitutes is widely praised as a brilliant and effective campaign, but it has not broken the company; it has embarrassed it, of course, and on occasions, it has dented the share price.[29 30] In that respect, the pursuit of food democracy has no final end-point but is a continual process of monitoring, educating, speaking out and holding the powerful to account. Among the skills that NGOs bring to the Food Wars are perhaps their speed and flexibility, their ability to challenge carefully crafted commercial images, a high level of public trust, international networks and media credibility, and quality and independence of research and policy proposals. The internet age has accentuated this mechanism.[22 31] All the NGO armoury can be amplified by skilful use of social media in addition to traditional campaigning tools.

NGOs by no means have public discourse to themselves. Vast armies of global public relations companies and lobbyists work to promote the interests of their corporate clients. This is the world of 'spin', which Miller and Dinan suggest ought really to be known as propaganda because 'it implies the unity of communication and action. It is communication for a purpose.'[32: 4] They are right. The distinction between journalism and PR is easily blurred. The growth of PR has been phenomenal. Whereas in 1980 there were 45 PR workers for every 36 journalists in the USA, by 2008 this had risen to 90 PR people for every 25 journalists.[33] PR and lobbies are not alone in promoting commercial interests. Advertising is the oldest form of exerting control. In 2011 global ad-spend was $466bn,[34] it grew by 50 per cent in 1995–2009.[35] Just as we can discern a spectrum of positions and roles among NGOs working on the world of food, so there is a spectrum in commercial lobbying and PR. Stung by health attacks on its industry, the fast-food industry spawned a Center for Consumer Freedom, which specialises in attacking critics, and singling out leading thinkers for personal critique.

One could argue that the creation of such ideological organisations defending the right to poor health is a backhanded compliment to NGO effectiveness in shaping food and health discourse, but it makes food governance messier. At the same time, it shows company strategies are taking their gloves off when dealing with their critics. Often a first line of defence is to deny links between consumption and ill health and to downplay food's role in eco-systems impact. This gets harder as evidence builds up. As a result, a second strategy comes to the fore. In this, more responsible food companies engage with NGOs and begin to take on board the case for transition. This then becomes the tricky terrain of judging whether there is assimilation and dissipation. Being invited into stakeholder consultation can blunt the health or environmental critics' edge. There is certainly some scepticism in social movements about public–private partnerships (PPPs) which have mixed results regarding their effectiveness to deliver public health benefits.[36] The next step is the final slug-out, when the gloves are off. That's when democracy really matters.

Building on existing commitments

Besides their national commitments and policy objectives, governments are often already signed up to existing commitments which signal the directions food and health policy could take. Table 8.3 gives a list of some key global commitments to nutrition, food safety and sustainable development. These go back to the founding of the UN system in the 1940s, with some governments placing a stronger

Table 8.3 List of global commitments, post-World War II

Occasion	Date	Nutrition	Safety	Sustainable food supply
Universal Declaration of Human Rights	1948	+	+	
UN Covenant on Economic, Social and Cultural Rights	1966	+		
Stockholm Environment & Development	1972			+
World Food Conference (Universal Declaration on the Eradication of Hunger and Malnutrition)	1974	+	+	
Convention on the Rights of Child	1989	+	+	
Innocenti Declaration on Breastfeeding	1991	+	+	
UN Conference on Environment and Development & Rio Declaration, UN Framework Convention on Climate Change and on Biological Diversity	1992			+
International Conference on Nutrition	1992	+	+	+
World Conference on Human Rights, Vienna & Vienna Declaration and Programme of Action	1993	+		+
UN 4th World Conference on Women & Beijing Declaration and Platform for Action	1995	+		+
World Food Summit	1996	+	+	+
UN Habitat 2 & Istanbul Declaration	1996			+
Global Convention on Climate Change (Kyoto Protocol)	1997 (2005)			
UN General Comment on the Right to Adequate Food	1999	+	+	+
World Health Assembly (Resolutions 53.15, 51.17, 53.18)	2000	+	+	+
UN Millennium Development Goals	2000	+	+	+
World Food Summit (Rome)	2002	+	+	+
World Summit on Sustainable Development (Johannesburg)	2002	+	+	+
WHO Global Strategy on Diet and Physical Activity (DPAS)	2004	+		+
L'Aquila Declaration on Global Food Security (G8 but signed by 28 countries)	2009	+	+	+
UN Conference of Sustainable Development (Rio+20)	2012	+		+
New Alliance for Food Security and Nutrition	2012	+	+	+
Rome Declaration and Framework for Action (International Conference on Nutrition – ICN2)	2014	+		+
UN Sustainable Development Goals	2015	+	+	+

emphasis on aspects of food and health than others. The point is that there is a long history of formal concern; but these attempts are far from being consistently integrated in practice.

Such global commitments are reminders there is no policy vacuum. On the contrary, food has a long presence in public health, environmental protection and sustainable development policy, but are these achieving the requisite grip on the undesirable features of the food system? And who decides what is undesirable? That's the democratic challenge. If these policies were in command, surely the indicators reviewed elsewhere in this book would not be as sobering as they are? Fine resolutions and commitments meet the harsher realities of food culture being driven in directions at variance with what a more sustainable food world requires.

Food rights

The notion of food rights is another lever on food system change. Slowly since the 1948 Universal Declaration on Human Rights,[37] there has been an accrual of international commitment to ensure that all humans receive a 'right to food'. This process of policy accumulation has been summarised elsewhere.[38 39] From the 1948 Universal Declaration to the 2004 Voluntary Guidelines on the realisation of the Right to Food, international human rights lawyers have built the case for each nation state to be held accountable for meeting a decency threshold for food. [40] In 2000, the UN created a new role of a Special Rapporteur on the Right to Food to audit progress. The first was Jean Zigler (2002–8), the second Olivier De Schutter (2008–14) and the third Hilal Elver (appointed in 2014). This office and its many country and topic reports have become a useful mirror on the state of food democracy.[41] The Rapporteur has powers to review country performance or sensitive issues such as land-grabs, the plight of smallholders and also inequalities of health and access to food. The Rapporteur reports to the UN Secretary General through the UN Office of the High Commissioner for Human Rights, and UN Committee on Economic, Social and Cultural Rights. This Office highlights an important political division; in one corner are those who argue rights are something which can or should be enforced and that ethics can (re)shape markets. [42] For them, rights are real. In the opposing corner are those who think this is a woolly liberal obfuscation, which delays and distorts the pursuit of the more effective mechanism of market discipline.

How institutions frame food and health: Codex and the EU

Trade has long been a key arena in which significant agricultural and food politics takes place.[43 44] At the great agriculture and food conferences of the 1940s at which Productionism was put in place – in 1942 at Hot Springs, 1945 in Quebec, 1947 in Washington, DC, and 1948 in Cairo[45]– the conviction that Productionism would deliver progress was embedded. These were optimistic times. The global consensus was to raise the quantity of food and to sell and transport it to the people as cheaply as possible. This is the era of the founding of the UN, symbolised by its striking new building in New York, away from Geneva and the war-zones of Europe where the League of Nations had been based (and where still some UN agencies reside).

It was not long before the optimistic glow began to fade. In 1964, a grouping of developing countries was created to argue that the new world order favoured the rich world. New bodies were created such as UNCTAD within the UN, and G77 outside. Anti-colonial politics articulated the view that aid and development were holding poor nations down,[46] and that the new world order 'rigged' the rules to benefit the rich, thus structuring a world with unequal access to food.[47] While these political fights found expression inside the UN and its food-related agencies such as the FAO, WHO and UNICEF, and newer programmes such as UNCTAD and UNEP, these bodies were being sidelined by existing Bretton Woods institutions such as the World Bank and IMF and newer institutions favouring neo-liberal political and economic approaches. Important among these was the creation of the WTO in 1994. This came out of long negotiations to bring agriculture and food into the General Agreement on Tariffs and Trade (GATT), the world trading rules.

Like the UN, the GATT came into existence in the 1940s and was signed in 1948 by two dozen countries committed to reducing tariffs and barriers to trade and to promote trade as the means to development and wealth-creation. Food issues under the provisions of the GATT were slow to be incorporated because of resistance from the developed countries, particularly the USA, which saw the GATT as a threat to its new food power, flexed in World War II and bolstering the Western side of the Iron Curtain through the Marshall Plan.[48] The GATT was slowly revised and extended almost as soon as the first rules were created, but food was only brought into the system in the protracted negotiations known as the Uruguay Round (1987–94). The GATT system is now policed and developed by the WTO secretariat from its headquarters in Geneva.

Besides its general commitment to trade liberalisation, the 1994 GATT contained a number of subsidiary agreements which transformed the world of food, such as the Agreement on Agriculture (AoA), on Technical Barriers to Trade (TBT) and Sanitary and Phytosanitary Standards (SPS), where food and health issues are encoded. An approach to standards was put in place which is being repeated in 21st-century bilateral deals such as the TTIP or TPP. The US and the EU fought hard in the Uruguay Round and thereafter to position their agricultures to advantage and to protect the interests of their food industries, in particular their positions on biotechnology and their definitions of risk. This saw the rich world protecting their trade interests against less-subsidised agricultural exporting countries, known as the Cairns Group (17 agriculture-exporting countries committed to a market-oriented trading system) and the informal 'Like Minded Group' of developing countries. NGOs from across the world were educated by these processes, and began to be sceptical about whose interests were triumphing. Hard-won safety and public health protection assumptions were swept away in the pursuit of harmonizing and reducing regulations to facilitate cross-border trade. Some developing country governments were also angry. At a meeting in September 2003 at Cancun, Mexico, to EU and US negotiators' astonishment, a group of 21 countries from the South led by Brazil, China and India emerged and confronted the food hyper-powers. The temporary G-21 Group took a hard line, demanding that the US and the EU reduce their barriers, cut subsidies and open up markets to their cheaper commodities,[49] without which conditions it would not entertain any kind of agreement.

It presented evidence of the poor being penalised, indicating that, in rich countries, agriculture represents less than 2 per cent of total national income and employment, whereas in middle-income countries, agriculture accounts for 17 per cent of GDP, rising to 35 per cent in the poorest countries; yet the interests of rich country agribusiness had predominated. G-21 said that enough was enough. As a result of this pressure, the entire Cancun negotiations went into abeyance, with the challenge from civil society groups for the West to deliver a more equitable food economy ringing loud. Pressure on the EU and US to reform their highly subsidised agricultures intensified; but their response was to threaten a return to bilateral trade agreements. This, we can note, is what has now happened. There has been no completion of a new WTO round since 1994, but bilateral deals are being done more frequently. The EU expanded its membership. Latin America created Mercosur. The USA created NAFTA. And now we await the completed TTIP and TPP, or will they be altered by public pressure?

Who sets global standards?

The SPS (Sanitary and Phytosanitary Standards) agreement within the 1994 GATT was a particularly important development in the long struggle over food democracy in relation to health. The 1994 GATT catapulted a previously low-key world body, the Codex Alimentarius Commission ('Codex'), into the hot seat of setting world food-safety standards as benchmarks for food trade. When Codex was set up as a UN body in 1965, with the FAO and WHO as its secretariat, it was to advise national governments. The 1994 GATT assigned to Codex a key and different role: arbitrating on world food standards in trade disputes. This was furiously contested, not least since it transpired that Codex meetings were full of industry representatives from sectors on which standards were to be set.[50] As a result, some efforts to improve 'declaration of interests' was instituted but in reality, two decades on from being given this significant new role, Codex has become part of the established policy architecture.

The critics concluded that Codex had become a facilitator of trade rather than a protector of ecological public health. Codex is a technocratic body, dealing with only single issues, and unable to provide the leadership in integrating food, environmental protection and public health. Stacked with industry advisors, it is not currently the body to halt the nutrition transition or food's environmental crisis.

Injecting health and environment into regional institutions: the EU

The European Union is another interesting battleground. Although its policy presence is dominated by the Common Agricultural Policy (CAP) – a bastion of Productionism and a bizarre mix of protectionism, internal liberalisation and external mercantilism – the EU has also had to bend to public pressure over food scandals and crises of confidence over food quality.[1 51]

The foundation of the CAP was a classic expression of the Productionist vision for health. With Europe wracked by World War II and experience of famines (such as that in the Netherlands in 1944) and food shortages, its priority was to rebuild agriculture as a home-grown policy for the reconstruction of post-war Europe.[52 53] Enshrined in 1957, it has grown vastly. Still in 2014 it accounted for about 40 per cent of the EU total budget, while shaping the actions of 28 Member States and a population of half a billion people.

Public health, by contrast, only formally entered the EU remit with the Amsterdam and Maastricht Treaties of the 1990s which revised the Treaty of Rome. Article 152 stated that the European Commission would take a strong health line in all its policies, but the public health group has a relatively small staff while the Agriculture Directorate has hundreds; size of budgets matters. To date, the only health audits of Commission food policies in general and the CAP in particular have been externally conducted and have not filtered into mainstream policy-making.[54] In 2003, the Swedish government conducted a health audit of the CAP, producing a radical, evidence-based set of policy recommendations to:[55]

- phase out all consumption aid to the manufacturers of dairy products with a high-fat content;
- limit the School Milk Measure to include only milk products with a low-fat content;
- introduce a school measure for fruit and vegetables;
- develop a plan to phase out tobacco subsidies within a reasonable time frame;
- redistribute agricultural support so that it favours the fruit and vegetable sector and encourages increased consumption.

Agriculture, subsidies and health: where the money goes

The issue of subsidies is a fierce battleground for food democracy. Subsidies are disproportionately the policy instrument of rich nations for the simple reason that they have deep pockets.[56 57] But subsidies have also come under fire from neo-liberal critics as distortions of 'free' markets.[58] The World Bank and IMF, for example, in the era of Structural Adjustment, used loans to developing countries to leverage reductions of existing governmental subsidies. This was applied to the developing world but not applied so ruthlessly in their homelands.[59 60]

Farmsubsidy.org, a campaigning NGO, monitors the subsidy situation from official sources. Table 8.4 gives its verdict that only one in ten beneficiaries of EU farm subsidies in 2012 was even published that year, but the total value of payments made was €22bn, 45 per cent of the CAP budget of €54bn. This suggests that many smaller subsidy recipients were not publicly known. Table 8.5 presents the different cultures of secrecy in the EU member states. Denmark, for example, withheld only 12 per cent of subsidy beneficiaries, accounting for only 4 per cent of total subsidies in that country. By contrast, Austria withheld information on

Table 8.4 EU subsidies, top 30 recipients, all years to 2012

Name	Amount (all years)
Various beneficiaries	€641,532,359
Tate & Lyle Europe (031583)	€594,270,084
Avebe B.A.	€433,774,989
Tereos	€355,862,808
Saint Louis Sucre	€287,490,301
Campina Melkunie Veghel	€260,114,283
Doux	€223,214,655
SC Fondul de Garantare a Creditului Rural – IFN SA	€220,000,000
Nestle Nederland BV	€193,279,415
Corman SA	€184,690,194
Tate & Lyle Europe	€170,957,385
Tiense Suikerraff. Raff. Tirlemontoise NV SA	€164,941,364
Hoogwegt International BV	€161,601,809
Navobi B.V.	€151,243,294
Corman SA	€149,455,603
Junta de Andalucia	€145,619,981
Italia Zuccheri SPA	€139,754,719
Krajowa Spółka Cukrowa Spółka Akcyjna	€138,806,128
Tiense Suikerraff. Raff. Tirlemontoise NV SA	€137,647,553
Hoogwegt International BV	€134,051,911
Meadow Foods Ltd	€127,223,714
Eridania Sadam SPA	€125,262,919
Azucarera Ebro, S. L.	€119,445,484
Nestle Nederland BV	€118,548,842
CSM Suiker B.V.	€118,406,046
Cristal Union	€114,464,451
Friesland Coberco Butter Products	€109,066,782
F.IN.A.F. First International Association Fruit Soc. Consortile A R.L.	€108,383,082
EDIA – Empresa de Desenvolvimento e Infra-estruturas do Alqueva, S.A.	€103,095,249
Landesumweltamt Brandenburg	€102,919,992

Source: farmsubsidy.org, 2014[61]

Table 8.5 EU secrecy about subsidies: the number of subsidy beneficiaries redacted (data kept secret), plus estimated funds redacted (data kept secret) by country, 2011

Country	Estimated beneficiaries redacted	Estimated funds redacted
Austria	97%	86%
Belgium	84%	69%
Bulgaria	94%	36%
Cyprus	No data	No data
Czech Republic	79%	27%
Denmark	12%	4%
Estonia	80%	15%
Finland	95%	86%
France	68%	35%
Germany	97%	72%
Greece	No data	No data
Hungary	0%	19%
Ireland	Extraction underway	Extraction underway
Italy	Data inaccessible	Data inaccessible
Latvia	82%	Unknown
Lithuania	99%	74%
Luxembourg	No data	No data
Malta	99%	84%
Netherlands	56%	20%
Poland	99%	71%
Portugal	96%	49%
Romania	95%	11%
Slovakia	79%	17%
Slovenia	98%	71%
Spain	94%	56%
Sweden	10%	3%
UK	94%	74%

Note: The estimates are made with reference to aggregate spending data in the 2010 budget and data on individual CAP payments in 2009, the last year of budget transparency.

Source: farmsubsidy.org, 2011[61]

97 per cent of beneficiaries, accounting for 86 per cent of total subsidies in that country. This variation suggests different levels of democratic accountability.

At the global level, the OECD has championed the critique of subsidies. For decades it has produced two now well-known estimates. One is the amount of subsidy going to producers, the so-called Producer Support Estimate (PSE), and the other is a calculation of how much consumers are subsidising producers through the price of foods they buy, the so-called Consumer Support Estimate (CSE). The OECD takes the long view, gently pushing for subsidy reduction by publishing annual figures. In 2011, the PSE in OECD countries was $252.5bn, nearly 19 per cent of production value. Countries varied. Australia's PSE was about $1.55bn, equivalent to around 3 per cent of total production whereas Canada's PSE was just over $7bn, 14 per cent of total production. By contrast the average EU PSE was $103bn accounting for 17.5 per cent of total production; and the US PSE was $30.6bn which was 7.7 per cent of total production.[62]

Conventional economists do not like subsidies, whether being given to the poor, as in welfare payments, or to the rich, as in the trading system. In the UK, an inquiry by the House of Lords (historically a bastion of landed interests) produced the interesting estimate that 20 per cent of UK farmers received 80 per cent of subsidies. A subsequent study found that 16 per cent of farm holdings received 69 per cent of subsidies, and a study by Oxford University for the UK government in 2003 showed that producers in six eastern British counties – the 'grain barons' – received more than £540m out of the more than £2bn in CAP aid which the UK received annually.[63] East Britain's grain farmers received £121 from each consumer, whereas other (often less profitable) farmers received £41 from each taxpayer. Three out of England's nine regions take half the total CAP subsidy received by England.

Food democracy requires open information. Information lifts the lid on how things actually work, and creates space where one can ask: what do people want?[64] Pressure groups have argued worldwide that sums currently benefiting particular wealthy interests could be spent supporting eco-systems protection or producing food products more in line with healthy guidelines. The problem, they say, is not subsidies, but who gets them and for what. Campaigning optimists point out that, after decades of pressure, the EU does now have a funding strand to support the environment, the so-called pillar 2 of the reformed CAP. This enabled farmers to be paid for eco-system services such as protecting biodiversity and water. The point is that there are democratic choices to be made about where money goes and how to spend it. Somehow, the food system's vast externalised costs need to be reined back. Food should provide public goods not public 'bads'.

Agencies: another response to the crisis of governance?

One response to difficulties in governance has been the setting up of agencies – government bodies wholly funded out of taxation, yet which can be presented as 'independent' and science-based, although how they are constituted and the remit they are given shapes their effectiveness. A key goal for food agencies is the maintenance of public trust in food, shaken as it can be by adulteration and safety scandals.[65] The US Food and Drug Administration's modern regulatory functions stem from the 1906 Pure Food and Drugs Act,[66] passed following the national outcry about meat adulteration and fraud exposed by Upton Sinclair's campaigning book *The Jungle*.[67] Nearly a century later, following a decade of scandals over safety, a new European Food Safety Authority (EFSA) was created, partially emulating the FDA.[51 68] But it was found to be floundering when, in 2013, the routine use of horsemeat in European sausages was exposed. Giant food companies were proven not to have known what was in their own-label products. A continental-wide web of subterfuge and crime was exposed. This was despite the supposed enshrinement of the principle of traceability for foodstuffs in EU food law since 2000. Some of the largest food companies in the world were found not to know what was in 'their' products, even though they could nominally identify the companies from which the products were sourced.[69]

Despite such failings, agencies have proliferated as states have de-layered and tried to soften their role. There are now many agencies in many countries covering environmental protection, medicines, food safety, trading standards and public health, all dealing with aspects of food. This can create another problem: how to integrate their findings into better policy and practice.

They can be a catalyst for change, but also a brake. In the UK, for example, a new Food Standards Agency was created in 2000 after the greatest period of food crisis since World War II. Its brief was to rebuild trust, but although its new chair was an ecologist, its remit excluded environmental matters. It did manage to extend its remit to include some nutrition matters. And its board was also chosen to be arms-length from any food industry involvement but over time, this weakened, and within a decade it was suffering 'agency capture', a term used to describe the process by which regulatory bodies can become close to and be influenced by the industries they were created to regulate.[70] A change of UK government in 2010 led to a narrowing of the FSA's remit. And in 2014 the FSA was exposed as having withheld the publication of campylobacter contamination rates on the casting vote of its then chair, a former president of the national farmers organisation.[71 72] Happily the ensuing scandal did lead to publication, but the damage to the agency's credibility was done.

The workings of such bodies indicate how institutions can themselves become political battlegrounds. Denmark, for example, created its food agency initially as an environmental agency in the 1970s following public concerns about pesticides. This evolved into a multi-sectoral body, first answering to the Environment Ministry, then to Health, and then it was reintegrated into a Ministry of Food in the 1990s, from the ashes of the old agriculture ministry.[73] Agencies can also be created for specific commercial purposes. For example, the National Food Authority of Australia, a food safety body set up in the late 1980s, was merged in the early 1990s with New Zealand's regulatory bodies to create the Australia–New Zealand Food Authority (ANZFA) and then relaunched in 2002 as Food Standards Australia New Zealand (FSANZ).[74] This was to enable two major food exporting countries to combine resources and meet their export pressures.

Governments like to change their own institutions, arguing that democratic votes give them the right to do so. Bodies are abolished as well as created; strengthened as well as marginalised. After World War II, the UK created a Committee on Medical Aspects of Food Policy (COMA) to give oversight to food's impact on public health. It issued important early warnings about food's impact on NCDs. For its pains, when the new FSA was created in 2000, COMA was replaced by a Standing Advisory Committee on Nutrition (SACN) and given a much narrower remit. Its traditions and wider view were dissipated. Expertise was lost.

Some countries, notably in Scandinavia, have created national Food Policy or Nutrition Councils to inject public health into national food policy. The UN's 1992 International Conference on Nutrition proposed the creation of such Policy Councils everywhere as mechanisms though which expert advice and evidence could be channelled to policy-makers. It drew on the Norwegian experience. Norway's was created in the 1960s when politicians were unhappy with the performance of its Inter-ministerial Council for failing to coordinate policy implementation. The National Nutrition Council was to champion health across wider food and social policies.[75] It won influence and respect. Former opponents of policy in the food and agriculture industries found themselves more willing to listen to arguments from nutritionists and health authorities. The Nutrition Council's functions were then reintegrated into the state machinery, locating it within the Directorate for Health and Social Welfare and integrating it with agricultural, fishery, price, consumer and trade policies as well as educational and research policy. Whether the champion role has been lost remains to be seen.

A particularly dynamic level for food democracy has been the emergence of local Food Policy Councils. These are often instigated by food activists. There are now decades of experience of such local and municipal bodies,[76] with much international sharing of experience.[77 78] One example of this was how Toronto City Council learned from the short-lived London Food Commission, a six-year body which advised the then Greater London Council in the 1980s, to create its own Food Policy Council. This was bedded closely into Toronto's municipal structure, and it has retained its leverage inside that structure, while building external citizens links.[79 80] North America has witnessed a powerful movement of such initiatives, summarised in a useful review of their strengths by Food First in California.[81] In the developing world, such thinking is most often built into public health bodies or economic development agencies. Food Policy Councils, however, remain an important institutional bridge between consumer-citizens, organised civil society, existing bodies and the food system. A review of 13 North American cities in a report by the City of Portland, Oregon,[76] concluded that the councils can be very effective in linking metrics, engaging high-profile political actors, but there is no 'one size fits all' ideal structure. The burgeoning City Sustainable Food movement around the world has spawned much experience on both what urban authorities can do and on their limitations.[82 83] Such councils need to fit the locality. They need national support too, as studies of Belo Horizonte's inspiring leadership in Brazil have suggested.[78 84 85] The policy space such Councils can inhabit is due to the continual evidence coming from science on the one hand, and citizens' pressure for better democratic control over local food systems, on the other hand.

Another vehicle for democratic engagement is the Parliamentary or Congressional Inquiry. These can be routine, such as when the Parliaments call for evidence on particular topics, or can be arms-length from government to address a particular crisis, as when a group of elected members hold hearings. Most democratic countries have such processes. These are parliaments holding themselves to account.

Food democracy versus food control: a permanent tension?

We have proposed in this chapter that the history of food governance can usefully be understood as a long struggle between two conflicting forces: 'food democracy' and 'food control'. The latter suggests a situation where relatively few sectors or bodies or companies exert power to shape the food supply to suit their interests. They favour *dirigiste* policy frameworks even if they speak the language of 'freedom' and 'choice'. Food control implies 'top-down' decisions and the mediation of views and interests to suit the controllers. There is limited dialogue, and few resources are allocated to investigate the range of policy options. The direction of policy travel is set in advance by vested interest. 'Food control' presupposes people, animals, plants and the environment being controlled in order to maintain a particular pattern of order, authority and predictability.

'Food democracy', on the other hand, gives scope for a more inclusive and participatory approach to food policy. Its ethos is, if anything, 'bottom-up', considering the diversity of views and interests in the mass of the population and food supply chains. The needs of the many are favoured over the few; mutuality and symbiosis are pursued. There is genuinely open debate between alternative and opposing views. The core assumption of 'food democracy' is that the public good – in this case the ecological and public health – will be improved by the democratic process.[23]

Some of the clearest and best studied examples of 'food control' are in times of war or under authoritarian political regimes. Forced farm collectivisation under the USSR,[86] or the imposition of tightly enforced rationing in World War II,[87] are examples. Under wartime circumstances, governments frequently impose tough regulations and rules, ordering the food supply chain and rationing access to food. [87] In today's architecture of food governance, governments are not the sole actors in the democratic struggle. If a regulation or proposal is made in the open, it can be fought over. Consultations occur. Delays result. That is why crises are so intriguing for policy analysts; they expose both the architecture and the vested interests.

Conclusion: human liberty and consumer choice

Central to any discussion of food democracy and food control is the question of human liberty. Is health an individual responsibility? Is food a matter for personal choice? If people want to eat out-of-season foods sourced from the four corners of the world and if the market will meet their needs, should they not be allowed to do so? Should they be even allowed to consume unfit food? Does it make sense to talk of people 'deciding' their health or environment? Neoliberals favour the argument that modern consumers have market power by 'voting' with their daily purchases of food and consumption patterns. It is true that developed economy consumers luxuriate in historically unparalleled ranges of foods. No one forces them to consume as they do, is the argument (while ignoring the distorting effects of marketing or PR). It is true that a visit to a hypermarket with its 30,000 or more items on sale is truly astonishing to those who have never experienced it. From a health and sustainability perspective, however, these modern cathedrals of choice offer contradictory experiences. The public consumes now but pays later. The 'consumer votes' theory may apply to the affluent in both the developed and developing worlds, but does not apply universally. Prices do not reflect full environmental and health costs. Consumers do not have full information about their food. Labels give some information but not all. They rarely convey the nature of production process or impacts. Although the demand for better food labels encourages manufacturers to declare product contents, the production processes can remain obscure, and one cannot assume they are in full control of their products, as the 2013 EU horsemeat scandal showed. Yet the food democratic cry to know the full story of what goes into food and what its full health and environmental consequences are remains central to the pursuit of food democracy.[88] It is one of the persistent themes in food policy.[39]

Is consumerism falling out of love with ever more choice? Beneath the patina of consumer choice there is some cultural homogenisation. Hypermarkets have given 30,000–40,000 items under one roof, but the number of food shops has declined, put out of business by economies of scale. About half of all the food and drink purchased in the UK at the turn of the millennium came from just 1,000 hypermarket outlets controlled by four dominant retail players. In many countries levels of supermarket concentration are very high. Giant companies dominate the food trades. On supermarket shelves, it is highly likely that consumers will find similar leading brands in particular product categories. As this book goes to press, an interesting tension within the food retail sector has emerged. Germany's massive retailers, Aldi and Lidl, offer far less choice but lower prices. The UK's giant Tesco, the third largest food retailer in the world, decided in 2015 to cut the amount of brands it offered. No longer would it offer choice of 69 coffees or multiple sardines. As one supplier told us in early 2015, that may sound fine if your product is the 'chosen one' to stay on their shelves, but it means you are out of business if you are 'de-listed'.

Where is food democracy headed? The modern food world is complex. There are multiple actors at multiple levels. That is ultimately why government is so important to food democracy. Active, sceptical and well-educated consumers are essential for good food systems, but they also need help and support not driven by commercial interests. Governments cannot be relied upon, and must themselves be scrutinised and held to account; but only some form of government and at all levels (from local to global) can be entrusted to set frameworks which have real legitimacy, and to shape the public interest.

References

1   Van Zwanenberg P, Millstone E. *BSE: Risk, Science, and Governance*. Oxford and New York: Oxford University Press, 2005.
2   Elliott C. *Elliott Review into the Integrity and Assurance of Food Supply Networks: Final Report – a food crime prevention framework*. <https://www.gov.uk/government/policy-advisory-groups/review-into-the-integrity-and-assurance-of-food-supply-networks> London: HM Government, 2014.
3   USTR. The Trans-Pacific Partnership. <https://ustr.gov/tpp> Washington, DC: Office of the United States Trade Representative, 2015.
4   Lilliston B. *Big Meat Swallows the Trans-Pacific Partnership*. Minneapolis, MN: Institute of Agriculture and Trade Policy, 2014.
5   TTIP. The Transatlantic Trade and Investment Partnership. <http://ec.europa.eu/trade/policy/in-focus/ttip/about-ttip> 2015.
6   Hansen-Kuhn K, Suppan S. *Promises and Perils of the TTIP: Negotiating a Transatlantic Agricultural Market*. Minneapolis, MN: Institute for Agriculture and Trade Policy, 2013.
7   Sharma S. *10 reasons TTIP is Bad for Good Food and Farming*. Minneapolis, MN: Institute for Agriculture and Trade Policy, 2014.
8   Raghavan C. *Recolonization: GATT, the Uruguay Round and the Third World*. London: Zed Press, 1990.
9   Watkins K. *Rigged Rules, Double Standards*. Oxford: Oxfam Publications, 2001.
10 Richards D, Smith MJ. *Governance and Public Policy in the United Kingdom*. Oxford: Oxford University Press, 2002.
11 Colebatch H. *Policy*. Maidenhead: Open University Press, 2009.
12 Wallace H, Pollack MA, Young AR. *Policy-Making in the European Union*. 6th edn. Oxford: Oxford University Press, 2010.
13 Hirst P, Thompson G. *Globalisation in Question: The International Economy and the Possibilities of Governance*. 2nd edn. Cambridge: Polity Press, 2001.
14 Deacon B, Ollila E, Koivuslao M, et al. *Global Social Governance, Globalism and Social Policy Programme*. Helsinki: STAKES, 2003.
15 Dobbs R, Sawers C, Thompson F, et al. *Overcoming Obesity: An Initial Economic Analysis*. New York: McKinsey Global Institute, 2014.
16 World Economic Forum, Accenture. *More with Less: Scaling Sustainable Consumption with Resource Efficiency*. Geneva: World Economic Forum, 2012.
17 World Economic Forum, McKinsey & Co. *Realizing a New Vision for Agriculture: A Roadmap for Stakeholders*. Davos: World Economic Forum, 2010.
18 World Resources Institute. *World Resources 2002/2004*. Washington DC: World Resources Institute, 2003.
19 Hawken P. *Blessed Unrest*. New York: Penguin, 2007.
20 Sustain. Sustain: the alliance for better food and farming. <http://www.sustainweb.org> London: Sustain, 2015.
21 Solnit D, Solnit R. *The Battle of the Story of the Battle of Seattle*. Oakland, CA: AK Press, 2009.
22 Gabriel Y, Lang T. *Unmanageable Consumer*. London: Sage, 2015.
23 Rayner G, Lang T. *Ecological Public Health: Reshaping the Conditions for Good Health*. Abingdon: Routledge/Earthscan, 2012.
24 Lang T. Going public: food campaigns during the 1980s and 1990s. In: Smith D, ed. *Nutrition Scientists and Nutrition Policy in the 20th Century*. London: Routledge, 1997: 238–60.

25   Consumers International, World Obesity Federation. *Recommendations towards a Global Convention to protect and Promote Healthy Diets: Proposal to the World Health Assembly.* London and Geneva: Consumers International and World Obesity Federation, 2014.

26   Lang T. *Food Fit for the World? How the GATT Food Trade Talks Challenge Public Health, the Environment and the Citizen; a Discussion Paper.* London: Public Health Alliance and Sustainable Agriculture Food and Environment Alliance, 1992.

27   Smith NC. *Morality and the Market: Consumer Pressure for Corporate Accountability.* London: Routledge, 1990.

28   Witkowski TH. Colonial consumers in revolt: buyer values and behavior during the nonimportation movement, 1764–1776. *Journal of Consumer Research*, 16(2) (1989): 216–26.

29   Allain A. Breastfeeding is politics: a personal view of the international baby milk campaign. *The Ecologist* 21(5) (1991): 206–13.

30   Palmer G. *The Politics of Breastfeeding.* 2nd edn. London: Pandora, 1993.

31   Klein N. *This Changes Everything: Capitalism vs the Climate.* London: Allen Lane, 2014.

32   Miller D, Dinan W. *A Century of Spin: How Public Relations Became the Cutting Edge of Corporate Power.* London: Pluto Press, 2008.

33   McChesney RW, Nichols J. *The Death and Life of American Journalism: The Media Revolution that will Begin the World Again.* Washington, DC: Nation Books, 2010.

34   MarketingCharts. Global web ad spend to rise 31% in 2 yrs. <http://www.marketingcharts.com/television/global-web-ad-spend-to-rise-31-in-2-yrs-18358> [accessed July 2011]. Thetford Center, VT: Marketing Data Box, 2011.

35   Shaw R. Streamlining advertising spend to do more with less. <https://www.som.cranfield.ac.uk/som/som_applications/somapps/contentpreview.aspx?pageid=13273&apptype=think&article=86> Cranfield: Cranfield University School of Management, 2009.

36   Kraak VI, Harrigan PB, Lawrence M, *et al.* Balancing the benefits and risks of public–private partnerships to address the global double burden of malnutrition. *Public Health Nutrition*, 2011: 1–15.

37   UN. *Universal Declaration of Human Rights.* Adopted and proclaimed by General Assembly resolution 217 A (III) of 10 December 1948. Geneva: United Nations, 1948.

38   Shaw DJ. *World Food Security: A History since 1945.* London: Palgrave Macmillan, 2007.

39   Lang T, Barling D, Caraher M. *Food Policy: Integrating Health, Environment and Society.* Oxford: Oxford University Press, 2009.

40   FAO. *Voluntary Guidelines to support the progressive realization of the right to adequate food in the context of national food security.* Adopted by the 127th Session of the FAO Council November 2004. Rome: Food and Agriculture Organisation, 2004.

41   De Schutter O. *Reports of the UN Special Rapporteur on the Right to Food.* <http://www.srfood.org> Louvain/Geneva: UN Economic and Social Council, 2014.

42   Eide A. *Right to Adequate Food as a Human Right.* UN Human Rights Study Series, 1. New York: United Nations, 1989.

43   Goldsmith E, Mander J. *The Case Against the Global Economy and for a Turn towards Localization.* New edn. London: Earthscan, 2001.

44   Hertz N. *The Silent Takeover: Global Capitalism and the Death of Democracy.* London: Heinemann, 2001.

45   Boyd Orr J. *As I Recall: The 1880s to the 1960s.* London: MacGibbon & Kee, 1966.

46 Amin S. *Unequal Development: An Essay on the Social Formations of Peripheral Capitalism,* trans. Brian Pearce. New York: Monthly Review Press, 1976.

47 George S. *How the Other Half Dies: The Real Reasons for World Hunger.* Harmondsworth: Penguin, 1976.

48 Kindleberger CP. *Marshall Plan Days.* London: Allen & Unwin, 1987.

49 Watkins K, von Braun J. *Time to Stop Dumping on the World's Poor. 2002–2003 Annual Report: Trade Policies and Food Security.* Washington, DC: International Food Policy Research Institute, 2002.

50 Avery N, Drake M, Lang T. *Cracking the Codex: A Report on the Codex Alimentarius Commission.* London: National Food Alliance, 1993.

51 Commission of the European Communities. *White Paper on Food Safety.* Brussels: Commission of the European Communities, 2000.

52 Tracy M. *Government and Agriculture in Western Europe, 1880–1988.* 3rd edn. New York and London: Harvester Wheatsheaf, 1989.

53 Grant W. *The Common Agricultural Policy.* Basingstoke: Macmillan Press, 1997.

54 Robertston A, Tirado C, Lobstein T, *et al. Food and Health in Europe: A New Basis for Action.* Copenhagen: World Health Organisation Regional Office for Europe, 2004.

55 Dahlgren G, Nordgren P, Whitehead M. *Health Impact Assessment of the EU Common Agricultural Policy.* Stockholm: National Institute of Public Health, 1997: 36.

56 Oxfam. *Rigged Rules and Double Standards: Trade, Globalization and the Fight Against Poverty.* Oxford: Oxfam, 2002.

57 Oxfam. *Make Trade Fair.* <http://www.maketradefair.com/en/index.htm>. Oxford: Oxfam, 2008.

58 Knudsen A-CL. *Farmers on Welfare: The Making of Europe's Common Agricultural Policy.* Ithaca, NY, and London: Cornell University Press, 2009.

59 Bello W, Cunningham S, Rau B. *Dark Victory: The United States, Structural Adjustment and Global Poverty.* London: Pluto Press with Food First and the Transnational Institute, 1994.

60 Bello WF. *The Food Wars.* London: Verso, 2009.

61 Farmsubsidy.org. EU farm subsidies for all countries, all years. <http://farmsubsidy. openspending.org/EU> [accessed June 2014]. London: Farmsubsidy.org, 2014.

62 OECD. StatExtracts, 2014. <www.stats.oecd.org/index.aspx?Datasetcode= MON20123_1>.

63 Nuffield College Oxford. *Identifying the Flow of Domestic and European Expenditure into the English Regions.* Report to the Office of the Deputy Prime Minister. Oxford Nuffield College University of Oxford, 2003.

64 Oxfam. *Working for the Few: Political Capture and Economic Inequality.* Oxford: Oxfam, 2014.

65 Wilson B. *Swindled: From Poison Sweets to Counterfeit Coffee – The Dark History of the Food Cheats.* London: John Murray, 2008.

66 Food and Drug Administration. History of the FDA. <http://www.fda.gov/aboutfda/ whatwedo/history/default.htm> [accessed June 2014]. Washington, DC: Food and Drug Administration, 2014.

67 Sinclair U. *The Jungle.* Harmondsworth: Penguin, 1906/1985.

68 Trichopoulou A, Millstone E, Lang T, *et al. European Policy on Food Safety.* Luxembourg: European Parliament Directorate General for Research Office for Science and Technology Options Assessment (STOA), 2000.

69   Elliott C. *Elliott Review into the Integrity and Assurance of Food Supply Networks – interim report.* <https://www.gov.uk/government/policy-advisory-groups/review-into-the-integrity-and-assurance-of-food-supply-networks> London: HM Government, 2013.

70   Millstone EP, Lang T. Risking regulatory capture at the UK's Food Standards Agency? *The Lancet,* 372(9633) (2008): 94–5.

71   Food Standards Agency. Reduction of campylobacter from poultry; publication of the retail survey results. Report by Steve Wearne to FSA Board, July 2014. <http://multimedia.food.gov.uk/multimedia/pdfs/board/board-papers2014/fsa-140707.pdf>. London: Food Standards Agency, 2014.

72   Lawrence F. Food poisoning scandal: how chicken spreads campylobacter. <http://www.theguardian.com/society/2014/jul/23/-sp-food-poisoning-scandal-how-chicken-spreads-campylobacter> *Guardian,* 23 July 2014.

73   Lang T, Millstone E, Rayner M. *Food Standards and the State: A Fresh Start.* London: Centre for Food Policy (then Thames Valley University, now City University London), 1997.

74   FSANZ. Food Standards Australia New Zealand. <http://www.foodstandards.gov.au/Pages/default.aspx> [accessed June 2014].

75   Lang T, Rayner G, Rayner M, *et al.* Policy Councils on Food, Nutrition and Physical Activity: the UK as a case study. *Public Health Nutrition,* 8(1) (2005): 11–19.

76   Hatfield MM. *City Food Policy and Programs: Lessons Harvested from an Emerging Field.* Portland, OR: City of Portland, Oregon Bureau of Planning and Sustainability, 2012.

77   Rocha C. *Brazil-Canada Partnership: Building Capacity in Food Security.* Toronto: Center for Studies in Food Security, Ryerson University, 2008.

78   Rocha C. Developments in national policies for food and nutrition security in Brazil. *Development Policy Review,* 27(1) (2009): 51–66.

79   MacRae R, Donahue K. *Municipal Food Policy Entrepreneurs: A Preliminary Analysis of How Canadian Cities and Regional Districts are Involved in Food System Change.* Toronto CA: Canadian Agri-food Policy Institute, Toronto Food Policy Council, Vancouver Food Policy Council, 2013.

80   Toronto Food Policy Council. About the TFPC. <http://tfpc.to> Toronto: Toronto City Council, 2015.

81   Harper A, Shattuck A, Holt-Giménez E, *et al. Food Policy Councils: Lessons Learned.* Oakland, CA: Food First Inc., 2009.

82   Jennings S, Cottee J, Curtis T, *et al. Food in an Urbanised World: The Role of City Region Food Systems in Resilience and Sustainable Development.* London The Prince of Wales's International Sustainability Unit and 3 Keel, 2015.

83   Birch E, Keating A. *Feeding: Food Security in a Rapidly Urbanizing World.* <www.feedingcities.com>. Philadelphia, PA: Penn Institute for Urban Research, 2014.

84   Rocha C. Small farms and sustainable rural development for food security: the Brazilian experience. *Development Southern Africa,* 29(4) (2012): 519–29.

85   Rocha C, Lessa I. Urban governance for food security: the alternative food system in Belo Horizonte, Brazil. *International Planning Studies,* 14(4) (2009): 389–400.

86   Dunman J. *Agriculture: Capitalist and Socialist.* London: Lawrence & Wishart, 1975.

87   Hammond RJ. *Food: The Growth of Policy.* London: HMSO/Longmans, Green & Co., 1951.

88   Lawrence F. *Not on the Label: What Really Goes into the Food on your Plate.* 2nd edn. London: Penguin, 2013.

9   The future

Lay then the axe to the root and teach governments humanity.

Thomas Paine, 1737–1809

Core arguments

The book has shown clear tensions in the policy framework that shapes food systems. These are the Food Wars. The Productionist paradigm that dominated the 20th century is under stress. It cannot cope with the complexity of demands for change in relation to both human and environmental health, and at the same time remain economically or societally viable. The two emerging paradigms – the Life Sciences Integrated and Ecologically Integrated paradigms – offer increasingly coherent versions of modernisation and vie for the attention of policy-makers, supply chains and the public. The terrain is fissured by the competing demands and evidence from business, government and civil society. As ever, the relationship between evidence and policy formulation is problematic and politicised. A period of experimentation is underway in which new solutions to the various crises outlined in the book are being proposed by the corporate sector, suppliers, NGOs and social movements, who vie to win the battle for consumer culture and state attention. They compete over where the public interest lies. A feature of the Food Wars is where and how to draw the line between individual, sectoral and public interest. Currently, the likely 'winner' in the paradigmatic struggle to replace Productionism is the Life Sciences Integrated paradigm. It offers a new vision of major change within the tradition that Productionists like. The Life Sciences Integrated paradigm is itself fractured, with some seeing this as a technical fix on a new scale, while others think eco-systems really do require completely new protection. Life Science dominance is also vulnerable to rising bodies of evidence about food's impact on society, not just health and eco-systems. Social divisions are massive within and between societies. Geopolitics is putting this under further strain. This and rising externalised costs, together with the realisation that their own lives and health are at stake, may alter the preparedness of consumers to act, not just think, like citizens with a long-term commitment to ecological sustainability, which is the core of a new ecological public health. The Food Wars look set to continue, but the actors, positions and evidence which may shape their outcome become ever clearer.

Charting the paradigms

At the conclusion of this book, we return to its starting point: the Food Wars are battles over food policy. They occur within and between states, corporate sectors and civil society; they frame food's relationship to health, environment and well-being, on an individual, domestic and societal or population level. We have shown that the food system is under pressure to address:

- what is meant by a good diet for the 21st century;
- improved quality of life and public health from food;
- social justice at national and international level;
- a less damaging relationship between production and eco-systems;
- more sustainable systems of growing, processing and selling food;
- a more accurate and fully costed approach to food pricing;
- a role for food in cultural life which is not distorted by powerful commercial forces;
- more democratic distribution of power and open governance across food supply chains;
- confidence of consumers and civil society in the food system.

These are reasonable requirements, yet the book has shown that there are very real conflicts of perspective and visions for the future – hence our metaphor of the Food Wars. In the first edition of this book (written in 2001–4), we outlined our notion of the three paradigms to help explain tensions we felt were emerging globally over food. Over the following decade we have noted how useful this theory was; tensions between and within the paradigms have become increasingly clear. Companies and governments who were resistant to the importance of eco-systems or who dismissed public health as a matter for individual choice have been slowly educated about the errors of such thinking. Hard-line productionists have realised that ecologically integrated thinking makes sense. Ecologically minded advocates have also realised they cannot turn the clock back on modern genetics; the question is who controls them and whether they are deviations. In short, the policy world has become more nuanced while the architecture of the policy options has become ever clearer.

It is hard to find many companies or governments who now deny food's impact on health and the environment, and the general role of food within sustainability. Yet they are locked into soft approaches to change: information, advice, labelling, 'nudge', roundtables. The harder types of policy measures and instruments such as regulation, laws, or taxes are still locked in policy filing cabinets and computers under 'only to be used in emergencies'.

Despite the welcome agreement that food sustainability has to be central to whichever version of food futures is conceived, the evidence suggests that the food system is still heading – or is that drifting? – into major trouble on many fronts. While hopeful that the language of sustainability and the recognition of the importance of trends such as obesity and climate change are widely shared, major tensions within the food system remain – between rich and poor, intensive and extensive approaches, more meat and plant-based diets, and so on. We do not yet detect signs that Food Peace has broken out. If anything, the Food Wars are intensifying. Table 9.1 gives a summary of how the paradigms differ in their approaches to the future. Apparently shared interpretations of food problems are in fact not interpreted in the same way.

These divergences are no grounds for pessimism about continuing Food Wars. We know that there are many people and organisations within all sectors of the food system – from farming to consumers – who share our analysis of the significant differences between visions for the future. Nor should Table 9.1 be taken to imply that there are no divergences within the paradigms. While the Life Sciences Integrated paradigm is currently in ascendancy, it is also fractured, with critics wary that this could become a mass technical fix. Choice-editing can be done up to a point, but consumers will ultimately have to change if planet earth is to adapt to climate change let alone reduce it, for instance. An initial rush by high level policy-makers to see the future of food as resolvable by a vast investment in life sciences to produce hi-tech products and low-impact processes is now being tempered by realisation of the enormity of the eco-systems crisis. Obesity, for example, cannot be resolved unless consumer culture changes. Nor can biodiversity loss. Much may depend on societal forces. Food is a sensitive issue; religious and other cultural mores are powerful. Also income inequalities and food prices are politically delicate. Social divisions are massive within and between societies. In that sense, there is good news and a consensus that big changes are inevitable; but, at the same time, there is an uneasy lock-in to the status quo. The default position and policy comfort zone is Productionism. Partly, we see this situation as the legacy of past public policy choices, and the initial success of Productionism. But there is also growing realisation that there are no single solutions to food system problems, only packages of solutions on multiple levels. This makes the food policy tasks ahead ever more complex.

Whether policy-makers address these issues and the choice of paradigm *explicitly* or leave them *implicit*, we remain convinced that new policy directions and leadership are needed. There are signs of 'big picture' frameworks emerging, at the UN, for example, with the 2015 Sustainable Development Goals, or at the EU with the espousal of the Green or Circular Economy. But much depends on how these wide-ranging frameworks are translated in the real world, and whether they just remain aspirations on paper. The challenge, as we have suggested throughout this book, is to optimise population and ecological health, and to assert the public interest. We think this is unlikely to happen unless there

Table 9.1 Different approaches to food futures, by paradigm

Policy focus	Productionist paradigm	Life Sciences Integration paradigm	Ecologically Integrated paradigm
Relationship to general economy	State in charge, shaping market solutions and addressing market failure	Corporation-led; large private sector science budgets; individualisation of solutions; competition within the marketplace	Population approach; social enterprise and fairness; emphasis on eco-systems shaping economic activity; cradle-to-grave approach
Approaches to diet, disease and health	Accepts societal burden of disease; main focus is on providing sufficient food	Individual choice is key driver; niche markets; nutrition is part of risk management and hazards control	The right to be well; aims for joint eco-systems and human health
Environment	Environment is there to be mined; costs are externalised; tendency towards mono-culture; industrial chemical dependency	Mono-cultural tendencies; hi-tech + low social knowledge approach; new bio-industrialisation; technocratic approach to sustainable intensification	Biodiversity at heart of food systems; ecological assumptions underpin sustainable intensification; social, environmental concerns shape technical knowledge
Food business	Commodity focus; industrial-scale ingredients and processing; pursuit of low-cost food; big budget marketing	Commodity focus with personalised niches; private industry dominates economic activity	Costs internalised where possible; tendencies to favour robust local food economies
Consumer culture	Original vision of mass markets for mass consumers is turning into differentiation by ability to pay	Appeal to hi-tech and gadget society; apparently personalised service; choice-editing 'beneath the radar' to preserve belief in consumer choice and cheap food culture	Societal responsibility based on a citizenship model; consumers become citizens; requires mass education to activate; heightened role for NGOs; price adjustment with cost internalisation
Role of the state	Paternalist state delivering markets and societal infrastructure	Balances of public and private sector; rhetoric of minimal state but accompanied by strong state action in some sectors	Sets common framework while protecting resources and eco-systems; encourages diversity and inclusiveness

is a noisy and informed civil society pressure for this more complex approach to the future of food.

What are the viable alternatives to the Productionist paradigm? As we have shown, evidence of the social, economic, environmental and health problems associated with current food trends has mounted since the turn of the millennium. Policy-making is not precise: it is a product of politics and the (im)balance of forces. It is a matter of timing, resources, who champions what, and which visions and imaginations are captured. In a food world dominated by supermarketisation and choice culture, the image of a future food system shaped by ecological considerations can be threatening. Meat eaters want to eat meat. Equally, the life sciences vision of the future can be scary, one shaped by scientists and technologists, responding to funders and investors rather than the public. Meat in this vision might be cloned meat. Who wants that? This battle for the hearts and minds of the consuming public will shape the outcome of the Food Wars.

Which paradigm will triumph?

Although Productionism is under threat and partly weakening in power and public appeal, the triumph of the other two paradigms is by no means certain. It is possible, as Thomas Kuhn originally argued when setting out his view of paradigms in scientific thought, for two or three paradigms to coexist. Social scientists have been studying the emergence of what they call 'alternative food networks', claimed by some as democratic experiments in future food systems.[1] Others see the vice-like grip of powerful corporate entities over food commodities and supply chains as blocks on more sustainable systems. A potential danger with both the Ecologically Integrated and Life Sciences Integrated paradigms, while they offer divergent scientific viewpoints on the worlds of food and health, is the polarisation within healthy and sustainable food production between rich and poor consumers – those who can afford such foods and those who cannot. We see such polarisation as socially unsustainable as well as unjust. A pool of ill health and inequality in communities acts as a drag upon society as a whole.[2-4]

The Food Wars' ultimate challenge will, therefore, be to clarify the organisational and practical aspects of the food system and to chart new directions. Childhood obesity, for instance, is an issue demanding answers about whether to control marketing and advertising and about moving food culture in a more health-enhancing direction. Equally, whether or not massively to expand GM crops for direct human consumption and whether individualised and personalised nutrition will improve population health profiles are very real issues which will help determine which paradigm becomes pre-eminent. Much will also depend on whether research into cutting-edge issues for each paradigm receives equal resources and funds. Commercial research is overwhelmingly dominated by investment in the Life Sciences paradigm, with little going into clarifying evidence on different interpretations of the biological sciences. Funding and resources are hopelessly skewed when it comes to public information and education. For every $1 spent by the World Health Organisation on trying to improve the nutrition of the world's population, $500 are spent by the food industry on promoting processed foods.[5] The deciding factors as to which paradigm triumphs overall, and which becomes pre-eminent in the food supply chain or whether there is a 'truce', include:

- the costs of ill health and healthcare for taxpayers;
- environmental pressures such as climate change, biodiversity loss, energy use and water stress on food system viability;
- whether the consuming public realises the fragility of resources used in food and this starts to affect their daily lives and food availability;
- whether opinion-formers grasp the enormity of the risks;
- the role of scientists and other specialists in articulating the urgency of the case for change;
- societal tolerance of food inequalities within and between societies;
- whether civil society creates coherent international alliances promoting food democracy;
- the level of public confidence and trust in the food supply chains and in science and technology;
- whether a coherent narrative about what the food problem is becomes accepted by the public;
- the funding available to competing notions of science and technology in each paradigm;
- how food commerce implements sustainable business practices;
- any unforeseen new crises arising from current food supply practices.

These are big 'ifs'. Much depends, of course, on the balance of forces and resources, but one thing is certain: the solutions to food system tensions will not be left to chance. It is highly political. Food crises always ultimately have to be responded to because if they are not, food availability for urbanised populations becomes at risk. When food prices rose in 2007–8, over 60 governments implemented some form of action to protect consumers from rocketing food prices. There were food riots in some countries. In a world which is supermarketised and where billions of people are accustomed to availability and affordability, and where they have in Amartya Sen's terms a sense of 'entitlement',[6] or in our terms there is more rather than less food democracy, the prospect of food anarchy is not a serious option for governments or the powerful. Societies can pressurise otherwise controlling forces to win progressive change. Public health history suggests that, if threatened by contagion, and if the rich are presented with evidence that prevention is possible but requires new policies or infrastructure, they will accept some form of taxation and discomfort if these deliver effective drains or good water systems or improved housing or transport or unadulterated but more expensive food. They grumble but accept. Even vested interests resistant to change will, when they see the inevitable, support the flow of funds. Such an era may be upon us in the Food Wars. In that sense, food politics fits the famous dictum of Clausewitz, the 19th-century Prussian military tactician, that war is politics conducted by other means.[7]

If the paradigmatic framework we have outlined is apt, it has profound implications for the future of the entire food supply chain. It is not that one paradigm is hostile to industry or to science and technology, and that the other favours it; on the contrary, the paradigms offer *different* approaches to industry, to consumers, to health, to science and technology, to managing the environment, and so on, as was illustrated in Table 9.1. Terms such as 'lifestyle' and 'choice' that current policy-makers hold dear will mean different things, according to which paradigm holds sway.

Within the still dominant Productionist paradigm, the policy emphasis has focused on unleashing the productive capacity in the food supply chain (particularly land and labour) and to aim for quantity and efficiency of output, defined in terms of yield, throughput and profitability. Its assumption is that the public good requires sufficiency and, in turn, that sufficiency will deliver the public good. Both before and after World War II, this policy made sense. Again, we stress that Productionism has been immensely successful in its own boundaries. Output rose dramatically and more mouths have been fed. Nevertheless it is now under

Table 9.2 Some contested features of the Productionist paradigm

Policy area	Contested feature
Technical	Transformation of nature (GM, seeds, land); impact of intensification; motives for investment and research (private/corporate financial returns versus public goods)
Commercial	The extent of corporate power; reliance on fossil fuels; trade liberalisation unleashing unequal market conditions; market concentration, monopolies and oligopolies suggesting competition policy failure; reliance on 'cheap' labour and low-cost production centres
Health	Individualisation; emphasis on safety rather than on degenerative diseases; costly healthcare services; externalised costs borne by society and families; continuing spread of non-communicable diseases such as diabetes and obesity
Culture	Globalisation, branding and marketing framing consumer choice; food mores affected by the consciousness and marketing industries; rapidity of change; artificial creation of choice and needs being shaped by marketing
Society	Food justice and fairtrade; widening income and dietary inequalities globally; the coexistence of under-, mal- and over-consumption; problems of access
Psychology	Neuroses associated with food (phobias, anorexia, etc.); media manipulation; targeting of children; concerns about mass marketing and soft 'psy-op' techniques such as 'big data' control to shape behaviour
Environment	Unsustainability; externalised costs; biodiversity loss; pollution, contamination and residues; water use and stress; soil damage; land use change
Governance	How food is regulated; the weakening of the state role; problems of multi-level governance; rise of NGOs and civil society; public trust; transparency and accountability

severe strain. Table 9.2 summarises some of the contested features covered in this book. Food policy today (as in the past) is highly contested space.

The new agenda

Through all this complexity, we detect a comparatively simple new agenda for food policy emerging. It can be summarised as: a food system operating to enable sustainable diets from sustainable supply and distribution. This actually requires that all sectors of the food world and governments:

*   integrate public policy *across* sectors (health, environment, trade, transport, regulation, welfare and education);
*   integrate *between* levels of governance (global, regional, national and sub-national/local);
*   reinternalise currently externalised costs so that consumer prices are real, fair and operate properly for market efficiencies;
*   ensure primary growers receive decent returns and living wages, in contrast to today's tendency for high profits and value-adding to be done off the land;
*   take a long-term approach to reform of food systems, and frame this to be delivered in incremental short- and medium-term actions;
*   link eco-systems and population health;
*   reform and revitalise policy institutions to make them fit for purpose, and use the full range of policy mechanisms to deliver the above.

If policy-makers were to take these aspirations seriously, this would have significant consequences. Longer term goals become clearer. New frameworks, targets and laws have to be created. New conventions and a reinvigorated democratic food governance are needed. How else are we to address the impact of supermarket concentration, the shift of rural populations to the cities in unprecedented urbanisation, the environmental crisis, the difficulty of regulating rapidly changing food markets and supply chains? These all add up to a highly complex challenge for food governance. It can be argued that in a transnationalising world, 'old' *dirigiste* models of food governance may not be appropriate any more, and that a dual system of food standards and 'rules-setting' has emerged, one created by the state, the other by companies setting regimes (such as Global-GAP the corporate standards-setting coalition).[8] Food companies have certainly been creating parallel systems of food governance with their own standards, conventions and policing, through contracts and specification. This dualism is not helpful and needs to be brought under more democratic scrutiny. That is not to say that all commercial governance is just protecting their interests; actually some companies have been conducting far-sighted experiments in reducing, for example, undesirable impacts. In their terms, what is now required is a new level playing field in which standards are ratcheted up, not held back by the lowest common denominator.

This may seem grand but what else is realistic in a food world undergoing a fundamental and global nutrition transition? No bland nutritional advice or labelling will get a grip of the vast cultural shift needed. In that sense, the ubiquity of obesity is a shocking reminder of the failure of public policy so far. It is time that ministers and ministries of agriculture, food, fisheries and health were challenged to give this agenda top priority as a coalition of NGOs and professionals argued in a resolution presented at the UN's International Conference on Nutrition in November 2014. [9] To date, no country's government has fully addressed the full range of problems outlined in this Food Wars thesis, but awareness is rising. If the scourge of tobacco can be confronted, why not obesity? They are different, of course, but the willingness to tackle the unacceptable is the same. The number and range of major reports cited in this book on different food challenges are testament to this pressure. How can the world go on producing food which destroys eco-systems after the weight of evidence from the 2005 Millennium Ecosystems Assessment?[10] How can it ignore the wide health gaps after the 2008 Commission on Social Determinants of Health?[11] How can it accept failure to meet the Millennium Development Goals faced with the annual FAO State of Food Insecurity reports?[12] Yet apparently it does. The pressure however is building up.

It is time to re-energise food policy and to reinvigorate what is meant by the public interest, and which strategic goals, instruments and measures ought to be deployed to achieve the simple goal of sustainable diets from sustainable food systems. The old 'top-down' model of public policy relied upon states deciding what they wanted to do, and upon the best, most efficient means for delivering what they set out to do. Taxes and fiscal measures? Advice, education, information? Regulations and laws? The patrician state would decide. We do not reject these instruments but are concerned about the ends to which they are used, and how they may best be deployed for effect. We worry that the policy reflex is currently to use soft, weak measures. Regulations, laws and fiscal measures should be core weapons in the Food Wars. It has been more fashionable and politically expedient since the 1980s, and global triumph of neo-liberal market economics, to see mechanisms such as codes of conduct, public–private partnerships, pledges, the issuing of advice, league tables of performance and 'naming and shaming' and the creation of voluntary approaches rather than the use of legislation and sanctions. Soft measures have their place, too, but they depend upon consensus among partners to work to the same end and are satisfactory mechanisms for resolving conflicts of interest.

In the 21st century, consumers may be more responsible but can they be the sole arbiters of societal change? Too often, governments and companies put the emphasis on consumers to look after their own health or to be responsible for environmental protection. This often is translated into a mundane choice of brands. This is too thin. Against that kind of consumerist logic, beloved of neo-liberals, choice-editing is more progressive. It recognises systemic change, but still keeps it beneath the radar of consumer consciousness. Choice-editing, too, cannot be left to the whim of corporate responsibility. That is why governmental frameworks are so important. They set the level playing field beloved of business.

These are potentially dangerous times for food policy. We do not see great advance unless there is more coherent public pressure on policy-makers to act.

No sector can do what is needed on their own, if humanity is to bridge public and ecosystems health through food supply for 9 billion people. The new era may be decided by the emergence of alliances determined to take food's impact on ill health, inequity and unsustainability as one of *the* challenges for humanity. To be sober about the situation, we see a number of broad policy scenarios:

- to do nothing and allow 'market forces' to run their course;
- to look to corporate solutions;
- to frame market conditions;
- to empower civil society to demand and consume differently;
- to wait for external shocks to change the food system.

It would be folly to favour the last option, although it looms large in many risk assessments, not least by the military. We see signs of all the above scenarios, but we hope that public pressure will build to pursue the goal of delivering human and ecological health through food, and of combining efficiency with authenticity and integrity of food systems.

Food is such an enormous sector of local, national and global economies, so vital to people's needs and existence, and so powerfully fought over by companies and other social forces wishing to capture markets, that conflict is almost inevitable. The Productionist paradigm is under stress. The Life Sciences Integrated paradigm has big commercial and institutional backing. More resources need to be put into policies from within the Ecologically Integrated paradigm which is under-funded but holds great promise. As we have sought to show, the data reveal that more needs to be done to tackle food's role in environmental and human health and societal divisions. The facts challenge each of the three paradigms. Concern about them is increasingly voiced by professions and movements. Clever modelling, wonderful experiments and declarations of intent are all emerging in the public policy arena. Yet, at the conclusion of this book, we must soberly conclude that these voices are still not coherent enough or working together enough to threaten business-as-usual tendencies. We would welcome a policy world in which the movements representing diverse interests such as public health, nutrition, environment, consumers, social justice, progressive business, development and citizenship could unite to reject the default position, and to press for fundamental modernisation of the food system in line with health, environmental and social justice. New overarching frameworks are needed to redirect the food system; that much is clear, but whether the social forces needed to achieve this are yet strong enough remains less clear. In this very practical sense, as well as in theory, the goal of food policy surely must be for humanity to be at one with nature. But culture, as ever, might turn out to be the defining arbiter. There is some way to go before the route to Food Peace is clear.

References

1   Goodman D, DuPuis EM, Goodman MK. *Alternative Food Networks: Knowledge, Place and Politics*. London: Routledge, 2012.
2   Wilkinson RG, Pickett K. *The Spirit Level: Why More Equal Societies Almost Always Do Better*. London: Allen Lane, 2009.
3   Commission on Macroeconomics and Health, *Macroeconomics and Health: Investing in Health for Economic Development*. Geneva: Harvard University/Center for International Development/World Health Organisation, 2002.
4   Marmot M. *Fair Society, Healthy Lives. Final Report of the Strategic Review of Health Inequalities in England Post 2010 (Marmot Review)*. London: University College London, 2010.
5   Millstone E, Lang T. *The Atlas of Food: Who Eats What, Where and Why*. London/Berkeley, CA: Earthscan/University of California Press, 2008.
6   Sen A. *Poverty and Famines: An Essay on Entitlement and Deprivation*. Oxford: Oxford University Press, 1982.
7   Von Clausewitz C, Howard M, Paret P. *On War*. Princeton, NJ: Princeton University Press, 1976 (1832).
8   GlobalGAP. What is GLOBALGAP? <http://www.globalgap.org> [accessed July 2014]. Cologne: Global-GAP, 2014.
9   World Obesity Federation, Consumers International, *et al*. Recommendations towards a convention to protect and promote healthy diets. <http://www.worldobesity.org/site_media/uploads/Convention_on_Healthy_Diets_FINAL.pdf> [accessed Dec. 2014]. London: World Obesity Federation, 2014.
10   Millennium Ecosystem Assessment (Program). *Ecosystems and Human Well-Being: Synthesis*. Washington, DC: Island Press, 2005.
11   Commission on Social Determinants of Health. *Closing the Gap in a Generation: Health Equity through Action on the Social Determinants of Health. Final Report of the Commission on Social Determinants of Health.* <http://www.who.int/social_determinants/final_report/en/index.html> Geneva: World Health Organisation, 2008.
12   FAO. *State of Food and Agriculture 2013*. Rome: Food and Agriculture Organisation of the United Nations, 2013.

Index